T0296321

CAMBRIDGE LIBRARY COLLECTION

Books of enduring scholarly value

Earth Sciences

In the nineteenth century, geology emerged as a distinct academic discipline. It pointed the way towards the theory of evolution, as scientists including Gideon Mantell, Adam Sedgwick, Charles Lyell and Roderick Murchison began to use the evidence of minerals, rock formations and fossils to demonstrate that the earth was older by millions of years than the conventional, Bible-based wisdom had supposed. They argued convincingly that the climate, flora and fauna of the distant past could be deduced from geological evidence. Volcanic activity, the formation of mountains, and the action of glaciers and rivers, tides and ocean currents also became better understood. This series includes landmark publications by pioneers of the modern earth sciences, who advanced the scientific understanding of our planet and the processes by which it is constantly re-shaped.

Siluria

The Scottish geologist Sir Roderick Impey Murchison (1792–1871) first proposed the Silurian period after studying ancient rocks in Wales in the 1830s. Naming the sequence after the Silures, a Celtic tribe, he believed that the fossils representing the origins of life could be attributed to this period. This assertion sparked a heated dispute with his contemporary Adam Sedgwick, ultimately ruining their friendship. First published in 1854, *Siluria* is a significant reworking of Murchison's earlier book, *The Silurian System*, which had appeared in 1839. Thorough in his approach, he combines his own findings with those of researchers around the world, touching also on the later Devonian, Carboniferous and Permian periods as well as questions of natural history. An important text in nineteenth-century geology and palaeontology, the work contains a valuable geological map of Wales along with detailed engravings of fossils, including crustaceans, cephalopods and fish.

Cambridge University Press has long been a pioneer in the reissuing of out-of-print titles from its own backlist, producing digital reprints of books that are still sought after by scholars and students but could not be reprinted economically using traditional technology. The Cambridge Library Collection extends this activity to a wider range of books which are still of importance to researchers and professionals, either for the source material they contain, or as landmarks in the history of their academic discipline.

Drawing from the world-renowned collections in the Cambridge University Library and other partner libraries, and guided by the advice of experts in each subject area, Cambridge University Press is using state-of-the-art scanning machines in its own Printing House to capture the content of each book selected for inclusion. The files are processed to give a consistently clear, crisp image, and the books finished to the high quality standard for which the Press is recognised around the world. The latest print-on-demand technology ensures that the books will remain available indefinitely, and that orders for single or multiple copies can quickly be supplied.

The Cambridge Library Collection brings back to life books of enduring scholarly value (including out-of-copyright works originally issued by other publishers) across a wide range of disciplines in the humanities and social sciences and in science and technology.

Siluria

*The History of the Oldest Known Rocks
Containing Organic Remains,
with a Brief Sketch of the
Distribution of Gold over the Earth*

RODERICK IMPEY MURCHISON

CAMBRIDGE
UNIVERSITY PRESS

CAMBRIDGE
UNIVERSITY PRESS

University Printing House, Cambridge, CB2 8BS, United Kingdom

Published in the United States of America by Cambridge University Press, New York

Cambridge University Press is part of the University of Cambridge.
It furthers the University's mission by disseminating knowledge in the pursuit of
education, learning and research at the highest international levels of excellence.

www.cambridge.org
Information on this title: www.cambridge.org/9781108067195

© in this compilation Cambridge University Press 2014

This edition first published 1854
This digitally printed version 2014

ISBN 978-1-108-06719-5 Paperback

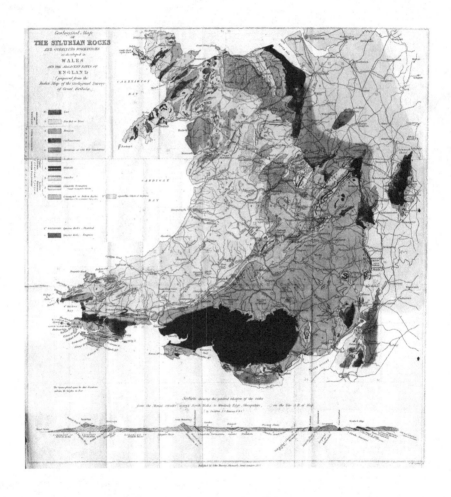

The material originally positioned here is too large for reproduction in this reissue. A PDF can be downloaded from the web address given on page iv of this book, by clicking on 'Resources Available'.

S I L U R I A.

THE

HISTORY OF THE OLDEST KNOWN ROCKS

CONTAINING

ORGANIC REMAINS,

WITH A BRIEF SKETCH OF THE DISTRIBUTION OF GOLD
OVER THE EARTH.

BY

SIR RODERICK IMPEY MURCHISON,

G.C.S^t.S.; D.C.L.; M.A.; F.R.S.; F.L.S.; HON. MEM. R.S. ED.; R.I.A.;

EX-PRESIDENT OF THE GEOLOGICAL AND ROYAL GEOGRAPHICAL SOCIETIES OF LONDON ; TRUSTEE OF
THE BRITISH MUSEUM ; HON. MEMBER OF THE ACADEMIES AND NAT. HIST. SOCIETIES
OF ST. PETERSBURG, MOSCOW, BERLIN, BRESLAU, COPENHAGEN, HOLLAND, SWITZERLAND, TURIN,
BOSTON, NEW YORK, PHILADELPHIA, ETC. ETC. ; AND
CORR. INSTITUTE OF FRANCE.

" Where were we when these grains of sand were assorted ? Compared with their date, the fall
of Babylon has just happened, and the Creation of man is an event of yesterday ! "
Geology, Rev. D. KING, 4th edi:. p. 116.

LONDON:
JOHN MURRAY, ALBEMARLE STREET.
1854.

SILURIA

THE

HISTORY OF THE OLDEST KNOWN ROCKS

OR

ORGANIC REMAINS

WITH A BRIEF SKETCH OF THE DISTRIBUTION OF GOLD
OVER THE EARTH

BY

SIR RODERICK IMPEY MURCHISON,

LONDON:
A. and G. A. SPOTTISWOODE,
New-street-Square.

Dedication.

TO

SIR HENRY THOMAS DE LA BECHE, C. B. F.R.S.

DIRECTOR GENERAL OF H. M. GEOLOGICAL SURVEY, &c. &c. &c.

To YOU, my dear De la Beche, who, by your labours and those of your associates, have demonstrated the wide extension of the Silurian Rocks in the British Isles, I dedicate this work; referring my readers to the instructive National Geological Museum of your foundation, for complete evidence of the truths I have endeavoured to sustain.

RODERICK IMPEY MURCHISON.

May 1. 1854.

PREFACE.

THE design of this work is sufficiently explained in the Introductory Chapter. I would here, however, make a few additional remarks, if only to express my great obligations to some of the authors referred to in the subsequent pages. And first to my friend, Mr. J. W. Salter, for his assistance and advice in describing, grouping, and comparing the fossils, all the most characteristic forms of which he has himself selected and drawn on wood. These woodcuts contain small figures either of species which have been discovered in the Silurian rocks of Britain since my former work was published, or of which better specimens have been obtained. The original typical forms so admirably delineated, according to their natural size, by Mr. James De C. Sowerby, in the 'Silurian System,' have been transferred from his etchings on copper to lithographic stones; and being classified and re-arranged, are presented in thirty-seven plates at the end of this volume. In regard to its illustrations, therefore, the 'Siluria' now offered to the public is a faithful outline of my previous labours and also of our present knowledge of the older palæozoic rocks, as registered in the noble series of organic life collected in the Government Museum of Practical Geology.

The chief deficiency in this part of the work, which my old friends will remark, is the absence of the beautiful lithographs of the Corals of the 'Silurian System,' drawn under the superintendence of my able associate, Mr. Lonsdale, and so lucidly described by him. These zoophytes not having been etched on copper, like the other organic remains, could not be transferred; and a selection has, therefore, been made of the most typical forms only, as represented in certain woodcuts. In naming and describing them, a few errata, alluded to in the volume, would have been avoided, had I

submitted the proofs, while going through the press, to the critical eye of
the same valued friend, whose assistance was of so much service to me in
preparing the original 'Silurian System,' and who, notwithstanding his state
of health and absence from London, has kindly enabled me to make those
corrections.

But passing over these and other defects, which might have been avoided
had I been less engaged in different occupations, I trust that my main
object will have been obtained, in presenting a clear, general view of the
succession of primeval life, and in rendering the earlier pages of geological
history accessible to many readers.

To render the work a *vade mecum* of geologists, on which foreigners, as
well as my countrymen, might depend, a coloured map is annexed, which
is simply a reduction of the geological map of Wales and the adjacent region
of Britain prepared by Professor Ramsay and the Government Surveyors,
under Sir H. T. De la Beche, and wherein the order of the Silurian Rocks
and their relations to overlying deposits are best displayed.   The reader
who may wish to examine the details of any one district, has only to look at
the large figures inscribed on this map, which refer him to each sheet,
illustrated in detail by the Government geologists.   The friends who supported me when I ventured to prepare my original map of the Silurian
region (at a time when a large portion of the country had not even been
represented in the Ordnance maps) will observe that the main features of
the range of the Silurian Rocks (Lower and Upper), and their relations
to *overlying* deposits, remain as I had traced them.   The fundamental
change made by my successors is, that nearly all the Welsh country
coloured in my original map as the Cambrian of Sedgwick, and supposed
to be occupied by rocks lower than those I described has been shown to
be composed of their exact equivalents.   In other words, the tract extending westwards from the Longmynd, which I long ago reduced to
order, as best exhibiting the Cambrian and Lower Silurian types of Shropshire and Montgomeryshire, *contains the same geological series as the
mountains of N. Wales;* the Cambrian rocks and the eqnivalent of the
Lingula flags, (lowest Silurian of the Survey), both inclusive.

It has truly been a subject of deep regret to me, that an old and
cherished friend, with whom I had long worked in foreign as well as
British lands, and whose powerful mind and brilliant eloquence have
thrown so much light on the science which we mutually cultivate, should,
of late years, have so strenuously objected to this application of the term
Lower Silurian.   But here the reader must remember, that the question

has been determined by many competent and independent authorities, all of them equally the friends of my distinguished associate as of myself. They have, in short, extended to the more complicated regions of Wales, the cession which I originally described in memoirs, map, and sections, illustrative of the adjacent Silurian counties.*

Indicating, as I have always done, a great difference in the organic remains of the Lower and Upper Silurian, I firmly adhere to my old view of the *union of these two groups in one system of life.* Aloof from their common *facies,* a careful revision of all the Silurian fossils in the Government Museum, collected from various parts of the British Isles, has led the palæontologists of that institution to the belief, that nearly one hundred species are common to the Lower and Upper divisions of the Silurian system; even excluding the Upper Caradoc, or intermediate zone, from the estimate. [See also the work, ' The Palæozoic Fossils of the Cambridge Museum,' and the Tabular List, with comments, in the Appendices A., B.]

Those persons who may wish to trace the historical evidences relating to the original researches, will do well to read a condensed sketch by that sound geologist, one of my first instructors, Dr. Fitton†, who clearly exposed the state of the subject in the year 1841, correctly noticing the effect also produced by other early labourers in the same field, among whom my most efficient coadjutor was the Rev. T. T. Lewis, of Aymestry. On the other hand, in addition to the various communications by Professor Sedgwick and myself, practical geologists will peruse with interest the memoirs of Bowman, Sharpe, and other authors published in the Proceedings and Journal of the Geological Society of London ; as well as the work of Professor Phillips ' On the Malvern and Abberley Hills,' a truly philosophical view of that Silurian region.

One of the most successful efforts to apply the true palæozoic succession to a distant part of Europe was made by my eminent and lamented friend Leopold von Buch ‡, who, simply by comparing fossils sent to him by General Tcheffkine with the types published in my former work, demonstrated that true Silurian rocks were developed in various parts of Russia. The work on Russia and the Ural Mountains, by my associates de Verneuil, von Keyserling, and myself, was, however (I hope I may say it

---

* See some good general recent observations bearing on this question, by J. B. Jukes, F.R.S., (Journ. Geol. Soc. Dublin, vol. vi., President's Discourse, p. 88.).

† Edinburgh Review, April, 1841, vol. cxlvii. p. 1.; see also Quarterly Review, 1839, vol. lxix., p. 102.

‡ Beitrage Gebirgs-formationen in Russland, Berlin, 1840.

without presumption), the first which, in extending those results to the north-western edge of Asia, developed a complete ascending series over the larger half of Europe, from the oldest known fossiliferous strata to the youngest tertiary deposits.

In perusing the fifteenth and sixteenth chapters, which indicate the parallelism of the palæozoic rocks of France, Spain, and the United States of America to those of Britain, geologists will at once recognize the vast amount of knowledge which has been contributed by my dear and enlightened companion, M. Edouard de Verneuil.

To my other numerous foreign contemporaries, and especially to M. Barrande, who have seen reason to apply to their own lands the classification and nomenclature first elaborated in the ancient kingdom of the Silures, I also tender my grateful acknowledgments. May the following pages not be without use in stimulating them to bring the older rocks of their respective countries into a closer comparison with our British types, than I have been able to effect in the present outline!

TABLE OF CONTENTS.

CHAPTER I.

CHAP. II.

BASE OF THE SILURIAN ROCKS, AND EARLIEST ZONE OF FORMER LIFE.

CHAP. III.

LOWER SILURIAN ROCKS — *continued.*

CHAP. IV.

LOWER SILURIAN ROCKS — *continued.*

THE CARADOC FORMATION.

CHAP. XII.

PERMIAN ROCKS.

CHAP. XIII.

CHAP. XIV.

PRIMEVAL SUCCESSION IN GERMANY AND BELGIUM.

CHAP. XV.

CHAP. XVI.

SUCCESSION OF PRIMEVAL ROCKS IN AMERICA.

CHAP. XVII.

CHAP. XVIII.

Conclusion.

APPENDIX.

LIST OF ILLUSTRATIONS.

WOODCUTS.

PLATES OF FOSSILS, ETC.

[*To be placed at the End of the Book.*]

ERRATA.

Preface, p. vii. l. 4. *for* cession, *read* succession.

Page
13. last line, *et passim, for* mezozoic *read* mesozoic.
43. last line but one, *omit,* The Olenus occurs in Bohemia.
44. l. 3. *omit* traces.
47. l. 7. from bottom, *for* has, *read* have.
54. l. 18. *read* (porphyries of Sedgwick), *after* igneous rocks.
68. l. 4. from bottom, *for* ch. 5., *read* ch. 8.
74. Note. *for* Pl. 3., *read* Pl. 11.; also ch. 8. instead of ch. 5.
  l. 20. *dele* Orthis alternata, Foss. 8.; and l. 22. *dele* Asaphus Powisii, pl. 2. See p. 98.
  l. 2. *for* B. ornatus, Conrad, *read* B. nodosus, Salter.
  l. 5. *for* next chapter, *read* chap. 8.
89. l. 21. *for* sixth, *read* fifth.
91. letters *under* Section near Builth, *for* c, *read* d.
103. l. 8. *for* girt, *read* grit.
105. l. 5. *for* amorphorus, *read* amorphous.
107. l. 5. *et passim, for* Petraia bina, Lonsdale, *read* Petraia bina, sp. Lonsd.
112. l. 2. *et passim, for* Rhynconella, *read* Rhynchonella.
116. l. 5. *for* extended, *read* extruded.
118. l. 4. *for* exterior nucleus, *read* exterior of the nucleus.
119. l. 7. from bottom, Fav. cristata, *for* Linn., *read* Blumenbach.
121. Foss. 12. f. 1. the name omitted is Diastopora? consimilis (Aulopora id. Sil. Syst.) The Author is indebted to Mr. Lonsdale for this correction, and also for a suggestion that f. 5. is a distinct species from the S. ramulosa, Goldfuss.
133. l. 10. *for* Avicula reticulata, *read* Pterinea Sowerbyi.
142. l. 5. from bottom *for* fossil, *read* seed-vessel of a plant.
  l. 10. from bottom, *for* Thelodus parvidens, *read* Onchus Murchisoni.
158. l 1. *for* Trenton, *read* Chazy.
161. l. 15. *for* sarcinulata, *read* lata.
  l. 16. *for* Avicular lineata, *read* Pterinea (Avicula) lineatula.
163. l. 7. from bottom, *for* comform, *read* conform.
164. note 2. *add* and vol. 5. 2nd ser. p. 1. *seq.*

Page
167. l. 5. dele the *. Note. l. 2. *for* vol. iv. old ser., *read* vol. v. 2nd ser. p. 1. *et seq.*
169. l. 13. *for* masses granular, *read* masses of granular.
  In the section, *for* letter o under Glenkeen, *read* c ; and a letter b should be placed under the dark rocks which rest on the granite (*). See descr. p. 172.
172. l. 10. *for* a, *read* c.
184. l. 8. *omit* O. flabellulum, f. 1.
187. l. 4, *for* last year, *read* in 1852 (the page was printed in 1853).
188. l. 6. from bottom *for* Pl. 9., *read* Pl. 5.
192. l. 2. *for* Foss. 10., *read* Foss. 23.
  l. 7. *for* Foss. 12. f. 7., *read* Foss. 7. f. 12.
195. l. 18. *dele* if, f. 7.——same species.
196. l. 6. *for* triated, *read* striated.
226. l. 8. *for* Mytilus antiquus, *read* Modiola antiqua.
231. l. 3. and 12. *for* Foss. 21., *read* Foss. 25.
240. l. 2. from bottom, *for* in which, *read* on which.
244. l. 19. *for* (g), *read* (f).
245. l. 7. *for* usher into, *read* usher in.
252. l. 8. *for* one sgeciesis, *read* one species is
256. l. 3. of description, place the (*) with 'Eruptive granite.'
  l. 5. *transfer* 'Club Mosses' to the Cryptogamæ above.
259. l. 6. *dele* in.
286. Note, l. 2, *for* tracts, *read* tracks.
304. l. 7. from bottom, *for* do not sto, *read* do not stop.
305. l. 17. *for* that, *read* those.
308. l. 2. *after* Polycœlia *dele* the semicolon.
310. l. 9. *for* Keratophoga, *read* Keratophaga.
315. P. S., end of, *for* Ruppendorf, *read* Ruppersdorf.
322. l. 19. *for* on, *read* in.
323. l. 4. from bottom, *for* Kutörga, *read* Kutorga.
327. l. 2. from bottom, *for* but the, *read* but to the.
  l. 5. from bottom, *for* Devonian rocks of Russia, *read* Devonian and Carboniferous rocks of Russia.
335. l. 2. *for* arise, *read* rise.
371. l. 10. *for* its, *read* the.
401. note *for* 1852, *read* 1853.
421. l. 11. *for* Shurman, *read* Shumard.
508. Index, *to* references to Dr. Buckland, *add* Alps, 500.

SILURIA.

CHAPTER I.

INTRODUCTION — ORIGINAL SILURIAN RESEARCHES. — DESIGN OF THIS WORK.

THE earliest condition of the earth is necessarily the darkest period of its geological history. The favourite hypothesis concerning the origin of the planet, founded on astronomical and physical analogies, is, that it assumed the form of a flattened spheroid from rotation on its axis when in a fluid state. Reasoning upon this idea, and looking to the structure of those rocks which either lie at great depths or have been extruded from beneath, the geologist has inferred that the crystalline masses, including granites which issued out from below all other rocks, and constitute possibly their existing substratum, were at one time in a molten state. The theory of a central heat, at first sufficiently intense to maintain the whole terrestrial mass in a state of fusion, but subsequently so far dissipated by radiation into space, as to allow the superficial portion to become solid, has been adopted by the greater number of philosophers who have grappled with the difficult problem of the first conditions of our planet. Most of them likewise have believed that all the great outbursts of igneous matter, by which the crust has been penetrated and its surface diversified, were merely outward signs of the continued internal activity of that primordial heat, now much repressed by the accumulations of ages, and of which our present volcanoes are feeble indications. If, then, the mathematician has correctly explained the causes of the shape of the globe, the geologist confirms his views when, examining into the nature of its oldest massive crystalline rocks, he sees in them clear proofs of the effects of intense heat. This original crust of the earth was subsequently, we may believe, broken up by protruded masses, which, issuing in a melted condition, constituted the axes and centres of mountain chains. Each great igneous eruption gave out substances that became, on cooling, solid rocks, which, when raised into the atmosphere, constituted lands that were exposed to innumerable wasting agencies; and thus afforded materials to be spread out as deposits upon the shores and bed of the ocean. In these hypothetical views concerning the production of the earliest sediments formed under water, we seem to reach a primary source; and once admitting that large superficial areas were originally occupied by igneous rocks, we have in them a basis from which the first sedimentary materials were obtained.

The earlier eruptions having necessarily occasioned elevations at some points and collapses or depressions at others, such changes of outline, aided by the grinding action of water, would occasion the formation of bands of sediment, which, adapting themselves to the inequalities of the surface, must have been of unequal dimensions in different parts of their range. In this way we may imagine how, by a repetition of the processes of elevation and denudation, the earliest exterior rugosities of the earth would be in some places increased, while in others they would be placed beyond the influence of sedimentary accumulation. May we not also infer, that the numerous molten rocks of great dimensions which were suddenly evolved from the interior at subsequent periods, must have made enormous additions to the solid crust of the earth, and have constituted grand sources for the augmentation of new strata?

Turning from the igneous rocks to crystalline stratified deposits, we now know that a great portion of the micaceous schists, chloritic and quartzose rocks, clay-slates, and limestones, once called primary, were of later origin. Many of these are nothing more than subaqueous sediments of various epochs, which have been altered and crystallized at periods long subsequent to their accumulation. This inference has been deduced from positive observation. Rocks, for example, have been tracked from the districts where they are crystalline, to spots where the mechanical and subaqueous origin of the beds is obvious, and from the latter to localities where the same strata are wholly unchanged, and contain organic remains. Transitions are thus seen from compact quartz rock, in which the grains of silica are scarcely discoverable with a powerful lens, to strata in which the sandy, gritty, and pebbly particles bespeak clearly that the whole range was originally accumulated under water. Other passages occur from crystalline, chloritic, and micaceous schists, to those clay-slates which are little more than consolidated mud, and from crystalline marble to common earthy limestone, in which organic remains abound. These and similar metamorphoses embrace the consideration of changes, like those, for example, by which ordinary limestone has been converted into dolomite and sulphate of lime or gypsum, or shale into mica-schist, as is seen in the secondary and tertiary rocks of the Alps.*

Elementary works will have, indeed, informed the student, that such changes of the original sediment have been generally accounted for by the influence of great heat proceeding from the interior of the earth, and which at different former periods manifested its power in the eruption of granites, syenites, porphyries, greenstones, and other substances formed by fusion. Let it, however, be understood, that the prodigious extent to which the metamorphism of the original strata has been carried in mountain-chains, and at different periods through all formations, though often

---

* See Alps, Apennines, &c. Quart. Journ. Geol. Soc. Lond. vol. v. p. 157. *et seq.*

probably connected with such igneous outbursts, must have resulted from a far mightier agency than that which was productive of the mere eruptions of molten matter or igneous rocks. The
latter are, in fact, but partial excrescences in the vast spread of
the stratified crystalline rocks, — symptoms only of the grand
changes which resulted from deep-seated causes ; probably from the
combination of heat, steam, and electricity, acting together with an
intensity very powerful in former periods.

Processes now going on in nature on a small scale, or imitated
artificially by man, may enable us to comprehend imperfectly in
what manner some of these infinitely grander ancient metamorphoses
were effected ; and the experimental science of chemistry, when more
extensively applied to the analysis of rocks, will, it is hoped, some
day reveal still more important truths in this, which is still one
of the most obscure points in the range of geological phenomena.

But speculations on such physical operations as those which have
affected the surface of the earth, are not here called for. At all
events, the earliest of the phenomena, with which alone we are at
present concerned, or the first formation of the known crust of the
planet, belongs to a period in which no definite order, — still less any
trace of life, — has been deciphered by human labour.*

The design of this work is much more attainable. Its aim is to
mark the most ancient strata in which the proofs of sedimentary
or aqueous action are still visible,—to note the geological position of
those beds which in various countries offer the first ascertained signs
of life, and to develop the succession of deposits, where not obscured
by metamorphism, that belong to such protozoic zones. In thus
adhering only to subjects capable of being investigated, it will

* The reader who desires to study the laws by which the superficial temperature of the earth has been regulated in the immensely long subsequent geological periods, will find them well explained in the profound essay of Mr. W.
Hopkins, " On the causes of changes of *climate* at different geological periods,"
Quart. Journ. Geol. Soc. Lond. vol. viii. p. 56.

be seen, that geology, modern as she is among the sciences, has revealed to us, that during cycles long anterior to the creation of the human race, and while the surface of the globe was passing from one condition to another, whole races of animals — each group adapted to the physical conditions in which they lived — were successively created and exterminated. It is to the first stages only of this grand and long series of former accumulations, and to the creatures entombed in them, that attention is now directed.

The convictions at which I have arrived being the result of many years of research, I have been urged by numerous friends to give a condensed, and, as far as is practicable, a popular view of the oldest sedimentary rocks and of their chief organic remains, and thus to throw into one moderate-sized volume the essence of my large works *, as sustained by the publications of many other authors.

Geologists are now pretty generally agreed, that the oldest organic remains which are traceable, pertain to the lower division of the rocks termed Silurian; but before any description of these ancient deposits, or of those preceding them, is given, a few words are required, in explanation of the researches by which our acquaintance with the earliest vestiges of life and order in the protozoic world has been attained.

One of the chief steps which led to the present classification, as admitted by my contemporaries, was the establishment of the 'Silurian System' of rocks and their imbedded fossils. Before the labours which terminated in the publication of the work so named, no one had unravelled the detailed sequence and characteristic fossils of any strata of a higher antiquity than the Old Red Sandstone; and even that formation was only known to be the natural base of the Carboniferous or Mountain limestone, and to contain a few undescribed fossil fishes. Not only were the relations and contents of all

* See Silurian System, Murchison, 1839; and Russia in Europe and the Ural Mountains, by Murchison, de Verneuil, and de Keyserling (J. Murray, 1845).

the inferior strata undefined, but even many rocks which are now known to be younger than the Silurian, were then considered to be of much more remote antiquity. No one had then surmised, that the great series of hard slates with limestones and fossils, which have since been termed Devonian, is an equivalent of the Old Red Sandstone, and younger than, as well as distinct from, the deposits of the still older Silurian era. On the contrary, British authorities believed (and I was myself so taught) that the schistose and subcrystalline rocks of Devonshire and Cornwall were about the most ancient of the vast undigested heaps of greywacke. In short, the best geologists * of my early days were accustomed to leave off with such rocks, as constituting obscure heaps of sediment, in and below which no succession of "strata as identified by their fossils" could be detected. The result of research, however, has been the elimination of several well-defined groups, all of which were formerly merged in the unmeaning German term ' grauwacke.' (See Chap. 14.)

Desirous of throwing light on this dark subject, I consulted my valued friend and instructor, Dr. Buckland, as to the region most likely to afford evidences of order, and by his advice I first explored, in 1831, the banks of the Wye between Hay and Builth. Discovering a considerable tract in Hereford, Radnor, and Shropshire, wherein large masses of grey-coloured strata rise out from beneath the Old Red Sandstone, and contain fossils differing from any which were known in the superior deposits, I began to classify these rocks. After four years of consecutive labour, I assigned to them (1835) the name Silurian, deriving it from the portion of England and Wales, in which the successive formations are clearly displayed, and wherein an ancient British people, the Silures, under their king Caradoc (Caractacus), had opposed a long and

* See those classical works, the first Geological Map of Mr. Greenough, and the Geology of England and Wales, by the Rev. W. D. Conybeare.

valorous resistance to the Romans. Having first, in the year 1833, separated these deposits* into four formations, and shown that each is characterized by peculiar organic remains, I next divided them (1834, 1835) into a lower and upper group, both of which I hoped would be found applicable to wide regions of the earth. After eight years of labour in the field and closet, the proofs of the truth of those views were more fully published in the work entitled the ' Silurian System ' (1839).

During my early researches, it was shown that the lowest of these (1833) fossil-bearing strata reposed, in the west of Shropshire, on a very thick accumulation of still older sediment, as exposed in the ridge of the Stiper Stones, and the Longmynd mountain; and the strata of the latter not offering a vestige of former life, they were consequently termed unfossiliferous greywacke.

At that time it was also supposed, that the contiguous slaty region of North Wales, then under the examination of Professor Sedgwick, consisted of rocks, in part fossiliferous, and of an enormous thickness, which rose up, according to my friend and fellow labourer, from beneath my Silurian types. Hence, another term, or that of Cambrian, was afterwards, or in the year 1836, applied to masses supposed to be inferior, before their true relations to the Silurian strata of Shropshire and Montgomeryshire had been ascertained. This assumed inferiority of position in the slaty rocks of North Wales being considered a fixed point, it was naturally thought, that such lower formations, the fossils of which were then undescribed, would be found to contain a set of organic remains,

* For the first tabular view of these four formations, the bottom one resting on the unfossiliferous greywacke of the Longmynd, see Proceedings Geol. Soc. Lond. vol. ii. p. 11., Jan. 1834. The characteristic fossil species were even then enumerated, and hence the classification which is now sustained is essentially twenty years old. It had even been previously stated by me, that the lowest known fossilbearing formation, or the ' *black trilobite schists and flags of Llandeilo, probably exceeded in thickness any of the superior groups.*'—Proc. Geol. Soc. vol. i. p. 476. 1833.

differing as a whole from those of my classified and published Silurian system. With other geologists, therefore, I waited for the production of the fossils which might typify such supposed older sediments; for in obtaining all the knowledge I had then acquired, by receding from upper strata whose contents were known, to lower and previously unknown rocks, I had invariably found that the latter were characterized by many distinct and new organisms. This fact, which had been first established in the tertiary and secondary deposits, was thus proved to be universally applicable by the occurrence of similar distinctions in the Carboniferous, Old Red, and Silurian rocks.

It was, however, in vain that we looked to the production of a peculiar type of life from the 'Cambrian' rocks. Silurian fossils were alone found in them; and the reason has since become manifest. The labours of many competent observers in the last fifteen years have proved that these rocks are not inferior in position, as they were supposed to be, to the lowest stratified rocks of my Silurian region of Shropshire and the adjacent parts of Montgomeryshire, *but are merely extensions of the same strata;* and hence the looked-for geological and zoological distinctions could never have been realized. In the following chapters it will be shown how Sir H. De la Beche, Professors Ramsay and E. Forbes, with Mr. Salter, and other geologists and palæontologists, have demonstrated, that the fossil-bearing rocks of North Wales are both in their order and contents the absolute equivalents of the chief mass of the strata which had been described and named by me 'Lower Silurian' in Shropshire and Montgomeryshire. These Government geologists have used my nomenclature in all their works relating to North Wales, and have, in short, determined the question physically as well as zoologically.*

But although in 1839, when my first work was completed, I held,

---

* See also Phillips, on the Malvern and Abberley Hills.—Memoirs, Geol. Surv. Vol. ii. Pt. 1. 1848.

in common with Professor Sedgwick, the erroneous idea of the infra-Silurian position of the rocks of North Wales, I soon saw reason to abandon that view, and to adopt (in the year 1841) the opinion which I have subsequently maintained. Thirteen years have elapsed since I was persuaded that the view I then took must be adhered to; first, because it had been ascertained, that in Scandinavia, Russia, Bohemia, and other countries, the oldest traces of former life were the same as the lower Silurian types of the British Isles; — and next, because many of the fossils figured in my work as Lower Silurian had been detected in the slates of Snowdon, which were then considered to lie near the bottom of the so-called 'Cambrian rocks.'

The leading object, therefore, of the present work is, I repeat, to bring out the 'Silurian System,' not as a mere abridgment of its original form; but such as it finally became in the year 1849, when it was honoured with the highest distinction which the Royal Society bestows*, and what it has proved to be, with the geographical and other additions made to it by the government surveyors at home, and by numerous geologists in other countries.

In extending my own researches to various distant lands, I found that as the true base of all rocks containing fossil remains was clear in Scandinavia, Russia, and Bohemia, and as the same fact was announced from North America, it was no longer difficult to describe the whole organic series *from a beginning*, and thus to record the succession of animals from their earliest known developments. In a word, as chroniclers of lost races, my associates and myself were enabled to register in our 'Russia and the Ural Mountains,' the types of former creatures from their apparent dawn. To the first chapters of that work, the reader is referred as fully explanatory of views which are here reiterated.† Then it was,

---

* The Copley Medal.

† The reader who desires to consult the documents which explain how my induction was arrived at, is referred to a memoir entitled, " On the meaning at-

that positive proofs, derived from a wide field of observation,
enabled us to commence geological history, with an account of
the entombment of the earliest animals recognizable in the crust
of the globe; and also to indicate the successive conditions which
prevailed upon the surface, in a long series of ages, and during the
many changes of outline which preceded the present state of the
planet. Then it was, that looking to the whole history of former
life, as exhibited in the strata, it was demonstrated from phenomena in one great empire alone (as had to a great extent been
shown in Britain), that during the formation of the sediments
which compose the crust of the earth, the animal kingdom had
been at least three times entirely renovated; the secondary and
tertiary periods having each been as clearly characterized by a
distinct fauna as the primeval series. In the work on Russia the
sequence was thus followed out truly, from the most ancient fossil-
bearing strata to the most recent stages in the geological series.

tached to the term Silurian during the last ten years," which will indicate to him
all my successive publications on this subject, including a geological map of
England and Wales, published by the Society for the Diffusion of Useful Knowledge, in 1843. (Journal of Geol. Soc. Lond. vol. viii. p. 173. See also the
memoir entitled, " On the meaning attached to the term ' Cambrian System,' and
on the evidence since obtained of its being geologically synonymous with the previously established term ' Lower Silurian.' " — Journ. Geol. Soc. Lond. vol. iii.
p. 165.) At the same time that I must protest against the recent proposal to absorb
my Lower Silurian into his Cambrian Rocks, let me record my high estima-
tion of the original memoirs of Professor Sedgwick, especially those on North
Wales, Cumberland, and the adjacent counties, which stand upon their own in-
trinsic merits. The publication on the palæozoic fossils of the Cambridge
Museum, which he is bringing out in conjunction with Professor M'Coy, will be,
I doubt not, a lasting monument in the history of geological science. If that
work had been published eighteen years ago, or in 1836, my friend, seeing that
his Bala and my Llandeilo rocks were identical, might have proposed (although my
fossils were first named and classified) that the Lower Silurian should be merged
in the Cambrian. But, now that the terms Lower and Upper Silurian have been
adopted in every country, the question is settled. My deep regret on the occasion of this difference of opinion has been expressed in the Preface; for in general
views, as in private friendship, we are cordially united.

In this volume attention is chiefly restricted to what has proved to be the protozoic, or first era of life. The plan, therefore, pursued will be, so far, similar to that which was adopted in the earlier chapters of the work on Russia; and these first leaves of geological history will be written from the clear traces of a beginning, — a plan which, for want of knowledge, was impracticable in Britain when the 'Silurian System' was published.

After a short sketch of the earliest and unfossiliferous sediments, full descriptions will be given of the Silurian rocks (Lower and Upper), followed by very brief accounts of the three overlying groups of palæozoic life, the Devonian, Carboniferous, and Permian.

The Devonian rocks were in previous years known only as the Old Red Sandstone, a name which has, indeed, become classical through the writings of Hugh Miller. These were termed Devonian, because the strata of that age in Devonshire, though very unlike the Old Red Sandstone of Scotland, Hereford, and the South Welsh counties, contain a much more copious and rich fossil fauna, and were demonstrated to occupy the same intermediate position between the previously described Silurian and Carboniferous rocks. At that time, however, none of the fossil fishes of the Scottish or English Old Red had been found in the sandstones, slates, schists, or limestones of Devonshire, or the Rhine, and objections might have been raised to the opinion formed of the age of the deposits. But the discovery made in Russia*, and afterwards extended to Belgium and the Rhenish provinces, of Scottish ichthyolites being associated with numerous mollusca of the Devonshire rocks, firmly established the truth of the comparison.

The Carboniferous rocks, so elaborately and usefully developed in the British Isles, have been already well investigated by many writers, particularly by Professor Phillips, and have been found

* See Russia in Europe and the Ural Mountains, vol. i. p. 64.

to extend, like the Silurian and Devonian, over immense regions in all quarters of the globe.

The great primeval or palæozoic series is now known to terminate upwards, in Europe, with certain deposits, for which, in the year 1841, I suggested the name of Permian. In the early days of geological science in England, this group was classed with the New Red Sandstone, of which it was supposed to form the base. But extended researches have shown from the character of its imbedded remains, that it is linked to the carboniferous deposit on which it rests, and is entirely distinct from the Trias, or New Red Sandstone, which, overlying it, forms the base of all the secondary rocks. The chief calcareous member of this Permian group was termed in England the Magnesian Limestone, in Germany the 'Zechstein;' but as magnesian limestones are of all ages, and as the German 'Zechstein' is but a part of a group, the other members of which are known as 'Kupfer Schiefer' (copper slate), 'Rothe todte liegende' (the Lower New Red of English geologists), &c., it was manifest that a single name for the whole was much needed. After showing how these variously named strata constituted one natural group, I therefore proposed to my fellow-labourers, de Verneuil and A. de Keyserling, that the vast Russian territory of Perm should furnish the required name. The term Permian* has, indeed, been adopted by several German authorities, and also by the Government Geological Surveyors of Britain.

In the opening chapter on the geology of Russia, we gave a general view of this palæozoic classification, as applied to Germany, France, Belgium, and North America; in all of which countries, as well as in Russia, it was shown, that a similar ascending order prevailed, from a base line of recognizable Silurian life, up through Devonian and Carboniferous deposits. In the nine years which have elapsed since the issue of that work, considerable addi-

---

* 'Penéen' of D'Omalius d'Halloy. (See Chap. 12.)

tions have been made to our knowledge, and all of them sustain the truth of our generalization. We then scarcely knew of the existence of true Silurian deposits in Germany; nearly all the grey-wacke of the Rhenish provinces and the Hartz having been assigned to the Devonian series. But since the opening out of the rich Silurian basin of Bohemia, which, in the hands of M. Barrande, has become the palæozoic centre of the continent, Thuringia and Saxony have been also found to contain Silurian rocks.

In Spain, several mountain chains have been shown by M. de Verneuil to consist of Silurian, followed by Devonian and Carboniferous rocks; whilst, in Portugal, Mr. Sharpe has described the first and last of these groups. Even Sardinia has exhibited, under the scrutiny of General A. della Marmora, her Silurian and super-jacent coal deposits. Again, as Devonian and Carboniferous strata overlie older rocks in North Africa, and Devonian fossils occur to-wards Central Africa* and at the Cape of Good Hope, there are already fair grounds for believing, that a similar order pervades the axial lines or ancient mountains of that vast continent.

In north-western Asia, the chief features of which are described by Humboldt and Rose, my colleagues and myself have explained, how the Silurian rocks of the Ural chain are succeeded by younger palæozoic deposits, and Pierre de Tchihatcheff has indicated a great extension of similar formations over large tracts of Southern Siberia and the Altai mountains; whilst in north-eastern Siberia, Adolf Erman has traced such rocks even to the Sea of Ochotsk.

In the giant Himalaya mountains, and in Hindostan, where till recently no systematic labours had been devoted to the older strata, we now know, that Silurian rocks, covered by secondary or mezozoic deposits, exist in those the highest mountains of the world;

---

* For North Africa, see Coquand, Bull. de la Soc. Géol. de France, 2nde Série, vol. iv. p. 1188. Some of the fossils collected by the enterprising traveller Overweg are also Devonian. For South Africa, the reader must consult a Memoir by Mr. Bain, not yet published in the Quart. Journ. Geol. Soc. Lond.

and that the Upper Punjaub contains a limestone charged with well-known carboniferous fossils, reposing, as in England, upon a red sandstone.* There is, indeed, every reason to believe that the mountain chains of Tartary and China are composed, to a great extent, of these older rocks; for whilst extensive coal-fields have been long worked in the environs of the capital, Pekin, Devonian fossils of the very same species as those of England and the continent have recently been sent from Kwangsi, far to the south of Shangai. Other fossils, identified by de Koninck as Devonian forms, were brought by M. Itier, from the Yuennan province, one hundred leagues north of Canton.†

In Australia, where a very short time since reference could be made only to rocks of the Carboniferous and Devonian age‡, we hear of true Silurian strata containing fossils like those of the British Isles. Some species seem undistinguishable. §

In South America, the lofty Cordilleras and plateaux, whose mineral characters had been so admirably described by Humboldt, are shown by Alcide d'Orbigny to consist in great part of such ancient sediments. Still more clearly has North America been found to contain a vast succession of these palæozoic rocks, and especially of their lower members. Numerous geologists of the United States have demonstrated, that their ancient strata followed the same order on a very grand and usually unbroken scale (particularly in the

* The Himalayan data are described, by Capt. R. Strachey; those of the Upper Punjaub, by Dr. A. Fleming. (Quart. Journ. Geol. Soc., vol. vii. p. 292., and vol. ix. p. 189.)

† See a description of the Chinese coal-field near Pekin, by Kovanko, Ann. des Mines de Russie, An. 1838, p. 191. No geologist can peruse Mr. Fortune's lively description of the Bohea mountains without suspecting, that a fine primeval succession may there be found. For the Chinese fossils, see Davidson, Q. Journ. Geol. Soc. Lond. vol. ix. p. 353.; and de Koninck, Bull. Acad. Roy. Sc. Belg. vol. xiii. pt. 2. p. 415.

‡ See Strzelecki's Australia, Foss. Fauna, Morris; M'Coy, Ann. Nat. Hist. 1847.

§ Memoir by the Rev. W. B. Clarke, Quart. Journ. Geol. Soc. Lond., vol. viii.; see also his collections, and those at the Government Museum.

western region); doubtless due to their having been exempted in such tracts from the intrusion of igneous rocks. Spread out in enormous sheets over the southern districts of Upper Canada, the Lower Silurian strata, invariably so called by every American geologist *, are there based on unfossiliferous slates, limestones and sandstones reposing on crystalline rocks, which, extending far northwards, are surmounted by other sedimentary masses similar to strata of the United States, and where Silurian fossils have been detected in limestones amid the polar ices. Adjacent to the southern end of this continent, similar remains have been collected by Darwin in the Falkland Islands.

In few of those regions, however, with the exception of North America (certainly not in the British Isles, where the strata are in many parts much obscured by igneous outbursts), is the sequence so undisturbed as in Scandinavia and European Russia. There, the successive primeval deposits extend over a large portion of the earth in regular sequence and in an unaltered state. Hence, though to the unskilled eye, Russia presents only a monotonous and undulating surface, chiefly occupied by accumulations of mud, sand, and erratic blocks, its framework, wherever it can be detected, exhibits a clear ascending series. The older sedimentary strata, deviating only slightly from horizontality, are there overlaid by widely-diffused masses of those Permian rocks which constitute the true termination of the long palæozoic period.

The following pages, as before said, will be chiefly devoted to the Silurian or first stages of this primeval series. They will be illustrated by woodcuts representing the most important organic remains, and certain typical, pictorial scenes, as well as vertical sections, chiefly taken from my original work. Faithful transfers from the original plates of the 'Silurian System' will also be given,

* See particularly the works of James Hall and Dale Owen, the Reports of Logan — the chief geologist of Canada — and the new general map of Marcou.

in a rearranged form, and with the modern nomenclature of the fossils.

If all the succeeding primeval rocks were to obtain the same amount of illustration as the Silurian, this work would be expanded far beyond the limits to which I must restrict it. The younger palæozoic, or the Devonian, Carboniferous, and Permian deposits, will therefore receive only such a description as may be sufficient to give the student a general view, and stimulate him to acquire a fuller acquaintance with them by consulting the various works wherein they are circumstantially described. But even the sketch of them in this volume will, it is hoped, suffice to show, that while the contiguous strata of two natural groups are intimately linked together by containing some species which are common to both, the principal fossils of each are certainly peculiar.

Although few mineral changes of the strata can be alluded to, an endeavour will be made to show, that gold, however it may now be spread over the surface, was originally accumulated in abundance in the older rocks only (especially in those which have been much altered), and in the associated eruptive masses.

Lastly, it is to be observed, that as the true sequence of the oldest fossiliferous strata was first detected in the British Isles, so the geological descriptions in this volume will be principally derived from our insular examples. At the same time, a general comparison will be instituted with the contemporaneous rocks of different quarters of the globe.

The importance of having, through patient surveys, mastered the obscurities which clouded the history of the earlier periods of animal life will thus, it is hoped, be rendered obvious, in showing that we have now obtained as correct an insight into the first fossil-bearing formations as we had previously acquired of the younger deposits.

CHAPTER II.

BASE OF THE SILURIAN ROCKS, AND EARLIEST ZONE OF FORMER LIFE.

OUTLINES, STRUCTURE, AND ORDER OF THE OLDER ROCKS. — EARLIEST CRYSTALLINE ROCKS. — THE LOWEST DEPOSITS IN WHICH A TRUE SEDIMENTARY ORIGIN IS SEEN ARE USUALLY VOID OF SIGNS OF FORMER LIFE. ORDER OF SUCCESSION FROM SUCH UNFOSSILIFEROUS STRATA UPWARDS TO THE LOWEST ZONE IN WHICH FOSSIL REMAINS HAVE BEEN DETECTED IN GREAT BRITAIN.

THE most ancient rocks in which traces of fossil animals have been detected vary much in structure and outline in the different countries where they have been recognized. Wherever such masses have remained in a state of comparative quiescence, from the period when they were raised up from beneath a primeval ocean and have since been only slightly modified by physical causes, they are necessarily unlike those rocks which, though formed at the same period and even composed of the same materials, have been penetrated by igneous matter, or subjected to alteration and dislocation through the action of heat, pressure, and other agencies.

In Russia, for example*, where some of the oldest deposits have been only partially hardened since they were accumulated at the bottom of the sea, and have been merely elevated into low plateaus that have undergone no great change or disruption, they have scarcely any resemblance to rocks of the same age in the British Isles. When, however, we follow these soft primeval strata of Russia

* Russia and the Ural Mountains, by Murchison, De Verneuil, and Von Keyserling, vol. i. p. 26*.

C

to the Ural Chain, in which there are numerous rocks of eruptive
character, we find that the beds, which on the west consist of mud
and sand, have there been converted into crystalline schists, limestones, and quartz rocks. Startling, therefore, as it may seem to
young geologists, the hillocks of slightly coherent mud, marl, and
sand near St. Petersburg are truly of the same age as some of the hard
slaty mountains of North Wales ; a fact which geological researches
have established by proving, that the deposits of the two countries
contain a similar group of organic remains, and occupy the same
relative place in the series of formations which compose the crust
of the globe. So also in North America, where strata of the same
period, as seen in the United States and the British provinces, are
usually in the state of ordinary sandstone, shale, and limestone, when
traced westwards over vast prairies to the sources of the Missouri,
are seen to have been also converted into crystalline rocks. This
phenomenon occurs along those numerous ancient cracks in the
crust of the earth through which the igneous materials were evolved
which specially mark the range of the Rocky Mountains and other
prolongations of the Andes.*

This is the change which geologists call metamorphism. For, as
the strata have been penetrated in many places by granites, porphyries, and other rocks which were once in a state of fusion, so
it is inferred that their transmutation has resulted from the action of
heat, and usually under the pressure of a former ocean. In such
examples the age of the original sediment is the same, though the
mineral aspect of the rocks in tracts distant from each other is so
very different.

* See Stansbury's Exploration of the Great Salt Lake of Utah. Philadelphia,
1852. The fossils collected by this enterprising engineer and geographer are
described by Professor James Hall, and prove the existence, if not of Silurian, at
all events of Devonian and carboniferous deposits, extending to the Mormon territory, where some of the lower strata, probably Silurian, are much metamorphosed
and mineralized.

In like manner it will presently be shown, that the comparatively low hills of western Shropshire, and the contiguous districts of Montgomery, parts of Herefordshire, and the adjacent tracts of South Wales, where the Silurian system was originally established, present to the student, in an intelligible manner, an arrangement and succession of strata little altered, which he will have much more trouble in deciphering amidst the hard, rugged, metamorphic, and crystalline formations of the same age in North Wales; even now that obscurities have been removed by the labours of the eminent persons who have laboured in the arduous task of bringing that region into order.* He will also see in it the very same phenomenon as in the Ural and Rocky mountains, — the conversion of the oldest of such strata into crystalline schists (Anglesea).

In Russia the Silurian rocks form either wide level plains, or low plateaus, whilst in other countries where they have been heaved up into mountains, they have a rounded outline, particularly

---

* See particularly the Memoirs of Professor Sedgwick, descriptive of the North Welsh mountains, in the Proceedings of the Geol. Soc. of Lond., vol. ii. p. 679. ; vol. iii. p. 548.; Quarterly Journ. Geol. Soc. Lond., vol. i. p. 5.; ib. p. 542.; vol. iii. p. 133.; also the Memoirs of Mr. Bowman, Reports of Brit. Association, 1841 ; Trans. Phil. Soc. Manch., vol. i. p. 194.; of Mr. Sharpe, Proceedings, vol. iii. p. 74.; ib. vol. iv. p. 10.; Quarterly Journ. Geol. Soc. Lond., vol. ii. p. 283.; Davis on Tremadoc, ib. vol. ii. p. 70.   See Ramsay and Aveline, Quarterly Journ. Geol. Soc., vol. iv. p. 294.   Independently of his works on the Silurian System, and Russia and the Ural Mountains, the author of this volume refers to his memoirs.   Lond. and Edin. Phil. Mag. 1835, p. 46.; Proceedings of the Geol. Soc., vol. i. p. 471.; ib. vol. ii. pp. 11. and 119.; ib. vol. iii. p. 640.; ib. vol. iv. p. 70.; Quarterly Journ. Geol. Soc. Lond., vol. i. p. 467.; ib. vol. iii. pp. 1. and 165.   But, above all, the geologist must consult the recently published maps and sections of the government surveyors, which have finally determined all questions at issue.   Of these, and particularly in reference to the complicated tracts of Wales, which he has so skilfully worked out with his associates. Professor Ramsay will soon publish a descriptive work in the volumes of the Government Survey, similar to that which has been done for the Malvern Hills and adjacent tracts by Professor Phillips.

where they consist of schists, originally composed of mud, the fine grains of which have given rise to equable atmospheric attrition. When, on the contrary, the shale and schist having been changed into hard slates, the sandstone into quartz rock, or the earthy limestone into crystalline marble, and particularly if the beds are highly inclined, and penetrated by igneous rocks, then striking peaks or abrupt cliffs and gorges are dominant.

Thus it is, that the Silurian rocks of different regions put on many external forms. In South Britain they are, necessarily, most varied in districts like western Shropshire, North Wales, and Cumberland, where igneous and intrusive rocks have diversified the outline by expanding, breaking through, and altering these sediments.

Observation has now taught us of what materials the fundamental rocks consist in the different countries where they occur. In Scandinavia, for example, as in parts of North America, the masses beneath all those formations in which there are the slightest vestiges of Silurian life, are primary rocks, including granitic gneiss, micaschist, as well as metalliferous schists and quartzose crystalline masses. In Bohemia, however, as in Great Britain, and other portions of North America *, the lowest zone containing fossil remains is underlaid by very thick buttresses of earlier sedimentary accumulations, whether sandstone, schist, or slate, which, though

* In North America this formation may also be termed azoic; for along the western portion of Lake Superior, and in the country extending southwards into the State of Michigan, rocks of this class rise from beneath the Potsdam sandstone, or lowest formation in which Silurian fossils are known. Citing the American authors, Foster and Whitney on the azoic rocks of Lake Superior, of King on those of the State of Missouri, of Logan on those of the British frontier, and of Engelmann on Texas, and placing them in relation to the excellent works of James Hall and Dale Owen, M. Desor has announced to the Geological Society of France, that in those regions there exists a vast formation of strata, in parts metamorphic, of silicious sandstones, chloritic and quartzose schists, characterized by the absence of organic remains, and a great abundance of iron ore. Bulletin de la Soc. Géol. de France, 2 Ser., tom. ix. p. 342.

often occasionally not more crystalline than the fossiliferous beds above them, have yet afforded no sign of former beings.

This is the important fact to which attention is first directed; for in such instances the geologist appeals to the book of nature where its leaves have undergone no great alteration. He sees before him an enormous pile or series of early subaqueous sediment, originally composed of mud, sand, or pebbles, the successive bottoms of a former sea, all of which have been derived from pre-existing rocks; and in these lower beds, even where they are little altered, he can detect no remains of former creatures. But lying upon them, and therefore evolved after them, other strata succeed, in which some few relics of a primeval ocean are discernible, and these again are every where succeeded by newer deposits in which many fossils occur. In this way, evidences have been fairly obtained to show, that the sediments which underlie the strata containing the lowest fossil remains constitute, in all countries which have been examined, the natural base or bottom rocks of the deposits termed Silurian.

The hypothesis that all the earliest sediments have been so altered as to have obliterated the traces of any relics of former life which may have been entombed in them, is therefore opposed by examples of enormously thick and varied deposits beneath the lowest fossiliferous rocks, and in which, if animal remains had ever existed, some traces of them would certainly be detected.

A very few words need here be said of those crystalline rocks which in numerous regions have been the nuclei out of whose materials the very oldest recognizable sediments have been composed. Passing over the consideration of the most ancient granites, porphyries, and other igneous rocks, there are doubtless also stratified crystalline rocks, which either lie under or have been raised from beneath all those deposits in which we can trace the signs of mechanical and subaqueous action.

Such rocks, formed, as I believe, at a period when the heat of the surface of the earth was antagonistic to the existence of living beings, unquestionably often form the axes or backbones of continents, upon the sides of which the oldest sedimentary formations have been deposited. But it is unnecessary here to dwell further upon them; whilst in the sequel we shall have to indicate the metamorphism of many formations of younger date. For the present work is, I repeat, simply intended as a popular sketch of the successive accumulation of those sedimentary strata which succeeded to the primary state of the planet, and in which the remains of the earliest known animals are entombed; the physical conditions which accompanied such deposits being duly noted.

Now, in some regions where the oldest of these deposits (all of them once sea-bottoms) have been exempted from such change or metamorphism, or very slightly subjected to it, we observe that the lowest sediments thus accumulated are void of traces of life. Proceeding in our researches, we next find that these, the most ancient greywacke grits and schists of geologists, which have been formed under water at the expense of previous rocks, are succeeded regularly, and without disorder, by other strata scarcely at all differing from the former in composition, in which, at various points of the surface of the globe, and at the same level or horizon, we here and there obtain the first glimpses of the same group of animal life. Once introduced to these primordial creatures, the inquirer into the subsequent operations of nature meets thenceforward with a never-failing storehouse

GENERAL ORDER OF THE PRIMEVAL ROCKS.

.x   y   1   2ª   2   3   4   5

Gneiss, &c. y; granite x.

of animal relics in all the formations which afterwards were successively accumulated. The accompanying diagram explains at a

glance this generalization, as established by an appeal to the structure of several countries, the rocks of which exhibit the first steps in
this long series.

In 1, the reader sees the older deposits formed out of the pre-existing rocks, consisting of gneiss, &c. y, or granite x, whilst 2^a represents
the earliest zone wherein any notable signs of life occur; and the
overlying masses, 2, 3, 4, 5, stand for the mass of the formations into
which the earliest primeval system of fossil animals is divided.

We will presently consider this ascending order of stratum upon
stratum, as it is presented to us in the region where the Silurian classification began, and in the adjacent country of N. Wales, to which
that classification has since been extended and applied by the
government geological surveyors. In the mean time, let me repeat,
that in Bohemia, as well as in N. America, there are crystalline
rocks, which, from their underlying position, are known to be of
higher antiquity than the azoic greywacke of the previous section.
So also it is believed that in the north Highlands of the British
Isles, and many other regions, there are equally ancient crystalline
rocks.

In no part of Siluria and Wales, however, are there any rocks of
more remote age than the oldest greywacke in which no fossils are
known. The pre-existing masses, out of whose materials the pebble-
beds, sandstones, and schists of the Longmynd in Shropshire, and of
the Llanberis and Harlech mountains in North Wales were formed,
have either subsided and disappeared, or have been covered over by the
strata under consideration. For, it is now ascertained that the schists
of Anglesea, which from their crystalline character were once supposed
to be more ancient than any other rocks of Wales, are simply an
altered part of the same greywacke which constitutes the base of the
Silurian series of deposits in the adjacent counties of Carnarvon and
Merioneth. In other words, the old slate and greywacke of Anglesea have been altered at one spot into chlorite and mica-schist,

in another into quartz rock. Agreeing with Sir H. De la Beche and the government geological surveyors, that this is the best explanation of the case, a small sketch is here annexed, to indicate the amount of curvature which these metamorphosed strata have undergone, as seen at the promontory of the South Stack Light-house, near Holyhead.

CONTORTED SCHISTS AT THE SOUTH STACK LIGHTHOUSE, ANGLESEA.

These contorted, crystalline rocks of Anglesea, with their intruded granites, are associated with stripes or patches, capriciously distributed as it were, of different palæozoic rocks, of Silurian, Devonian, and Carboniferous age, thus forming a sort of kaleidoscope, which the most experienced geologist might have difficulty in unravelling; the outlines of which, formerly laid down in Greenough's Map of England, are now very accurately detailed in the sheets of the Government Survey   The whole of the bands have a direction or strike from S. S. W. to N. N. E., as may be seen in the annexed map.

No sooner, however, do we pass the Menai Straits, and advance eastwards towards the west flank of the mountainous range of Snowdon, than we find huge buttresses of very ancient greywacke grit, schist, slate, and sandstone, having the same direction, from S. S. W. to N. N. E., in which, though their sedimentary character is obvious, and that they have not been altered as in Anglesea, no fossils have been detected through a thickness of many thousand feet. These are the true representatives of the Longmynd or bottom rocks of this work, and of which the crystalline schists of Anglesea are metamorphosed members. In other words, these unfossiliferous slaty rocks of North Wales, with their sandstones and grits, are now proved to be the equivalents of the strata of the Longmynd and Haughmond Hills, which were formerly described by me as types of that greywacke, which in Shropshire rises from beneath all the formations containing fossils of the Silurian system. (See section below the map.)

It is to these lowest known sedimentary rocks, whether in the Silurian region or in North Wales, that the government surveyors have restricted the term " Cambrian; " simply implying thereby, not that fossils may not be found in them, but that hitherto no traces of the Silurian forms have been detected in them. But such an application of Cambrian is not used in this work, because, when introduced by Professor Sedgwick, the term was employed, both by him and myself, as applying to a vast succession of *fossiliferous* strata, containing undescribed fossils, the whole of which were supposed to rise up from beneath the well-known Lower Silurian rocks. The government geologists have shown that this *supposed* order of infra-position was erroneous, and that all the fossiliferous rocks of North Wales, which had been called Cambrian before their included fossils were described, are physically the same strata as those which

had been laid down as Lower Silurian on the immediate west flank
of the fundamental rocks of the Longmynd. Naturally, therefore,
these North Welsh or so-called Cambrian rocks have been found
to be charged with Lower Silurian fossils.

As the same authorities have thus recurred to the original typical
tracts of Shropshire, a few words must here be said to explain the
structure of that northern portion of Siluria, as laid down in my original map and sections, with which the Cambria of Sedgwick has been
so clearly placed in parallel. The reader who desires to peruse a full
description of these ancient rocks of the Longmynd, which form the
mineral axis of Shropshire and Montgomeryshire, and are there seen
to rise out from beneath all the fossil-bearing strata, may consult the
21st chapter of the "Silurian System," which is exclusively devoted
to them and their associated igneous (trap) rocks as seen in the
hills of Longmynd, Haughmond, Ratlinghope, Linley, Pontesford,
Lyth, &c. To that description I have little or nothing to add,
nor am I aware of any correction required to be made in it. The rocks
remain still the unfossiliferous bottom beds they were then described
to be. Though attaining only a height of about 1600 feet above the

THE LONGMYND.

Caer Caradoc.   Wenlock Edge.   Clee Hills.

sea, they are of mountainous character, presenting a very ancient
aspect, and the woodcut annexed gives some idea of their forms.

In this view. the spectator is looking through one of the depressions of the Longmynd, locally called "gutters," to the tract diversified by the ridge of Caer Caradoc with the Wenlock Edge and Clee Hills in the distance.

In the Longmynd, and adjacent hills between that mountain and the Stiper Stones, these elder rocks are exposed in an enormous thickness of about 26,000 feet, of dark imperfect slates and grey schistose strata, with coarse, purplish, ferruginous grit or sandstone, and conglomerate.* According to the government surveyors, who have compared them, these Longmynd rocks are of much more than double the dimensions of strata of like age and position in North Wales; and hence they are truly *the* bottom rocks of the whole region.

Striking from S.S.W. to N.N.E., these strata have not afforded the trace of a fossil, after long and assiduous researches. They are vertical in the eastern or slaty portion, and in their western parts, or in the contiguous hills of Ratlinghope, &c., they dip at a very high angle to the W.N.W., under the perfectly parallel bands of schist and grit of the Stiper Stones, which are now classed in the government maps and sections as the lowest Silurian rocks of the region. This succession, which is delineated on a small scale in the section below the map, is sufficiently exhibited in the long

---

* The manner in which certain cavities of these rocks are filled with bitumen, and the veins of copper ores which they contain near the junction of the intrusive rocks, are facts worthy of attention; it being remembered, however, that these mineral changes were produced very long after the consolidation of such ancient greywacke strata. The same copper veins, *v*, also penetrate the sandstone and

3   *v*   1

limestone with Pentamerus oblongus, 3, which rest unconformably upon the older slates, 1, of the Longmynd, as in this diagram, taken from Sil. Syst., pl. 32. f. 4.

diagram on the opposite page (1). Then follow, in parallel masses, about 14,000 feet thick, the schists of the mining tract of Shelve and Corndon, in which Llandeilo flags, and true Lower Silurian strata, are replete with characteristic fossils (2 of section).

In this manner, the conformable succession from an unfossiliferous base, high up into the Llandeilo flags inclusive, is seen in this district of the region originally illustrated by myself. The accompanying woodcut is, indeed, merely a copy of coloured diagrams representing this order, which were made in the year 1833, and which, having been exhibited to the Geological Society of London, were afterwards published, 1839, in the " Silurian System." *

Now, the researches of the last few years, on the part of the government surveyors, have shown that the strata which occupy the loftier mountains of North Wales, ranging from Snowdon to Bala, are of the same age as these under consideration, which extend from the Shelve and Corndon tract to the Stiper Stones and there rest upon the Longmynd. In other words, the fossiliferous rocks originally described as Lower Silurian types, whether in Shropshire and Montgomeryshire, or in South Wales near Llandeilo, have been so clearly unfolded as to prove that they occupy the greater portion of North Wales.

The generalized coloured section at the bottom of the geological map annexed, which has been prepared for me by Professor Ramsay, who with his associates Aveline and Selwyn established this important conclusion, will at once render the subject intelligible to every reader; and my own original detailed section, as given on the next page, completes the comparison.

In the two buttresses of the earliest or unfossiliferous greywacke, of the Longmynd on the east, and of Llanberis and Harlech on the west, are represented, therefore, the oldest sedimentary rocks of England and Wales. (See map and coloured section.)

---

* Silurian System, pl. 32. fig. 1. et seq.

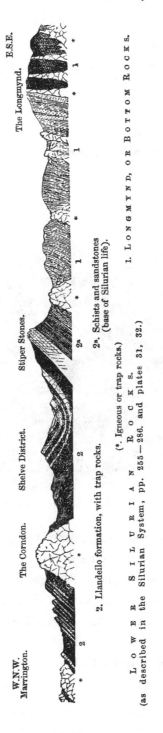

SECTION FROM THE LONGMYND ON THE E.S.E. ACROSS THE STIPER STONES TO THE TRACT OF SHELVE AND CORNDON ON THE W.N.W.

W.N.W.
Marrington.

The Corndon.

Shelve District.

Stiper Stones.

The Longmynd.

E.S.E.

2. Llandeilo formation, with trap rocks.

2. Schists and sandstones (base of Silurian life).

(*. Igneous or trap rocks.)

LOWER SILURIAN ROCKS.
(as described in the Silurian System, pp. 255—286, and plates 31, 32.)

1. LONGMYND, OR BOTTOM ROCKS.

N.B. The detailed horizontal section across a part of this tract, though prepared by the Government Surveyors, was not published when these sheets passed through the press, or I should have placed a reduction of it alongside of my old section of 1833, to show how in all essential points the two diagrams agree.

In fact, these masses constitute, in Wales as in Shropshire, the natural base of the lowest fossil-bearing strata, 2^a, to which in Britain they are every where conformable, and into which they gradually pass upwards.

These first steps in the geological series, which the reader is about to ascend, may be further illustrated by the accompanying pictorial illustration from North Wales.

PASS OF LLANBERIS FROM THE LOWER LAKE.

In this well known scene the spectator is supposed to be looking up to the Pass of Llanberis, from the lower lake of that name, the heights of Snowdon being in the distance on the right hand. In the foreground, and on both sides of the ruined castle of Dolbadarn, the faces of the cliffs exhibit flexures of the oldest strata here visible. These are the purplish and grey slaty rocks, 1, of the previous section, whose structure and relations in North Wales were formerly so well described by Professor Sedgwick. Containing the best roofing slates in the world*, and subordinate courses of greywacke grit, with rocks of igneous origin intermixed, they are seen to fold over and plunge to the east-south-east, so as to pass under the great and massive succession of schists which constitute the distant heights of the Snowdon range. The reader will observe that faint slanting lines have been introduced into the drawing to represent this ascending order, which the geologist can observe for himself, as he examines the sides of the valley in passing from the unfossiliferous slaty rocks of Llanberis, up to the overlying strata of Snowdon (2^a, 2 of section below the map), which, by their imbedded organic remains, are known to be of Llandeilo age. In short, the rocks in the foreground stand in the place of the Longmynd, whilst the overlying or distant masses are the equivalents of the rocks of the Stiper Stones, Shelve and Corndon of the previous section, p. 29.

A similar succession is visible at Barmouth and other places in North Wales, as will hereafter be noticed.

Whilst, however, the lowest rocks of Shropshire and Wales are void of fossils, it must here be noticed, that in Ireland and immediately opposite to Anglesea and Carnarvon, rocks of precisely the same mineral characters, and occupying the same place in the geological series, have afforded two species of a peculiar zoophyte.

The government surveyors have decided (and after two personal inspections of the ground, I quite agree with them,) that the purple and greenish greywacke of Bray Head, near Dublin, to which the

---

* The best slates lie in the masses to the left hand of the foreground, and are the property of my friend Mr. T. Assheton Smith. Thence they range to the quarries of Colonel Douglas Pennant, M.P., on the N.N.E. A very instructive diagram, representing the curves of the original strata, at Llanberis, and the manner in which they are traversed by planes of slaty cleavage, is given by Professor Sedgwick, Quart. Journ. Geol. Soc. Lond. vol. iii. p. 138. For a correct acquaintance with all the faults and fractures of these rocks, the geologist will, of course, consult the subsequently published maps and sections of the government surveyors.

quartz rock of the great and lesser Sugar Loaf is subordinate, corresponds exactly to the oldest buttresses of the opposite mountains in Wales, and which form the foreground in the preceding sketch. It is in this ancient Irish greywacke that a very peculiar fossil zoophyte was detected by Professor Oldham, and named after him by Professor E. Forbes. The Oldhamia is a polype of which two species have been detected, and one of them is here given.

FOSSILS, 1.

OLDHAMIA ANTIQUA, FROM BRAY HEAD, IRELAND.

The reader may look with reverence on this zoophyte; for notwithstanding the most assiduous researches, it is the only animal relic yet known in this very low stage of unequivocal sedimentary matter. The term azoic cannot therefore be applied to the Irish rocks of the same age as those which in England and Wales have afforded no such traces of life. A zoological reason is, however, added to the physical arguments above offered, for considering these very ancient deposits as simply the base of the oldest known series of former beings.

Returning then from Ireland, to the region of Siluria and Wales, in whose bottom rocks no fossils have been found, it will be seen by reference to the map, that of the strata containing Silurian organic remains, the lowest bands, or 2, 3, are most spread out on the west or Welsh side, the upper group, or 4, 5, on the English or original Silurian frontier.

In great part of South Wales, or in nearly all the hilly region to the S. of Cader Idris, as bounded by the sea on the west, Lower

Silurian rocks only prevail. There also, as in North Wales, they roll over in undulations with a general direction in some parts from N. N. E. to S. S. W., in others from N. E. to S. W.; the dip or inclination of the strata varying necessarily with each of their great folds. In the south-western parts of Montgomery and Radnor, as well as in Cardigan, and a large portion of Carmarthen, they are much less associated with igneous rocks than in Merioneth and Carnarvon on the one side, or in western Shropshire or eastern Radnorshire on the other. In the three last-mentioned tracts, Lower Silurian rocks are interstratified with felspathic trap; whilst projections of rock of irregular forms have burst through all the sediments at a later period. Viewed in this general sense, and excepting the few oases consisting of the older and unfossiliferous base, all the western region of Wales (excluding the old red sandstone and carboniferous deposits at its northern and southern extremities) consists of Silurian rocks. (See map.)

Let us now make a rapid traverse of the whole region from the higher Welsh mountains to Montgomeryshire and the border English counties of Salop and Hereford, where, as above stated, the Silurian types were first described.

The lowest rocks of Carnarvonshire, already adverted to, are much more slaty than those of the same age in the Longmynd of Shropshire. Though it is not compatible with the nature of this work to expatiate upon the structure in rocks called slaty cleavage, I may state that it is a crystalline determination and arrangement of the fine particles of mud of which the rocks were originally composed, into a countless number of thin plates precisely parallel to each other, and which usually cutting across the original lines of deposit, indicate (when observable) the true bedding or stratification.

The following woodcut will serve to explain the slaty cleavage of rocks as distinguished from the original layers or laminæ of deposit.

This remarkable physical feature of cleavage, which in former days was often mistaken for the true beds of the rocks, is often persistent

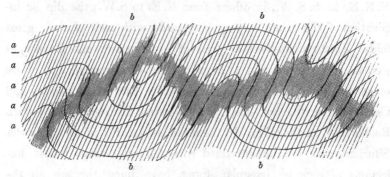

SLATY CLEAVAGE AND BEDDING.

a. Strata or beds in undulation. *b.* Lines of cleavage. (The dark tint represents the portions of a mountain in which the layers of deposit and lines of cleavage coincide.) From Sil. Syst. p. 400.

through great portions of a mountain-chain, and those who wish to be better acquainted with it, must consult the admirable memoir of Sedgwick *, who first threw a clear light upon the phenomena, and referred them to some great chemical change which heat and electricity probably combined to produce. The strata, however, I remind the reader, which are now found in this fine and crystalline state, were at one period nothing more than accumulations of mud; and there are still parts of the world, as in Russia, where, unaffected by any changes, they have remained almost in their primeval, muddy, and unsolidified condition up to the present day.

It is sufficient on this occasion to say, that their crystalline structure, as seen in Wales, and as indicated by the highly inclined lines, *b*, of the preceding diagram, was impressed upon the strata or undulating layers, *a*, very long after their accumulation as submarine mud, and even after they had been subjected to curvatures, as above indicated. Such a cleavage, prevailing at intervals from the lowest unfossiliferous greywacke or bottom rock through the Silurian system, is seen in

* See particularly Professor Sedgwick's Memoir on the Structure of large Mineral Masses, Trans. Geol. Soc. Lond., N. S. vol. iii. p. 461.

parts of other overlying palæozoic formations. In all the instances of
its occurrence in fossiliferous strata, the separation of the original
lines of bedding from the plates of cleavage is naturally much more
easy where fossil shells occur * along the layers of deposit, than in
those oldest, unfossiliferous, and finely levigated sediments, in which
the beds are often only distinguishable by the different colours in
the same slaty mass.

In all other diagrams or sections used in this work, young geologists must understand, that the divisional lines represent the layers or
laminæ of true beds as originally and successively formed by deposition under water. It is by long and patient labours in the mountains,
and by following the curvatures and breaks of these lines of deposit,
that the whole symmetry of the rocks has been determined, in
spite, as it were, of the obscurities produced by slaty cleavage
(which occurs more or less in all the North Welsh Silurian rocks),
and also by the interpolation of igneous rocks of various characters.

From this allusion to an important physical feature, by which many
ancient rocks of different epochs which were originally composed of
fine mud have been more or less affected, and particularly in Britain,
we must return to the natural order of succession as read off in the
earlier pages of primeval history.

Lowest zone of animal life in Great Britain. — Near the two extremities of the coloured section of the map, or in Shropshire, on the
one hand, and near the Menai Straits on the other, the lowest zone
in which any vestiges of former life are discoverable consists of the
schists and sandstones, 2ᵃ, which overlie the unfossiliferous strata, 1,
above described. On the west flank of the Longmynd of Shropshire, these schists, though of some thickness, are only imperfectly

* The reader who would well understand how fossil shells have been distorted
by the action which produced the cleavage, must consult an able memoir of Mr
D. Sharpe, Quarterly Journ. Geol. Soc., vol. iii. p. 74. See the example of a
distorted trilobite in this work, chap. 8. p. 201.

developed in a valley of denudation, and in the slopes between it
and the singular quartz rocks called the Stiper Stones. There the
lowest beds have only been found to contain fucoids and a rare
graptolite, a genus of zoophyte which, as will hereafter be seen,
characterizes the Silurian strata from their beginning to their close.
Although the section at page 29, which passes from the Longmynd on the E.S.E. to the mining country of Shelve on the W.N.W.
across the quartzose ridge of the Stiper Stones*, will convey a
general idea of the succession, the lover of nature†, as well as the
geologist, will recognize in the Stiper Stones, objects well worthy of
a visit and a close examination.

These rocks are made up of a number of broken and serrated
ledges jutting out from beneath the heath and turf†, like rugged
Cyclopean ruins, to form the summits of the hills which flank the
mining district of Shelve, at heights varying from 1500 to 1600 feet
above the sea. They are sandstones, which in the highest parts of
the ridge have been considerably altered, probably by the former
action of the heat evolved during the protrusion of those igneous
rocks which abound in the adjacent hills.

The first of these sketches represents the general range of these
protruding rocks, as seen on their western faces, where they dip at a

* It is also worthy of remark, that in the long transverse section below the map,
extending in a right line over 90 miles of country, the lowest rocks, near each extremity, are accompanied by much mineralization of the strata. In both, copper
and lead veinstones occur, and in parts of Merionethshire gold also occurs in
small quantities. These auriferous conditions of the Lower Silurian and other palæozoic rocks will be explained in chap. 16.

† The northern summit and slope of one of the elevations of the Stiper Stones
(called Bleak Hill in the Ordnance maps), is covered with a forest of holly trees of
great size. These hard, old, indigenous tenants of the soil, defying all cold blasts,
are appropriate emblems of the extreme antiquity of the rocks of this mountain.
The views of the Stiper Stones on the next page are taken from the Silurian System,
pp. 268, 282.

‡ See map and sections of the Silurian System, and p. 268. of that work.

high angle under the schists of the Llandeilo formation, the low
country of northern Shropshire being seen in the distance.

THE WESTERN FACE OF THE STIPER STONES.

In the following sketch, on the contrary, the spectator is supposed
to be looking southwards along the eastern escarpments of these
broken rocks, the hills of Linley being in the distance; on the left

THE EASTERN FACE OF THE STIPER STONES.
In this case the strata dip to the right hand of the spectator, whilst in the preceding sketch
they dip to the left.   In the upper view the spectator is looking to the N. N. E. ; in the lower
to the S. S. W.

hand is the valley which has been excavated in the shale which lies
between the Longmynd and these Silurian sandstones.

In following these quartzose rocks in their strike from the cul-
miating points of the Stiper Stones to the Nils Hill quarries near Pon-
tesbury, on the N.N.E., we find the strata gradually assuming the
ordinary character of a sandstone and pebbly grit, the beds of which
are ocasionally separated by shale occasionally greenish and steatitic.
On the surfaces of some of these beds are ripple-marks, so common in
all arenaceous deposits which have been formed under water, and also
branching raised casts, strongly indicative of organic origin. Some
of these are certainly the forms of fucoids which were deposited on
an ancient sea-shore or bottom. The rock so characterised is, indeed,
very analogous to the lowest sedimentary stratum in Sweden, which
there lies at the base of the fossil series of deposits, and which, from
its exposing similar forms, was, in another work, called the "fucoid
sandstone."* In Shropshire, small flakes of anthracite occur, though
rarely, in the joints and cavities of these silicious sandstones; and
this substance, we may very rationally infer, might have resulted
from the conversion into carbon of the original sea-weeds whose
casts are still visible in the sandstone. This remarkable ledge of the
Stiper Stones, with its underlying schists, a fuller description of
which must be sought in the "Silurian System," constitutes, as it
were, a divisionary band between the ferruginous, unfossiliferous
greywacke of the Longmynd, 1, with its copper veins on the
one hand, and the lead-bearing, fossiliferous Lower Silurian schists
and sandstones, 2ª, 2, of Shelve and Corndon on the other.

The strata that repose conformably on the silicious grits of the
Stiper Stones (see long section, p. 29.) are flagstones, which,
weathering to a rusty brown colour, exhibit an internal texture and
colour resembling usual forms of greywacke in many countries. They
are well exposed, in a vast succession, in highly inclined ledges, dipping about 55° to the W.N.W., on the north side of the deep
transverse combe of Mytton's Dingle. Their surfaces are also marked

* Russia in Europe and the Ural Mountains, vol. i. p. 16.; Sil. Syst. p. 285.

by casts of fucoids; but I have been unable to detect other fossils in them, though they pass up into and form an integral part of the Llandeilo formation.

The next overlying strata, which are spread out in undulations over the mining tract of Shelve, are black schists, with occasional bands of earthy sandstone, in parts calcareous and light coloured. These form the chief mass of the Llandeilo formation, with its numerous graptolites, trilobites, mollusks, and other fossils, and will be treated of hereafter. (See section, p. 29.)

Miniature as it is by comparison with North Wales, this tract of the Stiper Stones and Shelve, which, with a description of its fossils, was formerly described and mapped as Lower Silurian, and which extends from Shropshire into Montgomeryshire, is, I repeat, now determined to be the geological prototype of the grander undulations of the same rocks, which spread out over so large a portion of the principality of Wales.

But whilst the strata in the escarpment of the Stiper Stones have afforded little more than casts of fucoids, and traces of graptolites*, certain districts in North Wales contain remains which mark more definitely the lowest known shelly band in the British Isles.

On the lower flanks both of Snowdon and Cader Idris, where they dip away from the above-mentioned subjacent greywacke grits and slates (the No. 1 of the section), these strata consist of schists, with glossy surfaces of lightish and dark grey and black colours.

Occasionally they are much laden with crystallised iron pyrites, and interstratified with felspathic ash-beds or volcanic grit, sometimes calcareous, which proceeded from submarine volcanoes. In this region, such eruptions, of which more notice will be taken in the sequel, have considerably expanded the dimensions of the whole series of beds, whilst here and there irregular and projecting bosses

---

* The Cruziana of the Welsh Lingula beds, p. 42, has also just been detected at the Stiper Stones by Mr. Salter.

of eruptive igneous rock, that have since broken through the strata, have greatly diversified the surface of the country.   Still, there are many spots near Harlech, Tremadoc, Llanberis, &c., where the most perfectly conformable and equable passage from lower unfossiliferous schist and grit to the beds under consideration are seen.   This, for example, is the succession at Barmouth.

b. Lingula Schists with imperfect slaty cleavage.
a. Grits and Schists, or Longmynd and bottom rocks.

The fossil remains of this primeval zone of life in N. Wales consist notably of the flat bivalve shell, the Lingula, of which examples are here and there to be seen, as on the flanks of Cader Idris and near Tremadoc.

It is worthy of remark, that the covering of this animal, as represented f. 1. of the following woodcut, is very horny, and only slightly calcareous; showing that its inhabitant was suited to the conditions of a sea whose bottom was composed of mud and sand, and which contained little or no lime wherewith to supply the fabric of a thicker shell.   It is, indeed, a remarkable fact, that in the vast thicknesses of the inferior or unfossiliferous greywacke before adverted to, there is scarcely any lime, and not a trace of a shell; whilst in these the lowest strata in which calcareous matter is visible, it has only occurred in such small quantities as to afford a very thin testaceous covering to the few animals of this early zone of life.   We shall presently see that when the Silurian fauna became abundant, as in the next overlying strata, it was accompanied by a corresponding development of lime which served as the pabulum for the construction of the shells of the imbedded mollusks.

In ascending through the different stages of the Silurian system, it will also be seen, that different species of the genus Lingula occurred most frequently at those intervals in which there was a return to similar sedimentary conditions, *i. e.* whenever the original, muddy sea-bottom was only slightly impregnated with lime.

Reverting to the original typical district of Shropshire, it must be remarked, that the beds immediately underlying the sandstones and quartz rocks of the Stiper Stones, and which stand in the place of the Lingula flags of Wales, are similar in composition to many strata which pervade the whole Silurian system. They are dark grey, fine-grained schists, occasionally containing a little mica, which either exfoliate in rude concretions, or split up into shivery and irregular fragments; the whole mass rapidly decomposing by atmospheric action, and returning into its original state of mud. It is this prevalent character of large portions of the Silurian rocks, particularly where they have not been intruded upon by igneous matter, or subjected to a slaty cleavage, which led me to apply the term " mudstones " so generally to them. And it is therefore specially to be remembered, that at the foot of the escarpment of the Stiper Stones, or at the base of every thing which contains Silurian fossils, this class of rock prevails.

In the mean time it is to be noted, that a form of Lingula nearly identical with that of North Wales, is also characteristic of the lowest Silurian rock of North America (the Potsdam sandstone). The genus has indeed lived on from the Silurian or primeval days to the present time, though its former associates, the graptolites and trilobites, vanished long ago from the world. The latter, or the earliest crustaceans, abound infinitely more in the Silurian than in the next overlying system of rocks, and of them the earliest which the labours of geologists have brought to light in Britain are the little Olenus and Paradoxides*, one species of the first-mentioned

---

* One species of Paradoxides, p. 43. Foss. 4. (locality unknown) has been found in North Wales, and, although very imperfect, Mr. Salter believes it identical with P. Forchhammeri, of the alum slates of Andrarum in Scania.

genus, O. micrurus, Foss. 2. f. 2., having been found associated with the Lingula in this primordial zone as it appears in Wales.

FOSSILS, 2.

1. Lingula Davisii, M'Coy.
2. Olenus micrurus, Salter.

3. Cruziana semiplicata, Salter.

Lingula Flags, North Wales.

In addition to the Lingula and Olenus, the government surveyors have lately detected a new crustacean of the Phyllopod tribe near Dolgelly, — Hymenocaris vermicauda, Salter. In beds of the same age near Bangor, there have been also found two kinds of fucoid, one of which is the Chondrites acutangulus, M'Coy, and a new species of the curious genus Cruziana (Bilobites, D'Orbigny). The above-mentioned fucoid, Chondrites, is the same species as Professor Sedgwick found in the Skiddaw slate of Cumberland, where it is associated with the Palæochorda, M'Coy, another genus of fucoids, and with the Graptolites latus. A figure is here given of the

FOSSILS, 3.

Hymenocaris vermicauda, Salter.

Lingula Flags, Dolgelly.

Hymenocaris vermicauda, from specimens in the collection of the Geological Survey. It occurs plentifully at Dolgelly and Tremadoc, in North Wales, associated with Lingula Davisii, and is therefore a true primordial fossil.

FOSSILS, 4.

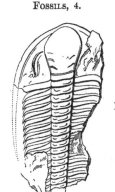

Paradoxides Forchhammeri, Angelin.?

Black Schists of North Wales.

The small Olenus is found in the lowest known fossil strata of another British tract. In the black schists on the western flanks of the Malvern Hills, strata which I had termed Lower Silurian, Professor Phillips detected three species of this genus, viz. O. humilis, W. 5. f. 1.; O. bisulcatus, f. 2.; O. scarabæoides (O. spinulosus, Ph.), f. 3. To these my friend Mr. Hugh Strickland has recently

FOSSILS, 5.

1. Olenus humilis, Ph.
2. O. bisulcatus, Ph.
3. O. scarabæoides, Wahl.?
4. Agnostus pisiformis, Brongn.

Trilobites from the black schists of the Malverns.

made the interesting addition of the Agnostus pisiformis, Brongn., Foss. 5. f. 4., a fossil only known in the oldest Silurian schists or alum slates of Sweden, of which hereafter. Associated with the Graptolite, the Olenus is also found in the same fossiliferous stratum of Scandinavia, which rests on a sandstone in which no other remains have been detected except fucoids, and is made up of the detritus of the subjacent crystalline rocks. Again, the Olenus occurs in Bohemia in the same or lowest horizon of life, and under similar conditions; and also associated, as in Sweden, with Agnostus, Paradoxides, and other trilobites, several of which have not yet been discovered traces in our island.* (See chap. 13.)

Let it not, however, be supposed, that fossil shelly remains occur in any notable quantity in this the lowest stratum in which they are known in Wales or England. On the contrary, the collections in the instructive public Museum of the Geological Survey teach us how very rare are the species which the labours of a series of years have been able to detect; and the student who may explore the North Welsh mountains, will duly appreciate the industry that has been required to gather together even these few primordial relics. The Lingulæ were first discovered by Mr. Davis in the year 1846, and termed by him Silurian fossils. † In Bohemia, Sweden, and America, as will hereafter be noticed, trilobites are, however, more plentiful in this zone.

In the following chapter the reader will be introduced to strata containing the chief mass of the Lower Silurian animals, some of which are similar in kind to the few crustaceans, mollusks, and graptolites already alluded to, but with very numerous additions of genera and species, in these, as well as in other classes; representing, in fact, nearly every known family of submarine beings except fishes.

---

* See the admirable work of M. Barrande, "Le Bassin Silurien de Bohême"— a monument of the talent and perseverance of the French author, in a foreign land, and one of the most remarkable monographs of our age. The first part only has appeared, in which the numerous trilobites are completely described.

† See Quart. Journ. Geol. Soc. Lond., vol. ii. p. 70. The rocks of that tract had, however, been described at a much earlier period by Professor Sedgwick, who, though he did not detect these fossils, has shown in a subsequent memoir the exact relations of these Tremadoc slates to the subjacent strata of Trawsfynydd and the overlying slates and igneous rocks of Festiniog and Moel-Wyn. Quart. Journ. Geol. Soc. Lond., vol. iii. p. 143.

CHAPTER III.

LOWER SILURIAN ROCKS — *continued.*

THE LLANDEILO FORMATION, ITS SLATES, SCHISTS, SANDSTONES, LIMESTONES, AND
INTERPOLATED IGNEOUS ROCKS. (2ᵃ AND 2 OF MAP AND COLOURED SECTIONS.)

THROUGHOUT the northern part of the Silurian region, and the
tracts of North Wales already mentioned, the rocks to be described
in this chapter have the same direction as the older schists and greywacke on which they rest. The direction of the Longmynd from
S. S.W. to N. N.E. has, in fact, been impressed on the whole of the
Llandeilo formation, the beds of which are perfectly conformable to and
graduate downwards into those of the older rocks. (See map and sect.
p. 29).

In the Silurian region, and in North Wales, as exhibited in the
general section below the map, 29, the Lingula zone, described in
the last chapter, is surmounted by thick masses of schist, 2, which,
with much interpolated igneous matter, pass up into and constitute
the lower part of the Llandeilo formation. These relations of superposition will be more clearly understood by referring to the sections
across the northern parts of the Silurian region, as formerly described
by myself.*

There, whether near Shelve in Shropshire, or at Llandrindod,
Llandegley, and Builth in Radnor, these strata are often interlaced
with conformable sheets of contemporaneous igneous matter, which
I termed volcanic grits.† They are also broken up by the intrusion

---

* See Silurian System, pp. 268. and 324. As early as the year 1833, I described
the Llandeilo formation, then termed black trilobite flagstones, as being of more
copious development than the overlying deposits. Proc. Geol. Soc. Lond., vol. i.
p. 476.

† The "volcanic ash" of Sir Henry de la Beche and the government surveyors.

of great bands of greenstone, porphyry, and other eruptive rocks.
In North Wales, the same deposits are not only to a greater extent
similarly affected by igneous matter, but are also, as before said, much
more slaty.

The mineral feature, however, by which the upper mass of the formation, whether slaty or not, is most distinguished from the strata
which preceded it, is in the occasional presence of lime, particularly
towards its central portion, where some of the schists, flagstones, and
shale become calcareous, and in parts pass into limestone, as at
Llandeilo in Carmarthenshire, Llampeter Felfrey in Pembrokeshire,
and Bala in Merionethshire.

In the lower or schistose parts, where very little calcareous matter
is visible, and nearly every bed is argillaceous, the fossils called
graptolites most prevail. This genus of zoophyte is supposed
by many naturalists to have been nearly allied to the living Vir-
gularia, a creature known only in deep seas. Others rather consider these extinct forms to belong to corallines or Sertularia. Be
this as it may, the geologist has observed that they are, for the most
part, found in beds of muddy origin, and *exclusively* in the Silurian
or earliest known system of fossil animal life. The whole group of

FOSSILS, 6.

LOWER SILURIAN GRAPTOLITES.

1. Rastrites peregrinus, Barr.
2. Graptolites Sedgwicki, Portl.
3. G. priodon, Bronn. (G. Ludensis, Sil. Syst.
4. Diplograpsus pristis, Hising.
5. D. nodosus, Harkness.
6. D. folium, Hising.
7. D. ramosus, Hall.
8. Didymograpsus sextans, Hall.
9. D. Murchisonæ Sil. Syst.

these little serrated creatures has been called Graptolithina, and has
been divided into about four genera, as represented in this woodcut.

One of these, the Diplograpsus, M'Coy, f. 4, 5, 6., is distinguished by having double sides like a leaf. A second, Graptolites of Linnæus, f. 2, 3., has teeth or serræ on one side only. A third, Rastrites, Barrande, f. 1., has teeth placed like the last, but not so crowded together. There are other branched forms, f. 8, 9., for which the name of Didymograpsus has been proposed, and one of which, the D. Murchisonæ, f. 9., is the most characteristic fossil of the black shales of the Llandeilo flags of Wales: it has also been found on the continent. Some of these forms are figured on our Pl. 12. f. 1.

The Diplograpsus, or doubly serrated forms, are chiefly characteristic of the Lower Silurian rocks; D. pristis, f. 8., and D. foliaceus, Pl. 12. f. 2., being the ordinary forms, whilst the one-sided species extend from the Lower to the Upper Silurian.

In no palæozoic rock younger than the Silurian is the true graptolite known, and hence, as types, they are most important to the practical geologist, who, in exploring many tracts of this age in Cumberland, the southern counties of Scotland, Ireland, Central Germany*, &c., meets with scarcely any other fossils. In Sweden, indeed, graptolites and fucoids so abound, as to have given a highly bituminous character to the lower strata, which, being also largely impregnated with iron pyrites, has afforded so much alum as to have procured for the strata the name of Alum slates. The graptolite also occurs copiously, in the Silurian region of Britain, and in Sweden, Norway, and Bohemia, in the shales of the Upper Silurian rocks.

Whilst, however, the mere presence of a true graptolite will at once decide that the rock in which it occurs is Silurian, it is only by finding the genera of these animals which display a double set

---

* Dr. Beck, of Copenhagen, assisted me in describing the few species of graptolite published in the Silurian System. M. Barrande has published a most elaborate and valuable treatise on the Graptolites of his Silurian Basin of Bohemia. Dr. Geinitz has systematised all the known forms of graptolites in a work, illustrated by clear and beautiful drawings, entiled, "Die Versteinerungen der Grauwacken Formation in Sachsen." Leipsic, 1852.

of serratures in the same species, that, in the absence of signs of the order of superposition, the field observer may presume he is examining the lower division of the system.

Deferring for a moment the consideration of the pseudo-volcanic or igneous rocks, which in some districts characterize the schists last spoken of, let us first take a glance at the great mass of the Lower Silurian rocks, as distinguished by very many organic forms, most of which will be treated of hereafter. (See chap. 8. and plates 1. to 11.) This great group was long ago described in the Silurian region, as made up of Llandeilo schists and flags beneath, and of Caradoc sandstone above; the names having been taken from the localities where each formation was, as it was thought, most characteristically developed. But, even in 1835, I specially cautioned geologists not to attach importance to the character implied by the names of flagstone or sandstone, for I well knew that such mineral features were partial and evanescent even in my own region, and must necessarily prove much more so in distant countries. The survey of Siluria alone had taught me, that the schistose and slaty beds of one tract became sandy in their extension, or vice versâ, and that the same courses which are so calcareous in some districts as to constitute limestone, were mere schists and flagstones in another. Hence I held to the rule, that in studying the primeval, as well as medieval or secondary and tertiary formations, the zoological contents of rocks, when coupled with their order of superposition, were the only safe criteria of their age.

In proceeding from the primordial zone described in the last chapter, to younger formations, we find that, in the British Isles at least, calcareous matter, doubtless formerly supplied from subterranean sources, increases upwards, and with it the number of fossil animals.*

* The observation in the text is only intended to apply to the number of animal remains found in the lowest formation in which there is any notable quantity of lime, by comparison with the subjacent strata where it is not perceptible. The fact, however, of the increase of limestone as the formations become newer, is generally true. Thus, no course of the limestone near Bala exceeds twenty or

At and near Llandeilo, the formation under review chiefly consists of calcareous flags, having a great thickness of schists below them, and sandy and gritty strata above them; but in following the deposit into Pembrokeshire, the calcareous portion is enriched, and becomes in some places, as at Llampeter Felfrey and Llandewi Felfrey, a copious mass of dark subcrystalline limestone with white veins. These rocks, which are there, as elsewhere, underlaid and overlaid by a great thickness of black schist, are thus represented in a section of the government surveyors, little differing from one I formerly published.

SECTION AT LLANDEWI FELFREY, PEMBROKESHIRE.

S.S.E.   N.N.W.

d c b^2 b^2 b b^2

b. Llandeilo schists and limestones, b^2.   *c.* Overlying Silurian rocks.   *d.* Old Red Sandstone. The contiguity of the Llandeilo formation to the Old Red Sandstone, and the tenuity of all the interjacent Silurian strata, will be explained hereafter.

Again, the very same formation as developed in Radnorshire (Builth to the Gelli Hills near Llandrindod), is chiefly schistose and flaggy, with comparatively little lime disseminated, and such also is its composition in the districts of Corndon and Shelve in Shropshire. The distribution of the limestone is equally irregular and uncertain in those tracts of North Wales which Professor Sedgwick so long explored. The limestone, for example, is visible near Bala, though, from its impure character, it is now no longer worked for use. But on its strike to the N. N. E. or to the S. S. W., it soon dwindles to so insignificant a course, that its position can only be detected by the

thirty feet in thickness, and even the same strata in Carmarthen and Pembroke are not more than double those dimensions. The Woolhope, or lowest Wenlock limestone (see chapter 6.), has, probably, about a maximum thickness of twenty-five feet. The chief Wenlock limestone is, in some places, more than a hundred feet thick, and the limestones of South Devon are, perhaps, twice that magnitude; whilst the carboniferous or mountain limestone is vastly thicker than any of the older calcareous rocks of Britain.

E

occasional presence of some of its fossils. A reference to the map
at once indicates how this limestone wedges out and disappears a
few miles to the south of Bala, thus leaving all the region between
North Wales and Carmarthenshire void of any such rock.

But besides this Llandeilo limestone, other shreds of calcareous
matter are discernible at intervals in the ascending order. One of these,
or the Hirnant limestone of Merioneth, a still thinner and more
partial course described by Professor Sedgwick, is, probably, on the
parallel with certain calcareous bands described by me as occurring
near Meifod in Montgomeryshire, and mapped as Llandeilo rocks by
the government geologists. To the north-west of Meifod these sandstones and schists, with partial limestones, seemed to me to repose on
flagstones with the Asaphus tyrannus, as exhibited on the banks of
the river Twrch, and described in my old sections. (See Sil. Syst.,
p. 307. and pl. 32. f. 9.)

From their mineral character, I formerly connected the Meifod
rocks with the next overlying formation. Their separation from the
Caradoc depends upon the researches of Professor Sedgwick and those
of the government surveyors. In some parts of Wales, and in the
Longmynd and Shelve districts, these rocks are unconformable to
the Caradoc sandstone. But the researches of the surveyors have
also shown, that this physical phenomenon of unconformity between these two divisions of Lower Silurian strata intimately connected by their organic remains is local, and that there are tracts
in North Wales, and even in the vicinity of Bala itself, where the
Llandeilo formation is conformably surmounted by the Caradoc sandstone, as represented in the coloured section appended to the map.
See also the diagrams in chapter 4.

In truth, Sir H. De la Beche and his associates have proceeded
on the axiom, that local unconformity of stratification does not constitute a ground for the establishment of a line of *geological* distinction and classification. They have even ascertained that in

South Wales, one portion of the carboniferous series is as uncon-formable to another and older part of the same coal-field, as the example which has been alluded to between the Llandeilo and the Caradoc of the Lower Silurian rocks of certain districts in North Wales and Shropshire.

In France, as in Bohemia (see chapters 13 and 14.) and elsewhere in foreign countries, there is always a perfect conformity between the lowest Silurian strata, and those which stand in the place of our Caradoc sandstone. The only great break which is there known occurs between the base of everything fossiliferous, and those great azoic masses of greywacke which represent the bottom rock, of the preceding chapters. In Britain, on the contrary, as already demon-strated, that fundamental rock is the regular and conformable base of all the Silurian strata. It is, indeed, manifest that the break or overlap in Wales, and along the south end of the Longmynd and Corndon tracts of Shropshire, is a local phenomenon.

I adhere, then, to my old divisions, under a belief, confirmed by subsequent extensive survey, that in many regions of the earth the geologist will find it impossible to classify by the means of such smaller divisions, and that he must simply view the whole as " Lower Silurian Rocks." Thus, however certain distinctions may be clear and good in typical tracts of Siluria and Wales, there are many parts even of the British Isles, and numberless foreign localities, where separation into the formations of Llandeilo and Caradoc is absolutely impracticable, either by mineral and zoological distinctions, or by dislocations.

In my original work it was stated that the so called Llandeilo flags and schists were often underlaid, or their place taken, by sandstones of considerable thickness. Such, for example, are the sandstones of the Stiper Stones, which as before stated (p. 29.), and as represented in the original sections, lie beneath the great mass of the strata now under consideration, and were from the first called " Lower Silurian."

On the other hand, the fact now generally admitted was always insisted on, that, however the mineral character might vary, very many of these strata were characterized by similar fossils, particularly by some of the commonest Lower Silurian types. In short, two groups or divisions only were dwelt upon, for the general scale of comparison, and these were Lower and Upper Silurian. (See Sil. Syst. *passim*.)

In tracts where it is unassociated with trap or igneous rocks, the Llandeilo formation above the Lingula schists consists thus in ascending order, first, of the dark schists with graptolites; next, of hard dark-coloured flags, sometimes slightly micaceous, often calcareous, with white veins of crystallized carbonate of lime, and which, here and there forming limestones, are specially distinguished by containing the large trilobites, Asaphus tyrannus, Ogygia Buchii, with several species of Trinuclei, as well as numerous shells, many of which are represented in the plates 1 to 10. The upper member is composed of strata of shale, schist, and sandstones, occasionally even conglomerates, in other tracts containing thin courses of calcareous or shelly sandstone. In the neighbourhood of Llandeilo, in Carmarthenshire, where, from the abundance of the above-mentioned fossils, the name was applied, the formation, consisting of schists, limestones, and sandstones, has been estimated by the government surveyors to have a thickness of about 4000 feet. In North Wales, where the rocks are much expanded by the intervention of igneous materials, the dimensions are vastly greater, and even to the west of the Longmynd and Stiper Stones this formation has a thickness of several thousand feet.

Igneous Rocks associated with the Lower Silurian.—In order to make the reader acquainted with the general aspect and nature of the British Lower Silurian rocks, it is essential to explain how in certain districts volcanic action affected the bottom of the sea during the deposition of the strata, and how they were dislocated by subsequent eruptions. In the vicinity of some igneous rocks, these schists and calcareous flagstones have been filled with mineral veins, and have undergone various changes. Besides ores of lead and copper, it will also be shown, in a separate chapter, that rocks of this age were rendered partially auriferous in Wales, and largely so in other countries. (See chap. 16.)

One of the tracts in the original Silurian region, where the Llandeilo formation is most metalliferous (*lead veins*), is that lofty ond rugged district of Shropshire, which lies around the village of Shelve and the Corndon mountain, and which extends from the west of the Stiper Stones into Montgomeryshire. (See Map.)

The student of natural phenomena is specially invited to examine this tract, because, together with the adjacent Longmynd and the Stiper Stones, it exhibits nearly the whole series of strata of North Wales in lower hills, and is therefore more easily accessible and comprehensible than the subcrystalline, contorted, and steep flanks of Snowdon and Cader Idris, which exhibit similar phenomena on a grand scale.

In reference to North Wales, I will merely say, that some of the finest examples of the stratified and contemporaneous beds of felspathic ash or volcanic grit are exhibited in the environs of Festiniog. There, the interval between the green and grey schists, of the age of the Lingula flags of Tremadoc, and certain black roofing slates of the Llandeilo formation, is occupied by syenite, satiny schists, and volcanic grits, the last being in parts coarse conglomerates. Again, to the west of Tan-y-bwlch, the latter in the form of ash-beds, which there are very finely levigated, differ chiefly from talc schist in being less quartzose, associating and alternating with hard greywacke grit, which is penetrated by large branching veins of white quartz.

Occasionally, this felspathic ash, which is separated into thick beds by slightly inclined layers of deposit, is traversed by numerous highly inclined laminæ of slaty cleavage, from which the associated coarse grits are exempt.

The repeated alternations of these pseudo-volcanic rocks with the schists and greywacke of North Wales can alone be well understood by reference to the maps and sections prepared by Ramsay and Selwyn. In the small map attached to this work, and which, as before explained, is simply a reduction of that of the government surveyors, numberless features are necessarily omitted.

An adequate acquaintance with such features can only be attained by traversing those mountains with the originals of these remarkable illustrations in hand. There, in the space of a few square miles, the geologist meets with numerous grand alternations of interstratified volcanic ashes and felspathic conglomerates, sometimes sandy and calcareous, which have been pierced by eruptive igneous rocks, whether syenite, felspar rock, or greenstone, here and there porphyritic. It is sometimes most difficult to separate the contemporaneous igneous rocks from the subsequently intruded masses (porphyries of Sedgwick); as, for example, at Cader Idris, where the succession is represented in a diagram, also taken from the sections of the Survey, which is given towards the conclusion of this chapter.

Then, again, let the student take any one of the closely worked quarter-sheet maps of this Survey (particularly those of Snowdon and Cader Idris), and, marking the faults or fractures as delineated by the great number of white lines, he may form some idea of the consummate skill and labour by which such results have been obtained.*

But to return to similar Silurian rocks which are seen in the tracts of Shelve, Llandrindod, and Builth, where they were first classified and placed in distinct relation to other deposits, and with which some of the copious masses of North Wales have been

* A fine example of this succession to the north of Festiniog was recently pointed out to M. de Verneuil and myself by Professor Ramsay.

identified by the government surveyors. Those districts were long ago shown by me to contain two classes of submarine igneous rocks; the one bedded and regularly interlaced with the strata, and therefore formed contemporaneously with them; the other not bedded, and which had broken through the others. Each of these is similarily divided into several varieties, for an acquaintance with which the reader must consult the original descriptions.*

Of the bedded pseudo-volcanic rocks in Shropshire and Montgomeryshire, some of the finest examples are to be seen in the picturesque defile of Marrington Dingle. There, the lowest strata, exposed at a high inclination in numerous quarries, are felspar rocks of a concretionary structure, which are surmounted by coarse mottled volcanic grits, consisting of a base of lightish and greenish grey granular felspar, mixed with sand and chlorite, and containing angular fragments of schists and porphyritic greenstones, the whole arranged in beds from two to four feet thick, and dipping at an angle of 45°, as expressed in this vignette.

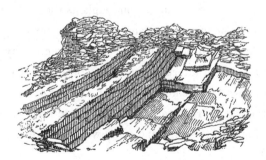

WHITTERY QUARRIES IN MARRINGTON DINGLE.
(From Sil. Syst., p. 270.)

In other parts of this district there are both felspathic agglomerates and ash-beds, or volcanic grits and slaty porphyries, with crystals of felspar. Some of these alternate in thick masses with the schist containing trilobites, others in courses of a few inches thick

* See Sil. Syst., pp. 269. 324. and *passim.*

only, and occasionally including fragments of the Ogygia (Asaphus) Buchii. Other organic remains are found in beds which are almost exclusively composed of igneous materials; thus showing that submarine volcanic action was rife in the sea-bottom in which these Lower Silurian strata were accumulated. The annexed woodcuts will convey some idea of the manner in which the volcanic breccia or ash, *b*, which alternates with the beds of schist, *a*, envelopes other fragments of schist and slate, *c*.

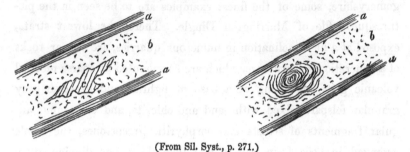

(From Sil. Syst., p. 271.)

Another diagram, taken from the coloured sections on the margin of my old map of the Silurian region, explains the manner in which, by the action of submarine volcanoes, such igneous dejections are supposed to have been accumulated, as represented by the light-coloured layers which alternate with the ordinary sediments, and further, how a volcanic cone, similar to that of Graham Island in the Mediterranean, as represented by the dotted line, may have disappeared and mixed its scoriæ and ashes with the ordinary submarine mud or sand.

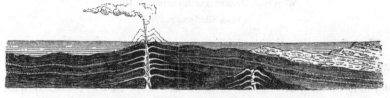

IDEAL REPRESENTATION OF THE MANNER IN WHICH SUBMARINE VOLCANIC DEJECTIONS WERE FORMED IN THE LOWER SILURIAN PERIOD.

(From bottom of Map, Sil. Syst.)

Some large masses of eruptive trap, which have penetrated the sedimentary and associated volcanic deposits at a subsequent period, are seen in the Corndon Hills as expressed in this view, the undulating grounds around them being composed of strata of the Llandeilo formation.

LOWER SILURIAN AND BEDDED TRAP, TRAVERSED BY ERUPTIVE ROCKS, THE CORNDON MOUNTAIN IN THE DISTANCE.
(From Sil. Syst., p. 271.)

Equally instructive examples of the alternations of bedded igneous rocks with Lower Silurian strata, and of posterior eruptions, are seen in the hills of Gelli, Gilwern, and Carneddau, near Llandegley, Llandrindod, and Builth in Radnorshire. Felspar porphyries are there regularly stratified, one example of which, as taken from the illustrations of my former researches, is here exhibited.

ALTERNATIONS OF LLANDEILO FLAGS AND SCHISTS WITH VOLCANIC GRITS.
(From Sil. Syst., p. 325.)

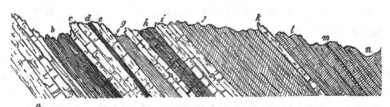

a. Coarse slaty felspar rock, slightly porphyritic and amygdaloidal, containing elongated concretions of green earth. This rock is regularly stratified in beds of three to four feet thick, and forms the mass of the hill, rising into the higher ground. *b.* Finely laminated greenish-

grey sandy flagstone, apparently hardened near the top. *c.* Fine-grained granular felspar
and courses of clay-stone, some of which are used as oven-stone. *d.* Altered flags, having
a conchoidal fracture, in parts almost Lydian-stone, with crystals of iron pyrites. *e.* Grey
felspar rock, the laminæ of deposit marked by feruginous streaks, probably due to the decomposition of some other mineral. *f.* Black shivery shale, containing a few concretions of
argillaceous limestone, with veins of calcareous spar. One of these, which fell under my notice,
was a septarium, two or three feet in diameter, containing many impressions of Graptolites.
This band of black shale was excavated to a considerable distance in search of *coal!* on the
strike of the beds, by the same individuals who have sought for lead ore. *g.* Hard thick-
bedded porphyritic felspar. *h.* Flagstone, with Ogygia Buchii, indurated in contact with
the trap; much iron pyrites. *i.* Grey porphyritic clay-stone. *j.* Black schist, with some
hard stone bands, in parts pyritous. *k.* Slaty porphyritic felspar. *l.* Black shale, with stone
bands and concretions of argillaceous limestone. *m.* Thin band of decomposed granular
felspar, weathering to a rusty colour, and looking like a coarse oolite. *n.* Black pyritous
schists, much contorted near the mouth of the gallery.

Thus, in a thickness of only 350 feet across the beds, this detailed
section exhibits bands of felspathic and porphyritic rocks, occasionally
both crystalline and amygdaloidal, with clay-stones, granular felspar,
and green earth; which materials alternate, six or seven times, with
finely laminated flagstones and black schists, one of which contains
graptolites, and another the Ogygia Buchii. (Pl. 3. f. 1.) In truth,
this little section exhibits an epitome of the structure of vast tracts
in North Wales.

Examples of such contemporaneous rocks, as well as those of
posterior intrusion, are well seen in Salop, Montgomery, and Radnor. In and around the Corndon mountain of Montgomeryshire,
(see view, p. 57.), in the volcanized and mining tracts of Shelve, the
eruptive and intrusive rocks are chiefly coarse-grained hornblendic
greenstones and felspar rocks passing into basalt. In many of these
cases, the shale or schist in contact with the eruptive rock has been
cemented into a complete porcellanite, with surfaces as smooth as the
finest lithographic stone. These altered shales are, in fact, the
" brand-erde," or burnt earth, of the Germans.*

* In England the most curious *actual* proofs of this conversion occur in the
South Staffordshire coal-field, particularly near the town of Dudley, where the
spontaneous and long-continued combustion of subterranean coal in abandoned
mines has produced in the shale and sandstones of the coal measures a variety
of burnt earths, which are of divers colours ; some of them resembling the riband
jasper of mineralogists.

The Breidden Hills, including the picturesque Moel-y-Golfa, on the right bank of the Severn, near Welsh Pool, also exhibit similar illustrations, both of contemporaneously bedded trap and of posterior or intrusive rocks, which have broken out along the same line at different periods, for a better acquaintance with which I must refer to former detailed descriptions.*

The annexed drawing, taken from the terrace of Powis Castle, shows these hills in the distance. They mark a line of eruption that separates the Lower Silurian rocks on the left hand or west, from the Upper Silurian of the Long Mountain on the right hand or east.

Rodney's Pillar.   Moel-y-Golfa.

VIEW OF THE BREIDDEN HILLS NEAR WELSHPOOL, FROM POWIS CASTLE.
(From a Drawing by Lady Murchison, Sil. Syst., p. 302.)

Now, it has been demonstrated that along this line of eruption some volcanic dejections were thrown down on the sea-bottom during the formation of the Lower Silurian strata; that other eruptions and dislocations afterwards took place along it, heaving those strata into highly inclined positions, like those on which Powis Castle stands;

* Silurian System, p. 287. *et seq.*

that on the edges of these beds carboniferous deposits were afterwards accumulated; and that again, after all these and other formations, including the new red sandstone, were completed, still more
recent eruptions occurred along this same line of fissure.*

The shortest method of explaining to the reader the structure of
the Breidden Hills is to exhibit this small diagram, taken from a
section just published by the Geological Survey, and which represents
a mass of porphyritic and amygdaloidal greenstone, which in its
protrusion has carried up included portions of slaty beds, and has
thrown off pebble beds and Upper Silurian rocks to the S.E., and
Lower Silurian to the N.W.

SECTION ACROSS THE BREIDDEN HILLS.

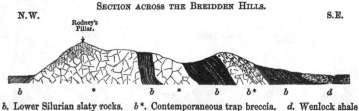

b. Lower Silurian slaty rocks.   *b*.* Contemporaneous trap breccia.   *d.* Wenlock shale
or base of Upper Silurian.   **.* Eruptive rocks.

One of the igneous rocks most easily examined on this line of
habitual outburst in ancient periods, is the largely excavated prismatic
mass at Welsh Pool, which seems to cut as a dike through the
Lower Silurian strata, which in contact with the igneous matter are
much altered. At a little distance, however, these beds contain organic
remains, some of which, particularly the Trinucleus concentricus,
were originally figured in the "Silurian System." (See pl. 4. f. 2—5.)

This prismatic rock, unquestionably of igneous origin, is largely
quarried as a building stone, like that of the Whittery quarries. It
is remarkable in having its light-green felspar matrix speckled with
minute nuts and kernels of calcareous spar, which, together with the
finely aggregated state of the felspar, renders the stone sectile and
valuable.

* For the proofs of this, see Sil. Syst., pp. 299. *et seq.*

Eruptive masses of greenstone and hypersthene rocks, described as cutting through Silurian strata, occur near Old Radnor *; but as these are associated with strata of posterior date to those we are now considering, they will be mentioned in the sixth chapter.

The Carneddau Hills, between Llandrindod and Builth, are, however, truly amorphous and eruptive masses of igneous rock, which have broken through and highly altered the Llandeilo flags and associated sandstones. Beautiful examples of this are seen on the banks of the Wye, west of Builth. (See Sil. Syst., p. 332. *et seq.*)

On this occasion one diagram only is selected, as taken from one of the coloured plates in my large work, to show, at one view, first,

SECTION ACROSS THE GELLI HILLS.
(From Sil. Syst., pl. 33. f. 5.)

*b**. Lower Silurian with interpolated contemporaneous trap.   *. Eruptive Rocks.

how the strata of the Llandeilo formation alternate with bedded igneous rocks, *b* *, and next, how they have been altered in contact with eruptive masses, *. It is in these black and hardened schists of contact that films of anthracite have been found, which have led credulous farmers to search for coal.†

Another frequent result, in this district, of the contact of the eruptive rock with the schists, is the production of much sulphuret of iron, the decomposition of which gives rise to the various mineral

* Sil. Syst., p. 318.

† A full exposure of this folly was given in the Silurian System, p. 328., with this diagram, representing an actual search after coal! at a spot called Tin-y-Coed, thus —

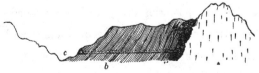

b, Lower Silurian schists traversed by a gallery, *c*, until the eruptive trap rock, *, was reached, where the enterprise was necessarily stopped.

sources of Llandegley, Landrindod, Builth, Llanwrtyd, &c., which
gush out around this volcanized tract.

Lower Silurian of Pembroke and Carmarthen.—Although some
allusions have been made, in the preceding pages, to parts of South
Wales, and to the general agreement of its structure with that of
North Wales, it is necessary to explain, particularly to all those
who may have consulted the " Silurian System," the general views
which I have long ago adopted, and which, though well known
to English geologists during the last ten years, may still be little
understood by foreigners and the public into whose hands the original
work may have fallen.

Whilst North Wales and the northern parts of Siluria have been
described as consisting of troughs, or undulating masses of Lower
Silurian rocks, resting to the west upon the unfossiliferous greywacke
of Carnarvon, and to the east on similar rocks of the Longmynd,
the same sedimentary base or bottom rock is again visible in South
Wales, but at one spot only. This is at St. David's, in Pembrokeshire, where the oldest greywacke, like the rocks of the Longmynd,
Llanberis, and Haerlech, is an intractable, hard, fine-grained, purple
and grey sandstone, with slaty schists, in which no organic remains
have been found.

In North Pembrokeshire, these bottom rocks are followed by
Lower Silurian strata, but with much irregularity, owing to the
association of vast masses of igneous rocks, which are interpolated in
a broken series of quartzose and felspathic conglomerates, grits, sandstones, and slates. Among the older igneous rocks of contemporaneous origin, are thick-bedded coarse conglomerates, and among
those of intrusive and posterior date, are large-grained greenstones
and compact felspar rocks, for a better acquaintance with which the
reader must consult other writings as well as my own.*

* See Kidd. Trans. Geol. Soc., old series, vol. ii. p. 791.; De la Beche, Trans.
Geol. Soc., new series, vol. ii. p. 1.; and Silurian System, p. 401.

With the exception then of the old greywacke of St. David's, all these rocks of sedimentary origin in North Pembroke, which were formerly termed Cambrian, have been found by the government surveyors to be, as in North Wales, mere undulations or expansions of what had been described as Lower Silurian strata in South Pembroke and Carmarthen. Many years have indeed elapsed since I publicly announced my belief in this result of the valuable labours of those geologists.*

Wherever, therefore (with the above exception), the reader of the original " Silurian System " meets with descriptions of schistose, slaty, or quartzose rocks of South Wales, which were called Cambrian, he must now consider that they merely refer to mineralogical variations of undulating strata of the Lower Silurian age. Of these, there are in Pembrokeshire, some sandstones and schists, which are on the parallel of the Lingula beds of North Wales, though the few fossils of that age have not yet been found in them any more than in Shropshire. And in truth, when the geologist examines North Pembroke (where alone in South Wales the bottom rocks are exposed), he will not be surprised that even the sedulous fossil collectors of the geological survey should have failed in procuring from such materials many traces of primordial life ; so much are the strata buried amidst masses of igneous rocks, both contemporaneous and eruptive, and so

* In the "Silurian System" it was indeed stated, as an apology (Notes, pp. 399. and 402.), that I had given little attention to North Pembroke, to which I only made one rapid excursion. Owing to its highly mineralized aspect, the vast profusion of associated igneous matter, and their highly crystalline nature, all these rocks were, at that day, considered to belong to a system inferior to the Silurian. That erroneous view was publicly renounced by me, many years ago, in two addresses to the Geological Society of London. I also endeavoured to give a wider currency to my views, by publishing, 1843, a general map of England and Wales, at the request of the Society for the Diffusion of Useful Knowledge. I have, since then, twice visited North Wales, and had all my views confirmed, by inspecting the ground with Professor Ramsay and Mr. Selwyn, in the company, on the one occasion, of Professor Nicol, on the other, of M. de Verneuil.

slaty and crystalline is their structure, accompanied, further, by
an absence of all calcareous matter.*

The lowest beds of that region in which fossils are visible are the
black schists with graptolites, formerly described by me either as the
base of the Silurian system, or as forming a downward passage into
what was then termed Cambrian, under the erroneous supposition
of the inferiority of position of another great mass or system with
which the schists were connected.

The graptolite schists are well exposed at Abereiddy Bay, where
the surfaces of slaty cleavage, as formerly pointed out, coincide in
one part absolutely with the laminæ along which the graptolites,
chiefly the G. Murchisoni (see pl. 12. f. 1.), have been deposited
as exhibited in this woodcut.

LOWER SILURIAN ROCKS IN ABEREIDDY BAY.
(From Sil. Syst., p. 400.)

Coincidence of Slaty Cleavage and Bedding.

In the Pembrokeshire section, the next strata in ascending order,
or the calcareous flags of Llandeilo, as defined by their trilobites
and other fossils, are rarely to be detected amidst the great intrusions
of igneous matter, and the enormous breaks, faults, and denudations,
by which the whole system has been obscured in this region. Some
depressions being filled with overlying old red sandstone and carboniferous limestone (of which hereafter), and others with highly
dislocated troughs of broken stone coal, or culm, the Llandeilo flags
are only visible in one locality, along the whole of the rugged coast

---

* See the tracts between the Precelly slates and Strumble Head, west of
Fishguard.

cliffs of this southern extremity of Wales, and just in the place where
I first described them.

This is at Musclewick Haven in St. Bride's Bay, where black
schists, in parts sandy, are traversed by dingy veins of black calcareous spar, and contain the usual trilobites (Asaphi and Trinuclei).*
The beds there dip at about 35° to the S. S.E., which inclination
would convey them under the great mass of the fossil-bearing
Silurian rocks, as exhibited in Marloes Bay. There is, however, no
visible connection, as the intervening promontory is much overlaid
by intrusive igneous rock. This insulated mass of Llandeilo schists
is, indeed, flanked by enormous faults or dislocations, by which, on
their north side, these Lower Silurian strata are thus brought
abruptly into contact with the old red sandstone.†

LLANDEILO SCHISTS IN MUSCLEWICK BAY.

(From Sil. Syst., pl. 35. f. 11.)

o. Old Red Sandstone.   b. Llandeilo Schists.   * Eruptive Rocks.

In tracing an ascending order in South Pembroke, the inferior
strata are seen to occupy the isles of Skomer, where, associated
with huge bands of trap rock, they plunge rapidly to the S.E., as
represented in the next drawing taken from Wooltack Park and the
mainland, and then dip under all the strata of Marloes Bay.

In pursuing the ascending order into Marloes Bay on the south-east,
we first meet with other bands of slaty greywacke, which also alternate
with igneous rocks, that seem to have cut out the schists with the fossils of the Llandeilo flags. For, we next pass over the inclined edges

* Silurian System, p. 394.
† Sir Henry de La Beche estimates this fault between the Llandeilo flags and
the old red sandstone, as an upcast of upwards of 2000 feet.

S.E.   N.W.

VIEW OF SKOMER ISLES, SEEN FROM THE MAINLAND.
(From Sil. Syst., p. 389.)

of a great thickness of quartzose sandstones, in which, though also
associated with much eruptive trap, remains of encrinites and other
fossils have been found; the whole dipping, like the inferior strata
of Skomer, to the S.E.   Then follow schists having a slaty
cleavage.   All these are overlaid in the cliffs (at Marloes Mire,
Rain Rock, and Gateholm), as expressed in this little view, by

Hook Point.

SILURIAN ROCKS OF MARLOES BAY, PEMBROKE, DIPPING UNDER THE OLD RED
SANDSTONE OF HOOK POINT.
(From Sil. Syst., p. 392.)

other sandy schists and sandstones of considerable dimensions. From
their position and mineral character, and not having found many
fossils, a large portion of the lower beds was classed by me with the
Caradoc sandstone; but in most of these the government surveyors

have since collected fossils which rather refer them to the bottom of the Wenlock, and consequently to the Upper Silurian rocks.

Such are the rocks represented in the foreground of the last vignette, and which continue conformably, layer over layer, to the dark-shaded headlands in the middle of the bay, where they plunge, as expressed, at a high angle, under other strata representing the mass of the Upper Silurian rocks, or Wenlock and Ludlow formations, which in their turn are surmounted by the old red sandstone of Hook Point. Reverting hereafter to the great discrepancy in lithological conditions which is here apparent in these Upper Silurian rocks when compared with the typical strata of the same age in Salop and Hereford, attention is specially called to this section, the only one in the British Isles, where a great succession of Silurian rocks is seen in clear relations to the old red sandstone on the sea-coast. We thus learn how difficult it is, even in these fully exposed coast cliffs, to distinguish the upper portion of the Lower Silurian from the lower part of the superior group by the structure of the rocks, and we are, therefore, compelled to revert to their fossil contents.

Here is a vast thickness of sandstone and sandy schists, which, dipping under strata of shale and calcareous beds with many fossils of the Wenlock limestone, has its surface marked by numerous brown impressions of the stems of encrinites (a character very common in the true Caradoc sandstone). These beds are also affected by a slaty cleavage, giving to them a very antique impress.

In truth, they are lithologically and stratigraphically the Caradoc of the Lower Silurian, to be described in the next chapter; whilst the palæontologist must range them with the base of the Upper Silurian on account of a prevalence of certain forms. And thus it ever must be in strata so intimately connected as the lower and upper members of this natural system, wherever such clear divisionary features are wanting as those to which attention will be called in the sequel. In fact, the same slaty impress, which affects all the lower strata in Marloes Bay, is continued upwards through shales and schists which represent the Upper Silurian rocks; nearly the whole of the Silurian system, up to its beds of passage into the old red sandstone, being here exhibited in the space of less than three miles.

The Lower Silurian rocks of Pembroke, composed in some parts of sandstones, in others of schists and flagstones, are much more calcareous as they trend eastward and north-eastward into Carmarthenshire. Rolling out in undulations from the complicated, disturbed, and slaty masses of North Pembroke, these strata become much more simple on a zone ranging from Haverfordwest by Sholeshook and the north of Narbeth to Llandewi Felfrey and Llampeter Felfrey, where the calcareous matter* so increases, that it forms, as before said, by far the finest masses of Llandeilo limestone in all Wales, as in this woodcut, which is here reproduced from p. 49.

SECTION AT LLANDEWI FELFREY, PEMBROKESHIRE.

S. S. E.   N. N. W.

d   c   b

d. Old red sandstone.   c. Thin zone of Upper Silurian rocks.
b. Llandeilo limestone, with underlying and overlying schists.

There the beds, as originally described, are charged with the prevailing fossils of the Llandeilo formation, including Ogygia Buchii, pl. 3. f. 1.; Asaphus tyrannus, pl. 1.; Leptæna sericea, pl. 6. f. 13.; Orthis elegantula, pl. 6. f. 5.; with the coral Halysites catenulatus, and also the Echinosphærites aurantium and E. Balticus, of the lower limestones of Sweden and Russia. (See chap. 5.)

When the limestone thins out, its position in the strata of schist and flag is often marked by bands of concretions only, and in other places by flagstones more or less calcareous. The latter type, or that

---

* Silurian System, p. 397.

of calcareous flagstones, prevails around the town of Llandeilo. There, and in other parts of Carmarthenshire, the Lower Silurian rocks are, on the whole, much less interfered with or expanded by igneous rocks than in Pembrokeshire or North Wales. Hence, though they do contain some small subordinate bands of greenstone and felspar rocks, analogous to their larger types elsewhere, and are thrown into undulations, their relations are more clearly exposed in a small compass, thus:—

VIEW FROM DYNEVOR PARK, LLANDEILO, LOOKING TO THE HILLS ABOVE GOLDEN GROVE.
(From Sil. Syst., p. 347.)

Here, all the Silurian rocks, from the Llandeilo flags up to the Ludlow rocks, are exhibited in the vale of the Towy or on the banks of that river; the old red sandstone forming the distant hills.

In reference to my original opinions, I beg the reader to compare the coloured sections in the "Silurian System," pl. 34., with those since published by the Government Survey, to see how they both represent the same succession of conformable strata of Lower and Upper Silurian rocks in the environs of Llandeilo.

Among the lowest visible strata of these tracts, are schists with an imperfect slaty cleavage, which contain irregular elliptic masses of silicious grit*, made up of pebbles of quartz and fragments of older greywacke slate; the former being seen in some places to have indented the latter as if they had been subjected to an enormous pressure.

It is in the lower schistose beds of these rocks, that the remains of annelids or red-blooded sea-worms, now so generally found in the Silurian rocks of various countries, were first discovered; and which were described for me by Mr. Macleay.† Occasionally, as between Llandeilo and Llandovery, the upper members of these slaty schists are also marked by other silicious grits, which, from their perfect conformity to the masses of Upper Silurian rocks above them, and to the calcareous flags and schists of Llandeilo beneath, I naturally considered to be thin and partial representatives of the intermediate Caradoc sandstone. These beds are now classed by the surveyors with the Llandeilo formation.

In its prolongation from Carmarthenshire and Cardiganshire, into Brecknockshire, this slaty greywacke group is again distinguished by included quartzose greywacke sandstones, which in Radnorshire swell out into important and continuous mountain masses.

Not unfrequent to the east of Aberystwith, these sandstones become of considerable thickness near Rhayader on the Wye, where they contain subordinate conglomerates, some of the finest examples of which are seen in the deep gorge of the river Elain, a tributary of the Wye (Craig-y-voel).

Similar rocks, whether silicious grits, pebbly conglomerates or hard intractable sandstones, range to the north-west, and occupy the ridges of Gwastaden, Rhiw Gwraidd, Dolfan, Devanner, and the Maen

* See Silurian System, vol. i. p. 359, 360. A similar phenomenon has been observed in other countries, as if the pebbles of quartz had indented each other by pressure and friction.

† See Sil. Syst., p. 699.

rocks in Radnor, and Bryn Din and Moel-ben-twrch in Montgomery. Owing to great dislocations on their eastern frontier, they are occasionally placed in unconformable relations to the Upper Silurian rocks of that region.*

To the west, however, they roll over in great undulations, on certain parallels, as laid down on the maps and horizontal sections of the government surveyors (some of the arenaceous masses being not less than 1000 feet thick), and with a strike which in South Wales is usually N. E. and S. W., the dips being both to the S. E. and N. W., but more frequently in the latter direction. In my former examination, I only detected casts of encrinites in these sandstones, but the government surveyors have found a sufficient number of forms to enable them to refer these rocks to that portion of the Lower Silurian (Llandeilo division), with which they are physically united. In my opinion, they may be simply considered the central sandstones of the great schistose Lower Silurian group, whose bottom rocks are seen in the Longmynd and at Harlech and Llanberis, and whose highest term is the Caradoc sandstone, to be described in the next chapter.

In districts where they have been affected by the intrusion of igneous rocks, the slaty masses of this age in South Wales are much mineralized, just as in Shropshire and North Wales. They contain the lead ores of Nant-y-Moen, and the old Roman gold mines of Gogofau, the one to the north the other to the west of Llandovery; both of which mines occur in quartzose bands subordinate to graptolite schists. The mineral sources (Llanwrtyd, Builth, &c.) also rise from points of junction between the intrusive and sedimentary rocks.

In some districts, where these South Welsh strata were formerly described by me as charged with characteristic Lower Silurian fossils,

---

* See Sil. Syst., p. 316., where these hard and slaty sandstones are described as being cut off from the Upper Silurian by great faults. See also original map of the Silurian region and sections, p. 90., in the next chapter.

they have the antique and slaty impress of the oldest rocks. Thus, in the wild and barren tract of " Noeth Grüg," north of Llandovery, the

Noeth Grüg.   (From the Government Survey.)

S.E.   N.W.

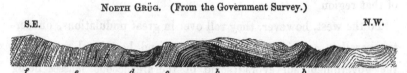

f   *e*   *d*   *c*   *b*   *b*

b. Lower Silurian schists, flags, and slates.   *c.* Lower Silurian grits and sandstone, classed by the surveyors with the schists above and below them.   *d.* Schists.   *e.* Upper Silurian shale and schist, capped by tilestone.   *f.* Old red sandstone.   (Compare with Section Sil. Syst. p. 352., and pl. 34. f. 3.)

strata are seen undulating, as represented by the curved lines, *b*, and are also traversed by cleavage planes (the highly inclined white lines). Consisting of coarse grits, sandstones, and schists, these strata are in parts replete with Lower Silurian fossils, among which are the Leptæna sericea, pl. 6. f. 14.; Pentamerus undatus, pl. 8. f. 5.*; P. oblongus, pl. 8. f. 3.; Atrypa crassa, pl. 5. f. 6.; Orthis Actoniæ, pl. 6. f. 11.; O. flabellulum, pl. 6. f. 12.; and the Lituites cornu-arietis, pl. 11. f. 2.

Though most of these species are found in the slates of Snowdon, the equivalent strata in which they are here imbedded are little more than two miles distant from the frontier of the Old Red Sandstone; all the intervening Silurian rocks, which elsewhere occupy vast tracts, being compressed in this district into the small horizontal space exhibited in the woodcut, as reduced from a section of the government surveyors. It is gratifying to me to observe that this diagram is quite in accordance with my old, published, coloured section across the same tract. (See Sil. Syst., p. 352. and pl. 34. f. 3.)

And here, in reference to the question whether there be or be not in this section, or in other similar traverses in Carmarthen-

* This fossil is termed Atrypa undata in the Silurian System, nearly all the so called Atrypæ of which work are now found to belong to the genus Pentamerus.

shire and Pembrokeshire, any small representative of the Caradoc
sandstone, as I thought, I must specially call attention to the fact,
that the correctness of my original sections is confirmed by the
detailed works of the geologists employed to survey the tract. For,

SECTION NEAR LLANGADOCK. FROM THE LOWER SILURIAN TO THE OLD RED SANDSTONE.
(The Spectator is looking to the south-west.)

b. Undulations of Llandeilo schists, flags, and limestone with interstratified trap *.   *c*. Sandstones, &c.   *d, e*. Upper Silurian.   *f*. Old red sandstone.

all these lower strata, in parts coarse sandstones, which overlie the
black and grey slaty schists and flags of Llandeilo, are now admitted
by those who followed me to be conformable to the Upper Silurian
strata or Wenlock and Ludlow formations, as exhibited in these
woodcuts, taken from the published section of the surveyors.*

SECTION NEAR LLANDEILO. FROM THE LOWER SILURIAN TO THE EDGE OF THE GREAT
SOUTH-WELSH COAL-FIELD.

(In this diagram the spectator is supposed to be placed a little south of Llandeilo, looking
north-east, or up the Vale of the Towy. Hence the points of the compass are reversed in
reference to the foregoing diagram.)

b. Llandeilo flags and limestone.   *c*. Sandstone, &c.   *d, e*. Upper Silurian rocks.   *f*. Old
red sandstone.   *g*. Carboniferous limestone.   *h*. Millstone grit.

Such facts indicate the propriety of adhering to two groups only
for all general purposes of comparison. In truth, the only Silurian

* I beg geologists to compare the several coloured sections of the Silurian
System, pl. 34., which represent the succession from the Llandeilo flags to the old
red sandstone in Carmarthenshire, with the horizontal sections of the surveyors
from which the above diagrams are faithful reductions.

formation which is fully developed in Carmarthenshire is the lower
one of Llandeilo with its associated schists and slates; all that
overlie it being feebly displayed. Thus, though the Upper Silurian
rocks here may be spoken of as a group, they can scarcely be
identified with the copious formations of Wenlock and Ludlow,
as seen in Shropshire and Herefordshire. Even the chief Lower
Silurian formation, or that of Llandeilo on which we have been
dilating, is at this spot compressed into the horizontal space of a
mile, whilst in North Wales it extends, as has been shown, through
undulations and expansions, over a whole region. We also see how
in this same tract, and in the longitudinal space of a few miles,
the very strata which are so calcareous at Llandeilo, have become
slaty sandstones and schists at Noeth Grüg, north of Llandovery;
a character indeed which they put on before they reach that town,
even at Mandinam and Goleugoed.* With this absence of lime the
prevalent trilobites of the Llandeilo flags disappear; the whole of
the lower division assuming so different a lithological aspect, that
except by rigid observation of other imbedded fossils the subforma-
tions cannot be distinguished.

In most parts, indeed, of South Wales, the only broad and clear
distinctions are those of lower and upper groups. I dwell, therefore, upon the fact, that in their range through this part of South
Wales, the Silurian rocks, both lower and upper, with the partial
exceptions mentioned, are entirely deprived of any subdividing
bands of limestone. Hence it is, that in Carmarthen, Brecknock,
and Radnor in the interior, or at Marloes Bay on the coast, the lines

* This is the only locality in Britain where the rare species Lituites (Nautilus),
pl. 3. f. 3., has been found, as well as the unique Diploceras (Orthoceras) bisiphonatum. (See chap. 5.) I cannot allude to these forms without expressing my
deep obligation to my deceased kind friend Mr. W. Williams of Llandovery, and
his son, now the Rev. Stewart Williams, for their aid in collecting many of my
best fossils, as acknowledged, Sil. Syst. p. 351.

of demarcation between the formations can with great difficulty be drawn. In truth, they are often so slaty and crystalline, that the highly inclined cleavage of the slates, as represented in the foreground of the following sketch, is the only feature visible to the unpractised eye; the real strata undulating or dipping at a much less angle, as represented in the sloping bank on the left of the foreground of this drawing.

LLANWRTYD WELLS.
(From Sil. Syst., p. 336.)

S.E.   N.W.

Lower Silurian slaty rocks.
View in a combe below the Baths of Llanwrtyd. (From a drawing by Mrs. Traherne.)

When viewed, however, on the grand scale, there is no district more explanatory of the general succession than this very tract between Llandovery and Builth, of which the above vignette represents a part. There, if the spectator stands on the slaty Lower Silurian rocks of the mountain of Esgair Davydd, above the baths of Llanwrtyd, he overlooks a wide area to the south-east, and has beneath his feet, and for a certain distance before him, a mass of the lower slaty rocks now under consideration; whilst in the dull round hills of the middle ground, are spread out the Upper Silurian of the Mynidd Epynt and Mynidd Bwlch-y-Groes. In the background, the Old Red Sandstone is seen rising to the Black Mountain of Hereford, and thence into the Fans of Brecon and Carmarthen, the highest mountains in South Wales.

The sketch of this grand geological scene being a proper introduction to the description of the Old Red or Devonian rocks, is therefore given in the tenth chapter, which treats of them; for in no one district of Wales and England are the Silurian and Devonian rocks so well brought together in a bird's-eye view : nor is the superposition of their groups anywhere, as far as I know, more apparent in the interior of the country and on so grand a scale. (See also large coloured view, Sil. Syst. p. 346.)

In the mean time, this long chapter may be concluded by reminding the reader, that the low western headlands of Pembrokeshire are of the same age as the loftier and grander mountains of North Wales; both of them having been identified with the rocks of the original Lower Silurian districts of Shropshire, Herefordshire, Montgomeryshire, Radnorshire, Brecknockshire, and Carmarthenshire. In many tracts, the copious evolutions of volcanic ashes and dejections have swollen out the earliest formed Silurian strata, and the finer muddy sediments of a primeval sea have since been changed into a crystalline and slaty condition. But no sooner do we pass from such hard and rugged centres (including the sandstones and conglomerates of the South Welsh slates), than we find that many of the intervening and central tracts in South Montgomery, Radnor, Cardigan, &c., consist of an undulating series of schists and shale, in which bands of sandstone are rare, and wherein limestones have altogether disappeared; a country, in fact, almost entirely deficient in anything like a stony framework.

No more striking example of this change from hard and bold rocks, to those of much less coherent nature and tamer outline, can be given, than by traversing from the low fossil rocks of Cader Idris and its magnificent bosses and depressions, to the soft hills and schistose tracts on the banks of the Dyfi, as in this diagram.

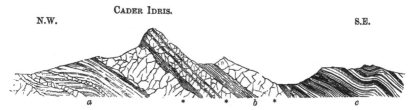

a. Summit of the Lingula schists with porphyry.   *. Grand masses of porphyry and other trap rocks, with courses of hard slaty Lower Silurian schists.   c. Lower Silurian, assuming its ordinary characters of shale when removed from the igneous rocks. ◄

In so doing, the field observer will see that the strata which, on the flanks of the mountain, are crystalline and slaty, are succeeded by other masses, which in proportion as they recede from the igneous rocks partake more and more of the character of "mudstone" — a character which prevails, as before said, in this region, from the base of the system with Silurian fossils, to the very youngest of its rocks.

To form some idea of the great development of volcanic or igneous materials in the Lower Silurian rocks of North Wales, the reader will understand, that the porphyries and trap rocks represented in the above diagram, belong exclusively to a set of phenomena which commenced about the close of the deposition of the Lingula schists. Thick as these lower igneous accumulations are, they were followed by others of scarcely less dimensions, which swell out the overlying series of the Snowdon slates*, as explained in the little coloured section beneath the map.

The next chapter will be dedicated to the consideration of a formation, in which, through some typical British tracts, sandstone so abounds, as to have induced me to term it the Caradoc sandstone— a rock which forms, in many places, the upper member of the natural group which has been under consideration.

* The large horizontal section across Snowdon by the government surveyors has just appeared, and shows in detail how the Lingula flags and certain igneous rocks rise up to the N.W. and S.E. from beneath the elevated trough of Snowdon. Professor Ramsay, who, with his associate, Mr. Selwyn, accomplished this arduous task, has just explained the results to the Geological Society. April 10. 1853.

CHAPTER IV.

LOWER SILURIAN ROCKS — *continued*.

THE CARADOC FORMATION.

CARADOC SANDSTONE AS DEVELOPED IN NORTH WALES AND RADNORSHIRE.   TYPES OF
THE FORMATION IN SHROPSHIRE, HEREFORDSHIRE, ETC.
THE CARADOC STRICTLY UNITED WITH THE LOWER DEPOSITS BY ITS STRUCTURE AND
FOSSIL CONTENTS.

THE superior member of the Lower Silurian rocks, or middle Silurian
of the government surveyors, is of varied composition in the different tracts in which it occurs. In the interior of North Wales
(Merioneth and Montgomery, and thence extending into West
Radnor, &c.) certain sandy masses, overlying the rocks described
in the last chapter, have been identified with the Caradoc sandstone
of the original Silurian region, and laid down as such in the coloured sheets of the Government Survey. In these tracts, like the
older sandstones and conglomerates of the Llandeilo division, the Caradoc is usually a hard, intractable, silicious rock, with many courses
of shale or schist, but with few or no traces of calcareous matter, and
seldom containing well preserved fossils. In fact, it frequently
bears a close resemblance to the very oldest greywacke grits and
sandstones in the sedimentary series, or the bottom rocks of the
Longmynd, Harlech, Llanberis, and St. David's; a lithological fact
which shows how all these rocks, from their base to the Caradoc
inclusive, are included in one natural group.

In portions of Wales and Shropshire this formation, as before
stated, is either transgressive, or unconformable to the inferior schists

and slates.* In many other parts, on the contrary, as in the wild tracts east of Bala, it reposes conformably for many miles in lofty escarpments on the upper Llandeilo rocks, composed of slaty schists with feeble courses of limestone.† A reference to the general section across North Wales, prepared for this work by Professor Ramsay (see bottom of map) completely establishes this position. In following that line of section to the S. E. from the environs of Bala, we see how the Llandeilo formation, rolling over in domes or anticlinal masses, supports basins of Caradoc sandstone and its overlying formation of Wenlock shale. In this way even the Upper Silurian rocks are brought by conformable folds into the proximity of the Llandeilo formation, as in South Wales. In the environs of Bala itself, indeed, the Caradoc formation of sandstone and shale occupies a trough flanked and underlaid conformably on the south-east, as well as on the north-west, by the upper division of the Llandeilo formation, as in this section, which has been reduced from one of the plates of the Government Survey prepared by Mr. Jukes.

LLANDEILO SCHISTS SUPPORTING CARADOC SANDSTONE.

b. Llandeilo formation, with its thin courses of Bala limestone on the N. W.   *c.* Caradoc sandstone, seen at Sir Watkin's Monument to be lying in perfect conformity on the Upper Llandeilo schist, which, rising out from beneath the sandstone in the Vale of Edeirnion, thence ascends to the S. E. to form the north-western flanks of the Berwyn Range. The powerful faults which affect the tract are represented by the dark traversing lines.

Now, it is the opinion of the surveyors, that in pursuing this section to the S. E. of the Berwyns, and in traversing the vales watered by the

* See the memoir of Professor Ramsay and Mr. Aveline, in which this feature is pointed at (Quart. Jour. Geol. Soc. Lond., vol. iv. p. 294.) The map annexed to this work shows the fact, in the wrapping round of the Caradoc sandstone upon the edges of the Llandeilo flags of the Shelve and Corndon districts, Shropshire.

† I examined this tract in 1851, in company with Professor Nicol.

Tannat and Fyrnwy rivers, the same series of Llandeilo or Bala beds as is represented in the *b* of the above section is repeated in obscure undulations until the Breidden Hills are reached (see p. 60.), where the Caradoc sandstone seems to be omitted, and the lower series is at once surmounted by the Wenlock shale. In that locality of powerful eruption the suppression of the Caradoc may, indeed, be due to dismemberments, of which hereafter. But there are tracts where the sandstones of the lower group, which are so expanded in parts of Wales (see p. 70.), naturally thin out and disappear ; and hence I adhere to my original opinion, that the presence or absence of such sandy rocks is a local feature in a group for which, in many cases, there is no safe term except " Lower Silurian."

The Caradoc formation is also seen, in the western and mountainous parts of Radnorshire, to fold over in undulations similar to those near Bala, *resting conformably on the Llandeilo schists, and overlaid in equal conformity by the Wenlock shale.* The next diagram represents a portion of the flexures which these rocks exhibit, as taken from a part of the large and detailed horizontal sections of the Geological Survey, relating to the country along the boundary between the counties of Radnor and Montgomery, in a section drawn from Cader Idris across the Severn and its tributaries to the river Ithon, and the sandstone ridges to the west of Radnor Forest. On this line the same conformability is seen in numerous localities, of which this is one example.

PEGWN'S FAWR, BORDERS OF RADNOR AND MONTGOMERY.

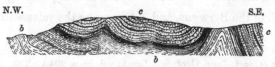

Junction Beds of the Llandeilo and Caradoc Formations.   (From a Section by Mr. Aveline.)

b. Llandeilo formation of grey, blue, and purple slates and shale.   *c.* Caradoc sandstone, composed of hard grey grits, with grey and blue schists, &c.

These facts, as well as those which are generalized in the section below the map, being the record made by independent and competent observers since the time of my personal researches, demonstrate that the break or dislocation between the Caradoc and Llandeilo formations (which together constitute the Lower Silurian rocks of the original classification) is purely a local phenomenon. And even in the tracts where these two formations have been physically disunited, we shall soon see that they are mutually so characterized by the same organic remains that they cannot be separated in any geological arrangement.

In the typical Silurian tract of Shropshire, around the eruptive trappean rocks of Caer Caradoc, from whence the formation was named, it is indeed cut off from those inferior deposits, with which it is elsewhere so intimately associated, by a great mass of igneous rocks, and by a very powerful displacement or fault, which brings it against the Longmynd or bottom rocks. The annexed view of the bold ridge of the Caradoc hills is taken from the slope at the escarpment of the Longmynd, a little to the south of Church Stretton.

| The Lawley. | Caer Caradoc. | Hope Bowdler. | Broccard's Castle. | Ragleath. |

THE CARADOC RANGE.

(Sketched by Mrs. S. Acton, from the eastern slopes of the Longmynd, overlooking the vale of Stretton, along which ran the old Roman road.)

G

The Caradoc formation is unconformable, at least in its uppermost or calcareous member, to the Llandeilo flags and schists of the Corndon and Shelve district, around the ends of which it wraps in the manner laid down by the government surveyors. (See Map.) On the other hand, however, or towards the east, the formation expands and gradually passes up into the Upper Silurian rocks of Wenlock and Ludlow.

In the lithographic view, p. 26., taken from the eastern slopes of the Longmynd, the spectator overlooks this range of eruptive hills of Caer Caradoc, &c., which, flanked on the east by sandstones, are followed by distant ridges of overlying Upper Silurian rocks and old red sandstone, as in this section. Here it is of much more complex

RELATIONS OF CARADOC SANDSTONE TO THE UPPER SILURIAN ROCKS IN SHROPSHIRE.

N.W.   S.E.
Caer Caradoc.   Wenlock Edge.

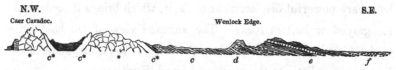

* Eruptive rocks. c*. Caradoc sandstone altered by eruptive rocks. c. Caradoc sandstone, &c.
d. Wenlock rocks.   e. Ludlow rocks.   f. Old red sandstone.

mineral structure than in North Wales and the mountains of Radnorshire. For the Caradoc of Shropshire contains good freestone, flagstone, grit, and conglomerate, with courses of limestone and shelly sandstone, and with this composite stony character the formation is much more highly fossiliferous than in any district of Wales with which I am acquainted.

The observer who has little time to spare may study the leading features of the deposit, in ascending order, by following the banks of the river Onny from the south-east flank of the Longmynd to Cheney Longville and the Craven Arms; but he who would endeavour to recognize all its variable, lithological features, must examine the masses as they spread out on their strike to the north-east, and expand in their course from the environs of Acton Scott, by Soudley, Church

Preen, Chatwall, and Acton Burnell, to the tract on the Severn, between the Wrekin and Coalbrook Dale.   In so doing, he will see how the upper portion, chiefly calcareous, dips everywhere conformably under the Wenlock shale, or base of the Upper Silurian. Along this line he will also see how the lower beds have in many places been fused and changed into quartz rock, and their schistose parts converted into dark brittle clay-stones, wherever they are in contact with the igneous rocks of Caer Caradoc, the Lawley, &c.*

As seen in the gorge of the river Onny, or in the various quarries of Soudley, Frodesley, Kenley, &c., the unaltered Caradoc sandstones are of considerable thickness.   The lowest masses of the formation which are visible, consist of grits and fine-grained sandstones of brownish, yellowish, red, purple, and even white colours, sometimes striped dull red and yellowish brown, sometimes green, and in other instances they are variegated with spots and concretions.   They thus resemble portions of the old red sandstone, carboniferous rocks, and even the newer red sandstones.   Some of the beds are as easily worked into building stones, troughs, and tombstones as the freestones of those younger formations; and, when remote from eruptive rocks, are, therefore, very dissimilar to the strata of precisely the same age which have been just mentioned as composing the hard rocks of Wales.

The typical Caradoc sandstones of Shropshire are, indeed, remarkable in being the oldest strata of the British Isles which present such a comparatively modern aspect; it having been formerly the sterile privilege of the old " greywacke " of this age to be considered useless except for the construction of rude buildings and stone walls.   Such truly is the character of the same sandstones in North Wales, which have been adverted to in the opening pages of this chapter.   And yet, notwithstanding this modern aspect, these rocks, as will presently

---

* For a detailed account of the mineral structure of this eruptive range, and of the Wrekin, in which felspathic rocks abound, but in parts of which syenite and greenstone also occur, see Sil. Syst., p. 225.

be shown, are charged with some of the oldest fossil types of the formation which lies far beneath them.

The remains of fossils occur at intervals throughout the deposit in Shropshire, and often so abound as to render some of the courses shelly limestones, which have occasionally been burnt for use; such local shelly developments being known to the workmen of the neighbourhood as " Jacob's Stones." Where the calcareous matter is more diffused in small particles through the mass, the rock becomes a hard calciferous grit, usually of a whitish drab colour, and breaking under the hammer with a conchoidal fracture, and a " chatoyant" lustre.

On the banks of the Onny, between Horderly and Wistanstow, the calcareous matter so augments as to form a band of limestone, which is, however, chiefly a pebbly quartzose conglomerate in a cement of lime. Another course of limestone is found at the eastern flank or upper end of the formation. In their north-eastern prolongation to Church Preen and Chatwall, the pebbly and conglomerate beds are not calcareous.

The recent cuttings of the Ludlow and Shrewsbury railroad have laid open several hillocks of the Caradoc sandstone, between Church Stretton and the Marsh-Brook station, the rocks of which consist of thick and thin bedded, dull, purple and red earthy sandstones, in parts finely micaceous, with usual partings of green earth, some courses of the shelly beds or " Jacob's Stones," and brown calciferous grits. In the cutting nearest to Church Stretton, the rock is very earthy and incoherent, and in it I found Diplograpsus pristis, Foss. 7. f. 14., a common Lower Silurian species which occurs in the Llandeilo schists of South Wales and the Snowdon group of North Wales. This variety of the Caradoc sandstone is very concretionary, exfoliates on weathering into balls, and is full of irregular joints, whose surfaces are coated with a film of the dark indigo-coloured oxide of iron which is so common in the Ludlow rocks. Mineralogically, indeed, the formation has already begun to assume the structure

of the Upper Silurian rocks, to the base of which its upper band is
everywhere conformable in this tract.

Besides the prevalent building-stones of Soudley and other places,
where thick-bedded sandstones are separated by courses of shale,
often of a dull greenish colour, surmounted by flags, there is an
inferior mass of shale often laden with fossils. Another variety of
the Caradoc sandstone, which assimilates the formation to the inferior one of Llandeilo, consists of stratified bands, along the outer
flanks of the Caradoc and Wrekin hills, which resemble the volcanic
grits described in the last chapter. These sandstones are allied to
greenstones of igneous origin, in being made up of green earth and
felspar, with only a few fine scales of mica. Like certain strata of
the inferior formation, they seem to indicate either a contemporaneous
volcanic disturbance in the bottom of the sea, or, that they have been
derived from pre-existing igneous rocks. My belief, however, is,
that some of these beds were really the result of submarine volcanic
dejections, thrown down in the sea-bottom whilst the Caradoc sandstones were accumulating, and were the precursors of those subsequent copious outbursts of the felspar and other trap rocks of the
Caradoc and Wrekin.

The superior masses of the Caradoc formation in Shropshire,
which are well displayed at Cheney Longville, and on the banks of
the Onny where that river flows across the strata, are chiefly made
up of fine-grained, sandy, thin flagstones, of gosling green and dingy
olive colours, amidst which some beds of shale occur, with courses
of grey-coloured shelly limestone.* Others of these calciferous sands
and earthy greenish flagstones begin to assume a lithological aspect,
frequent in the Upper Silurian rocks, and particularly in the Ludlow
formation, to which other varieties of the rock above alluded to (to
the south of Church Stretton) have also an affinity ; they are earthy ;
with great spheroidal concretions, and with a general tendency to

* The Asaphus Powisii was first found here. Pl. 2. f. 2.

exfoliate and decompose to mud. Thus, in composition, some of the varieties of the Caradoc sandstone connect it with the volcanic grit or ash of the Llandeilo formation, and others with the ordinary aqueous deposits of Upper Silurian age.

The highest calcareous band of the Caradoc formation is visible on the Onny, and is traceable, to the N. E., by Acton Scott and the Hollies to Cressage and Buildwas.* In this range the portion of the rock which is calcareous, with beds of sandy shale and pipe-clay, dipping under the Wenlock shale, passes conformably upwards into that lowest member of the Upper Silurian rocks, as formerly so defined. (Sil. Syst., p. 217.)

The fossils which most frequently occur in the heart of the Caradoc formation are as follow:

CARADOC FOSSILS, 7.

1. Calymene Blumenbachii. 2. Homalonotus bisulcatus, Salter. 3. Phacops truncato-caudatus, Portl. 4. Tentaculites annulatus. 5. Lingula crumena, Phill. 6. Orthis testudinaria, Dalm.

---

* In a minute re-examination of this district, made by the Government Surveyors whilst these pages are passing through the press, I learn that they have detected, on the banks of the Onny, a small amount of unconformity between the uppermost of the limestones containing the Pentamerus oblongus and the fossils figured on the opposite page, and the subjacent beds of flagstone shale and impure limestone in which occur the fossils given on this page. That this transgression is local is manifest from the fact, that in other tracts the Caradoc Sandstone and Wenlock Shale are represented by the surveyors to undulate in perfect parallelism. Believing that the limestone with Pentamerus oblongus is truly the bed of transition from the lower to the upper group, it is still my conviction that it must, from its fossils, be classed with the former. For, although the *Orthis alternata*, Foss. 8. f. 6, (which is a fossil of the beds subjacent), has been introduced, by mistake, into that woodcut, I must always consider that a band into which Trinucleus concentricus and Pentamerus oblongus range upwards from the Llandeilo formation is truly Lower Silurian. I willingly admit that several of the other species of fossils in this bed are common to the Wenlock and lower deposits; a fact which strengthens my view of the unity of the Silurian system of life.—R. I. M., 28th May, 1853.

7. O. vespertilio. 8. Strophomena tenuistriata. 9. S. grandis. 10. Bellerophon bilobatus.
11. B. ornatus, Conrad. 12. Orthonota nasuta, Conrad. 13. Nebulipora lens, M'Coy.
14. Diplograpsus pristis, His. 15. Graptolites priodon.
In addition to the above, the following characteristic fossils are figured in plates 4 to 10., or in
the woodcuts of the next chapter: — Orthis elegantula. O. flabellulum. O. Actoniæ. O.
calligramma (virgata, Sil. Syst.). Strophomena expansa. S. spiriferoides, M'Coy. Lep·
tæna sericea. Modiolopsis orbicularis (Avicula, Sil. Syst.). M. modiolaris, Conrad. M.
obliqua. Bellerophon acutus. Phacops conophthalmus, Eichw. (Asaphus Powisii, head of,
Sil. Syst.).
The species with no author's name attached are published in the " Silurian System."

After this mention of the chief fossils of this formation, the attention of the reader is again called to the fact that, notwithstanding the differences between the lithological character of these strata of Caradoc sandstone and those of the Llandeilo formation, *all the above-mentioned species of fossils prevail in both deposits;* thus absolutely uniting them in one natural group.

Even in the very uppermost stratum of the Caradoc formation, which here graduates into the Wenlock shale and true Upper Silurian Rocks, there are several species, such as the Petraia elongata, Foss. 8. f. 11.; Leptæna sericea, pl. 6. f. 13.; Orthis calligramma, Foss. 14. f. 3.; Trinucleus concentricus (T. Caractaci, Sil. Syst.), pl. 4. f. 2 —5.; Pentamerus undatus, pl. 8. f. 5.; Tentaculites annulatus, Foss. 7. f. 4., &c., which are known to occur on the flanks of Snowdon, or in beds of the Llandeilo age in North and South Wales.

UPPER CARADOC FOSSILS, 8.

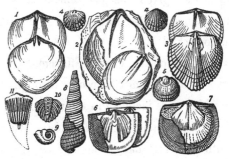

1. Pentamerus lens. 2. P. oblongus. 3. P. liratus, Sow. 4. Atrypa hemisphærica. 5. A. reticularis. 6. Orthis alternata. 7. Strophomena compressa. 8. Holopella cancellata. 9. Bellerophon trilobatus. 10. Encrinurus punctatus. 11. Petraia elongata, Phill.

The above woodcut, Foss. 8., exhibits the chief characteristic
fossils of this upper band, some of them, as above stated, being common in the inferior strata; whilst others, such as the Atrypa hemisphærica, f. 4.; A. reticularis, f. 5.; Pentamerus liratus, f. 3.; P.
lens, f. 1.; and Holopella cancellata, f. 8., are more characteristic of
this horizon. On the other hand, Encrinurus punctatus, f. 10.,
and Bellerophon trilobatus, f. 9., are of frequent occurrence in the
Upper Silurian group, of which hereafter.

It is therefore manifest that, to whatever extent dislocations may
have acted in producing to the west of the Longmynd in Shropshire,
or in some parts of Wales, a break between the Llandeilo and the
Caradoc formations, as before explained, or may have excluded the
mass of the Caradoc, as is seen along the south end of the Longmynd,
the operations were purely local, and had no influence whatever in
destroying the races of animals that inhabited the seas of that long
period. Those marine creatures lived on; some of them (judging
from their remains) becoming more abundant when the upper Caradoc beds were accumulated, than when the much older mud and sand
which afterwards formed the slates of Snowdon were deposited.
With such clear and strong zoological proofs, and the evidences of
conformable transition and succession in some portions of this region,
which have been already given, the government surveyors, who observed the transgressions alluded to, have decided that my general
classification was correct.

In defining the course of the Lower Silurian Rocks along their
eastern or superior frontier, from Shropshire through those counties
of South Wales which were included in the region of my former examination, or from the eastern parts of Radnor, Brecon, Carmarthen, to the south of Pembroke, I announced that the Caradoc
formation there became unusually thin, and was with difficulty
separable from the Llandeilo flags.

Strictly, however, considered as the Caradoc of Shropshire, the formation is seen to protrude along a line of fault as an isolated patch* among the Wenlock shale of Brampton Bryan Park, and is again visible further to the S. W. at two points in Radnorshire, the one in a dome to the south of the town of Presteign, the other on the western slopes of the hill of Old Radnor. (See Map.) In these examples it is a quartzose, pebbly conglomerate, passing into a hard greywacke grit (Corton), in which casts of the Pentamerus oblongus and other characteristic fossils occur, such as Atrypa hemisphærica, and very abundantly Petraia elongata and P. bina. Along this line it is also obviously the uppermost bed of the formation as brought up in domes and arches by a line of igneous eruption, which in evolving the adjacent crystalline hypersthene and felspathic rocks of Stanner, Hanter, and Ousel, has so altered the contiguous sand-stones, schists, and grits (just as around the Wrekin, and the Caer Caradoc) that in Old Radnor Hill and Yat Hill it is difficult to determine which parts of it are metamorphosed rocks resulting from the effect of olcanic eruption, and which have really been in a molten state. The diagrams illustrating this phenomenon, including the metamorphosis of the limestone of Nash Scar, will be given in the next chapter, which treats of the Wenlock formation.

In that district of Radnorshire, the uppermost Caradoc is a silicious grit and conglomerate, and is at once conformably surmounted by shale and limestone, which, from their fossils, are grouped with the Wenlock formation of the Upper Silurian rocks, as will be hereafter explained (pp. 102. et seq.).

At Corton, near Presteign, where the order is clear, the old exca-vations or quarries exhibit the strata thus; — the lower beds of the Wenlock formation dipping away from the side of a dome of coarse and pebbly Caradoc sandstone, laden with casts of the Pentamerus oblongus.

* Discovered by my friend the Rev. T. T. Lewis.

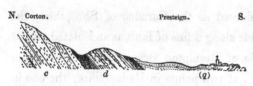

c. Caradoc sandstone. d. Lower Wenlock shale with lime-stone (Woolhope), see Chap. 6. (q.) Gravel of the Valley of Pres-teign.

But whilst we have in this north-eastern part of Radnor, as in Shropshire, so clear and conformable a succession from the Caradoc to the Wenlock formation, we behold a very different state of things by merely following the strata to the south-west into another part of the same county. There, between Llandegley and Builth, the Llandeilo formation is brought at once into unconformable contact with the Upper Silurian Rocks; the Caradoc formation being entirely omitted. Nothing short of a most powerful dismemberment will account for such contrasting positions of the strata, as are here ex-hibited in diagrams taken from the large horizontal sections of the survey. In fact, the Llandeilo Rocks must have been upheaved and removed from the influence of those waters under which the Caradoc sand was accumulated, to be afterwards submerged and covered by Upper Silurian strata.

Both of the following transverse parallel sections pass across a tract in which the Lower Silurian rocks (Llandeilo), with their Asaphi, Trinuclei, and interposed trap, and also the Upper Silurian Rocks, were formerly described; but I did not then represent what the government surveyors have since made out, the distinct overlap of one set of beds upon the other, and which shows how the equivalent of the Caradoc is entirely omitted.

UNCONFORMABLE RELATIONS OF LLANDEILO FLAGS AND UPPER SILURIAN. NEAR BUILTH.

(Taken from Horizontal Sections of the Government Survey.)

S.E.   N.W.
Gwaun Ceste.   Tyn-y-Coed (near to).

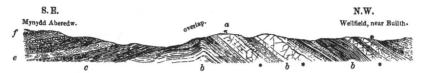

S. E.
Mynydd Aberedw.

N.W.
Wellfield, near Builth.

b. Llandeilo formation with Ogygia Buchii, &c.   * Trap rocks.   *d.* Wenlock shale resting on the uplifted edges of *b*, the Caradoc sandstone (or *c*) being entirely omitted.   *e.* Lower Ludlow.   *f.* Upper Ludlow.

The relations of the silicious strata in the neighbourhood of Llandovery and Llandeilo have been previously explained, and the reasons assigned for having formerly considered such beds to be the representatives, though in a very diminished form, of the Caradoc of Shropshire.

Seeing in that tract the comparative tenuity of all the formations above the Llandeilo, I simply termed the whole of the inferior mass " Lower Silurian," as distinguished from another and overlying mass of small thickness, which, being wanting in subdivisionary characters, was called " Upper Silurian " (see sections, p. 72, 73.). It was only in those examples, where strata between the lowest beds with true Llandeilo fossils, and others with Upper Silurian fossils, were of a sandy or silicious nature, that I applied the word Caradoc. And there is, in truth, so perfectly an apparent conformity between all these Silurian rocks in Carmarthenshire (the whole being within the range of a bomb-shell,) that if the phenomena of a disturbance and overlap had not been worked out in another part of the region, any field geologist, looking simply to the order of the beds, the character of the imbedded fossils, and the structure of the rocks, must have come to the same conclusion as myself, and have viewed the sandy strata of Carmarthenshire which lie above the black flags of Llandeilo, as the meagre extensions of a formation which, as Caradoc sandstone, is fully developed only in the more northern tracts of this region. (See coloured sections, Sil. Syst., Pl. 34.) The dwindling out of the masses of sandstone in their range from north to south, as laid down in the map (No. 3.), sufficiently indicate the phenomenon.

Lower Silurian of the Malvern Hills, Woolhope, May Hill, &c. —
Though less copious in dimensions than in North Wales and Shropshire, the Lower Silurian rocks have, to a considerable extent, their
representatives on the western flank of the crystalline and syenitic
range of the Malvern Hills. The little sketch here annexed represents the northern half of the Malvern Hills, or the North Hill, with
the Worcester, and Hereford Beacons, as seen from the undulating
grounds below Mathon Lodge, or near the outward flank of the
Silurian rocks, the mass of which occupy the intermediate and undulating low hills.

North Hill.   Worcester Beacon.   Hereford Beacon.

VIEW OF THE MALVERN HILLS FROM THE WEST.
(From a Sketch by Lady Murchison, Sil. Syst., p. 409.)

The black schists which rise up to the surface in the elevated grounds
between the higher syenitic hills and Eastnor Castle, and which lie
to the south-east of the foreground of the preceding sketch, were
formerly described by me as probably of the age of the Llandeilo formation.* A subsequent discovery made in them by Professor Phillips
of the small trilobite Olenus, and a recent one by Mr. Strickland of the
Agnostus pisiformis, fossils which are peculiar to the lowest Silurian
zone of other countries, have extended that view, and have led me
to believe that the black Malvernian schists, however diminutive,
may be as old even as the Lingula beds of North Wales. In another
chapter, and when treating of the Silurian rocks of Norway and

* Silurian System, p. 416.

Sweden, it will be shown, that mere thickness of strata is no true test of their age nor of the time which may have elapsed in their deposition. The hundreds of feet in the beds on the west flank of the Malverns may, in short, be the equivalents in time of thousands of feet of strata in North Wales. I also showed, in my former work, the existence of sandstones which, overlying these black schists, and charged with fossils, truly represented the Caradoc sandstone in structure and in fossils. The geologist must, however, consider my first publication as an outline descriptive only of the chief features of these remarkable hills, the detailed features of which have since been more elaborately and ably worked out by Professor J. Phillips, to another of whose discoveries attention will presently be called.

On this occasion I must chiefly restrict myself to the notice of a few data which have been brought to light since the publication of the " Silurian System." In a recent excursion* I observed, what had not before been detected; that the same black schists and sandy flagstones, or a portion of them, which constitute the lowest fossil-bearing sediment of the tract, are thrown off to the east as well as to the west of the eruptive or syenitic chain, at Midsummer Hill to the S. of the Hereford Beacon. This is seen in traversing the ridge near its south end, from Keys End Street on the east, by the White-leaved Oak, to the first arable ground on the west. The Malverns are therefore represented in the following diagram, as exhibiting a geological saddle, having a thin and partially metamorphosed flap only on the east side, and several thick flaps on the west, the lower part only of one of which is in an altered condition. Among the oldest strata which are visible on the east flank are the same black schists subordinate to sandy flags, which on the western slope of the ridge of syenite have given to the ground the name of Coal Hill†,

* This excursion was made in company with Mr. H. E. Strickland and the Rev. W. P. Brodie.

† Absurd trials for coal have been made by ignorant persons in this black Lower Silurian shale *on both sides* of the eruptive axis.

and in which Professor Phillips has detected the three species of Olenus already mentioned.

The reader of my former work will therefore be pleased to look at the woodcut (Sil. Syst. p. 418., and coloured section, pl. 36. f. 8., of that work) as somewhat incomplete, and will understand that the facts are more correctly expressed in this little diagram.

SECTION FROM THE MALVERN HILLS TO LEDBURY.

This woodcut, as taken from a coloured section in the "Silurian System," and slightly modified (pl. 36. f. 8.), explains the general order and the undulations on the west side of the Malvern Hills, in the parallel of Midsummer Hill, Eastnor Park, and Ledbury. The eruptive rocks * of the Malvern ridge on the right are seen to be flanked, on both sides, by altered Lower Silurian sandstones and schists, *b* *   To the west these are followed by other schists with Olenus, *b*, and Caradoc sandstones and conglomerates, *c*, the upper portion of which dip down from the Obelisk Hill of Eastnor, and pass under the Woolhope, or Lower Wenlock limestone, *d*. The latter is followed by the Wenlock shale, *e*, and the Wenlock lime-stone, *f*, which last, bending under the Lower Ludlow, *g*, reappears in a dome that throws off towards Ledbury the whole of the Ludlow formation, *h i*, after a flexure in Wellington Heath under the old red sandstone, *j*.

Although it is not within the scope of this volume to describe in detail the former effects of the eruptive molten matter in transmuting the character and condition of the sediments through which it burst forth, still I would beg any geological student who has it in his power to examine this natural section.   He will then see that as the strata which flank the ridge on either side approach to its centre, which is composed of igneous rocks, they gradually become more hard and brittle, and that, their flaglike character gradually ceasing, the beds pass into thick and subcrystalline amorphous masses, in which the lines of bedding are lost, and the substance is much altered and fissured by many devious joints with serpentinous coatings.   The central or eruptive mass, as exposed in the extensive quarries of Mid-summer Hill, near the White-leaved Oak, is a compact felspar rock, passing into syenite, but, as formerly noted, the northern extensions of this same eruptive ridge, as seen in the Worcestershire Beacon and the North Hill, contain many other varieties of igneous rocks, such as syenite having a very granitic aspect, greenstone, epidote rock, &c.

But returning to the Silurian succession. The Upper and greater portion of the Lower Silurian rocks of the Malvern district were first classed by myself, and subsequently by Professor Phillips, with the Caradoc formation. In my own work they were described as sandstones and conglomerates, often of purplish, sometimes of green and brownish tints, in which many of the same species of shells occur as in Shropshire and other localities.* Assigning to the Lower Silurian of the Malverns a thickness of 2000 feet, Professor Phillips correctly states, that the Holly-bush sandstone, which contains no fossils except fucoids, is the bottom of the whole series, and even underlies the black shale with certain small points of interposed (intruded?) trap rock or greenstone. Above the black Olenus shale commence the purple and grey conglomerate and sandstone, called Caradoc sandstone, here of about 1000 feet thick, as seen at Howler's Heath, Eastnor Obelisk, &c.

In delineating with great precision the order of the strata with their flexures and breaks, Professor Phillips discovered, on the west side of the Worcestershire Beacon, that one of the local conglomerates of the Caradoc age, and charged with several of the characteristic fossils, contained also pebbles and fragments of the syenite and other eruptive rocks of the Malvern chain. Hence it was evident, that along this line there had been eruptions which had left a shore or island of hard rock from which the Caradoc conglomerate in question had been derived. It is, however, to be inferred that the eruption, to which I have called attention, at the White-leaved Oak and Keys End Hill, took place at a more remote period, and even that the erupted matter traversed the oldest Silurian strata of the tract, or the Holly-bush sandstones. This agency having ceased, a low ridge

---

* This is the green sandstone described, Sil. Syst. pp. 416—418.

seems to have been left, to be acted upon by the sea of the subsequent epoch, so as to furnish the fragments of a younger portion of the Caradoc formation.

The posterior elevation of the Malvern Hills in a solid state, at a subsequent period, together with the folding back of the strata, so as partially to invert the order and place the Caradoc grits *over* the Wenlock shale (a phenomenon first pointed out by Mr. Leonard Horner as due to the uprise of the syenite), is one of great interest, and which occurs on a vast scale in the Alps and other countries.*

At present, however, I must pass from the consideration of this physical feature, to other natural appearances on the flank of the Malvern Hills, described by Professor Phillips and myself, to show how difficult it is to draw any definite line between the uppermost member of the Caradoc formation and the lowest portion of the overlying Wenlock shale. To the west of the Worcestershire Beacon, or near the north end of the chain, there occurs an impure limestone, occupying precisely the same position at the summit of the Caradoc sandstone, properly so called, as the Hollies limestone of Shropshire (p. 86.). Now, although this rock contains some fossils which truly belong to the lower group, such as Leptæna sericea, Atrypa hemisphærica, and Orthis calligramma, it possesses others which are characteristic of the Upper Silurian division. At Woolhope, however, as will hereafter be shown, a limestone occupying nearly the same position is, through every one of its imbedded fossils, an indisputable member of the Wenlock formation.

In Shropshire, on the contrary, the calcareous zone which separates the Lower from the Upper Silurian rocks pertains preeminently, as we have seen, through its organic remains, to the lower, whilst in

---

* This overthrow of the strata, together with an excellent account of the mineral structure of the Malvern Hills, was first given by Mr. Leonard Horner, Geol. Trans., old series, vol. i. p. 281. The nature and succession of the overthrown beds were illustrated in detail by myself, Silurian System, p. 423., and lastly by Professor Phillips, Memoirs Geol. Survey, vol. ii. p. 67. *et seq.*

Radnorshire, a limestone apparently or very nearly in the same position is altogether Upper Silurian in its fossils. The truth is, that in some tracts we must expect to meet with links, both lithological and zoological, which thus bind together the inferior and superior divisions of the system ; whilst in other districts, their separation will be more neatly marked.

The reader may compare my original description of this intermediary band of limestone, as it ranges by Stumps Wood and other places west of the Malverns, with the more precise details of Professor Phillips.*

For although it is quite clear that transitions like those above mentioned must be looked for, and have in fact been found to exist, over large areas of the earth's surface, it is well known that a mass of sediment which in one tract is calcareous often becomes sandy and argillaceous in another; and thus in such cases very close examination of the fossils can alone decide the exact line of demarcation. In reference to the original Silurian region, the uppermost limb of the Caradoc formation, if fossils decide the case, is an argillaceous limestone in Shropshire, and in Radnorshire a pure siliceous grit; though the localities before alluded to are only a few miles distant from each other. These local extensions of limestone will be still more instructively explained in describing the order of the Upper Silurian rocks.

In May and Huntley Hills in Gloucestershire, the Caradoc formation, consisting of conglomerate and sandstone, with certain characteristic fossils, indicates a mineral transition upwards into the Wenlock formation through alternating beds of sandstone with shale and impure limestone, which are of considerable thickness. It is

* In his memoir of the Malvern and Abberley hills, and their extension to Woolhope, May Hill, &c., Professor Phillips has skilfully and elaborately developed all the physical contortions of the strata in this district, and the mineral and palæontological characters of the rocks. He has further imparted great value to his work by the philosophical reflections with which it is interspersed. It is a monograph which reflects high credit on the Geological Survey of Great Britain.

H

to some of these upper and transition beds of sandstone at May
Hill, that Professor Sedgwick has recently directed attention* as being
almost exclusively charged with fossils, which Professor M'Coy refers
to the Wenlock age. The government surveyors, as well as myself,
have, however, collected Pentamerus lens, if not P. oblongus, from the
same strata; both species being in my opinion typical of the lower
division. Such transitions and fossiliferous changes must, I repeat,
be looked for in the range of all sedimentary systems, and this case,
serves only, in my opinion, better to link together the Lower and
Upper Silurian rocks.† This subject will be specially reverted to in
the eighth and ninth Chapters.

* On the separation of the Caradoc sandstone, into two distinct groups. Geol.
Soc. Nov. 1852.

† In a minute re-examination of the Caradoc district of Shropshire, made by
the Government Surveyors whilst these pages are passing through the press, I
learn that they have detected on the banks of the Onny a certain amount of
unconformity between the uppermost of the limestones (containing Pentamerus
oblongus, and the other fossils, figured p. 88.), and those subjacent beds of flagstone,
shale, and sandstone, in which, with a few of the same species, the great mass of
fossils are those figured in the woodcut Foss. 7. That this physical break is a
local phenomenon only, is manifest from the fact, that in some other tracts all the
strata of Llandeilo, Caradoc, and Wenlock age are represented by the same
surveyors to undulate in perfect parallelism. (Borders of Radnor and Montgomery.) It is still my conviction, that whilst the limestone with the Pentamerus
oblongus is what I always considered it — the bed of transition from the lower
to the upper group,—it ought still, from its fossils, to be rather classed with the
former than the latter. For, although the *Orthis alternata*, which is found in the
subjacent sandstone beds, has been introduced by mistake into the woodcut Foss. 8.,
I cannot admit that a band into which *Pentamerus oblongus* and *Trinucleus concentricus* range upwards from the Llandeilo formation, can be otherwise than
Lower Silurian. And if several of the other species in this bed are common to
the Wenlock, as well as to the inferior deposits, of which there is no doubt, this
circumstance only strengthens my view respecting the unity of the Silurian
System. In this survey with Mr. Aveline, Mr. Salter observed some new quarries
of Caradoc sandstone, to one of which the collector is particularly directed. It is
at Gretton, near Cardington. All the characteristic Lower Silurian fossils may
be there obtained in a very brief space of time in true and original Caradoc
sandstone.

The most eastern tract in England where the Caradoc sandstone has been protruded to the surface is at the lower Lickey Hills in Worcestershire, and again near Barr in Staffordshire, at both of which localities it supports the base of the Upper Silurian rocks of the adjacent tracts of Dudley and Walsall. At the Lickey, low heathy hills, which are in truth miniature mountains, are chiefly composed of quartz rocks, identical with those masses on the flanks of the Caradoc and Wrekin which have been formed by the fusion of the same sandstone. The Lickey quartz probably owes its lithological character to a similar cause; though in this case the internal heat has not found issue to the surface in the form of trap rock as in Shropshire. In other words, the Lickey quartz rock is supposed to be on a line of former eruption. On its flank this rock graduates into an ordinary grit or sandstone, in which the Pentamerus oblongus occurs, and thus all doubt of its age is removed; whilst just as in the other tracts above mentioned, this sandstone, particularly on its eastern and southern parts (Colmers and Kendal End), is conformably surmounted by a calcareous shale with a thin course of limestone which, as at Woolhope, Corton near Presteign, and Old Radnor, contains Upper Silurian fossils.

In concluding this chapter, we remind the reader that the Caradoc formation has been shown by stratigraphical, lithological, and zoological proofs to be an integral part of the Lower Silurian rocks, and that its most characteristic fossils are species which also occur in the inferior Llandeilo rocks. It was at the summit of this Caradoc sandstone that I long ago drew the line of demarcation between the inferior and superior masses of the Silurian system; and observations, since extended to many distant regions, have confirmed the general truthfulness of that division.

CHAPTER V.

UPPER SILURIAN ROCKS.

GENERAL CHARACTER OF THE UPPER SILURIAN ROCKS, AS DIVIDED INTO THE WEN-
LOCK AND LUDLOW FORMATIONS.

THE WENLOCK FORMATION OF SHALE AND LIMESTONE, WITH ITS CHIEF FOSSILS,
DESCRIBED IN ASCENDING ORDER.

In many parts of the region of England and Wales now under consideration, wherever the lower strata of the age of the Wenlock
formation are seen to rest upon the older rocks which have been last
described, they are linked together in conformable positions; the
upper bands of the lower group graduating insensibly into the deposits we are now about to consider. Of this succession several
examples have, indeed, been already cited. (See section below Map).

As the older schists and slates of Wales were assuredly at one
period nothing more than finely laminated marine mud, so is it still
more apparent that such was the former state of the greater portion of
the Upper Silurian; for even at the present day it is an accumulation
of similar materials, though in a softer and less coherent state.

Whether these argillaceous masses be examined in the wilds of
Radnor Forest and the eastern parts of Montgomery, in the western
parts of Shropshire (Long Mountain), or in many tracts of North
and South Wales (see Map), they present the same "facies" of a
thick, yet finely laminated, dark, dull grey shale, in which hard stone
of any strength or persistence is the rare exception. Their dominant
character is that of " mudstone."

Ranging chiefly from S. W. to N. E., and resting conformably upon

Caradoc and lower rocks, in numerous undulations*, these mudstones assume, however, in Denbighshire a much harder, flaglike character, and there they have even been so subjected to the influence of a crystalline transverse cleavage as occasionally to form roofing slates.

On the whole, however, the Upper Silurian rocks of North Wales have very much the lithological structure of large masses of the Lower Silurian of South Wales, where slaty cleavage has only so far operated as to leave the mass in that disjointed, incoherent state of mudstone, the "rotch" of the natives, so useless to the mason and the miner, and so cold and profitless to the agriculturist. In all such tracts, where the subdividing limestones are absent, or very feebly indicated, it is only by close observation of the imbedded fossils, that the formations, so clear and typical in other parts, can be recognized.

When, on the contrary, we follow the same deposits from North and South Wales to the exemplar tracts of Herefordshire and Shropshire, where the Upper Silurian rocks were first classified and described, we find them diversified by interpolated courses of limestone; much calcareous matter being also disseminated, both in nodules and in flagstones.

With additions like these to the richness and variety of the subsoil, which are so welcome to the proprietor and farmer, the geologist usually discovers a much greater abundance of fossil animal remains than in the same strata of the sterile western tracts. By observing the order of superposition, and by tracing the divisionary limestones, he reads off the order of the strata, and chronicles with precision the succession of their respective fossils.

* The strike of the same deposits varies in different districts. Thus, in North Wales, the prevalent strike is nearly N.N.E. to S.S.W. In Shropshire the prevalent direction is nearly N.E., S.W.; whilst in S. Pembroke all the Silurian rocks range from W. by N. to E. by S., in conformity with the direction of the great South Welsh coal-field (see Map). The *conformable undulations of the Caradoc and Wenlock formations* in various parts of Wales are represented in several of the published horizontal sections of the Government Survey.

In this way the Upper Silurian rocks are seen to consist, as a whole, of the two formations to which I assigned the names of Wenlock and Ludlow; each of these being subdivided in the manner expressed in this woodcut.

GENERAL ORDER OF THE UPPER SILURIAN ROCKS BETWEEN THE CARADOC SANDSTONE BENEATH, AND THE OLD RED SANDSTONE ABOVE THEM.

c. Pentamerus limestone on the summit of the Caradoc sandstone.* d^1. Wenlock shale with Lower or Woolhope limestone. d^2. Wenlock shale. d^3. Wenlock limestone. e^1. Lower Ludlow. e^2. Middle Ludlow or Aymestry limestone. e^3. Upper Ludlow. f. Bottom of Old Red Sandstone.

The inferior member of the Wenlock formation which rests on the Caradoc sandstone, as seen in Shropshire and parts of Wales, is chiefly a mass of dull argillaceous dark grey shale, rarely if ever micaceous.

Strictly speaking, there is no band of limestone (like d^1) subordinate to the lower shale below the escarpment of Wenlock Edge in Shropshire; the rock to which attention is now called being there merely represented, as also in some other tracts, by calcareous nodules.

In following the formation from Shropshire into Herefordshire, the great limestone, d^3, above the shale, d^2, is seen on the banks of the Lugg, west of Aymestry, to be diminished to a thin irregular stratum chiefly concretionary. Further to the south-west, and in Radnorshire, that rock, still diminishing, disappears entirely amid the mudstones above alluded to; but, to the south of Presteign, an inferior course of limestone sets in in the lowest part of the Wenlock shale. This rock (d^1 of the diagram), which has been alluded to in the last chapter, merits special attention.

* It has been explained, pp. 86. 98., that the limestone with Pentamerus oblongus, which rises from beneath the Wenlock shale, is still grouped by me with the Lower Silurian of Britain. For Russia and N. America, see chaps. 13. 15.

An examination of quarries now abandoned at Corton, one mile south of Presteign, shows indeed, that this limestone is fairly subordinate to a shale (there of black colour), which rests on the Caradoc grit and conglomerate before described; as expressed in this diagram, which is reproduced from p. 90.

LOWER WENLOCK AT CORTON, NEAR PRESTEIGN.

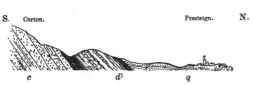

c. Caradoc sandstone, or coarse Corton girt, with Pentamerus oblongus, &c. d^1 Lower Wenlock shale and Woolhope or Lower Wenlock limestone. q. Gravel of the Valley of Presteign.

And again, at Haxwell, a similar arrangement was visible some years ago. But between these two upcasts of the Caradoc sandstone, which exhibit the true relations of the strata, lies the large and loftier rock of Nash Scar, in which the same limestones, whether thick-bedded or nodular, have been run together into one amorphous mass, in which the stratified character has been destroyed, and the shale driven off, as expressed in this diagram.*

ALTERED AMORPHOUS LIMESTONE OF NASH SCAR.
(From Sil. Syst., p. 313.)

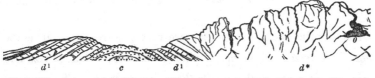

c. Arch of Caradoc sandstone throwing off limestone and shale (Woolhope), d^1, followed by d^*, or amorphous altered limestone. o. Caverns.

In tracing the strata southwards along this axis, other masses of limestone more or less amorphous are seen near Old Radnor, which, in proportion as they approach the eruptive masses of Stanner and

* As the rocks of Nash Scar and Old Radnor are the only great masses of limestone in this region, there being no other between this district and the sea-coast, they are of high value to the Welsh proprietors; the lime being transported to very great distances westwards.

Hanter rocks, and Ousel Hill, or the highly metamorphosed rock of Old Radnor and Yat Hill, are themselves subcrystalline, and unbedded, with coatings of serpentine upon the surfaces of the joints. On the other hand, on receding westwards from that line of eruption and metamorphism, into the Vale of Radnor, to the south-east of Harpton Court *, the limestone begins to resume its bedded character, resting on the pebbly Caradoc conglomerates which range by Old Radnor church and Yat Hill. Whilst this section, taken from one of the coloured sections of the Government Survey, ex-hibits the syenite and greenstone of Hanter Hill throwing off the Ludlow rocks to the south-east, it is also suggestive of the fact, that another body of igneous rock lies subjacent to the Caradoc con-glomerate and crystalline limestone of Old Radnor and Yat Hill, where the coatings of serpentine and brecciated and altered features of the stratified rocks are, in the eye of the geologist, conclusive evidences in favour of such relations.

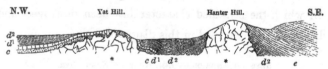

N.W.   Yat Hill.   Hanter Hill.   S.E.

e. Ludlow rocks. d^2. Wenlock shale. d^1. Woolhope or Lower Wenlock limestone (partially altered).   c. Caradoc sandstone (in parts altered).   *. Eruptive rocks. (Syenite, Greenstone, and Hypersthene rock.)

The eruptive rocks of this tract are very picturesque, and as they offer, in a very small compass, phenomena which characterize large mountain masses, a sketch of them is annexed. The spectator is placed near the south-eastern foot of Stanner rocks, near Kington, which are charged with hypersthene, a mineral which, though common in some foreign countries, and abounding at Loch Scavig, in Skye, has hitherto been found in one other British locality only.

* The seat of my esteemed friend the Right Hon. Sir T. Frankland Lewis, who, with his son, Mr. George Lewis, first urged me to put together all my geological documents respecting this region, and thus form the work afterwards called the " Silurian System."

Hanter.   Ousel.   Stanner Rocks.

VIEW FROM STANNER ROCKS (OUSAL WOOD, HANTER HILL, AND HERGEST RIDGE BEING
SUCCESSIVELY SEEN IN THE DISTANCE).

(From Sil. Syst., p. 311.)

Although the greater expansion in Radnorshire of the lower
Wenlock limestone, as compared with the few nodular beds of
Shropshire, must have been in a great measure due to the original
deposit, the form of which, in an unaltered state, is well seen
near Presteign, its amorphorus and massive condition at Nash
Scar and Old Radnor was doubtless caused by the action of heat
issuing along a line of former linear fissure, which, in some
places, evolved the hypersthenic rocks and greenstones of Stanner,
Ousel, and Hanter, upon the line in question, fused the stratified
limestone and the calcareous nodules of the shale into amorphous
and heterogeneous masses, more or less crystalline, leaving coatings
and films of serpentine on their faces and joints.

All the fossils which have been found in the limestone of this
tract, whether by Mr. Edward Davis, who discovered most of them,
and specially assisted me in studying them, or subsequently by
the government surveyors, are truly Wenlock and Upper Silurian
forms.

Some of these, such as the trilobites Bumastus Barriensis, Phacops caudatus, Encrinurus variolaris, with the shells Leptæna lævigata, and Nerita prototypa, large encrinite stems, and several corals, are well known published fossils of the limestones of Wenlock and Dudley*; and in addition to these, even the Pentamerus Knightii has also been found, which, as will hereafter be stated, is a very marked fossil of the Ludlow formation.

Wenlock formation of Woolhope, the Malvern Hills, May Hill, &c. — In describing the upward development of the Silurian rocks on the west flank of the Malvern Hills, a limestone in the same position was described by me as containing a mixture of Caradoc and Wenlock fossils, and it was shown how somewhat further to the south the same stratum becomes an integral part of the Wenlock formation. It is, however, on the outside of the inner dome of the symmetrical valley of elevation of Woolhope, that the lower Wenlock zone is most clearly exposed. Having been raised equably on the outward face of a central dome or nucleus of sandstone, the nature of the calcareous deposit can be seen in detail on the sides of the roads, whilst the best limestones, which have been largely opened out since I described the tract, afford numerous fossils which were unknown when the " Silurian System." was published.†

SECTION ACROSS THE ELEVATED VALLEY OF WOOLHOPE.
(From Sil. Syst.)

W. S. W.　　　　　　　　　　　　　　　　　　　　　　　E. N. E.

c. Caradoc sandstone. *d*¹. Woolhope or Lower Wenlock limestone and shale. *d*². Wenlock shale. *d*³. Wenlock limestone. *e*¹. Lower Ludlow. *e*². Middle Ludlow, or Aymestry limestone. *e*³. Upper Ludlow. *f*. Old Red Sandstone (base of).

* See Sil. Syst. p. 313.

† It was my want of acquaintance with any of these fossils in the limestone of Woolhope (for they were unknown when I wrote), that led me to class this rock with the upper part of the Caradoc; for if fossils should be detected in it, I naturally inferred that they would be the same as those of a limestone in a similar position in Shropshire, which, from its containing Pentamerus oblongus, was classed by me as the Uppermost Caradoc. (See pp. 86. 98.)

In ascending order, the beds may be thus described. The top of the Caradoc, or the central dome of Woolhope, *c*, with its transition beds, is a greenish earthy sandstone, containing the following fossils: Pentamerus lens (Atrypa, Sil. Syst.), Atrypa reticularis, Linn. (A. affinis, Sil. Syst.), Petraia bina, Lonsdale.

This course is followed by shale with calcareous, gritty flagstones, containing the large Pentamerus liratus (Atrypa lirata, Sil. Syst.), Foss. 8. f. 3., and in the more nodular portions abundance of Leptæna transversalis and Atrypa reticularis. Then succeed layers of an impure earthy limestone, containing Strophomena depressa, a species which, though occurring in the Lower Silurian rocks of North Wales, becomes much more abundant in the Wenlock formation. In the next overlying bands of shale, calcareous matter so increases as to form strong beds of dark, indigo-coloured, argillaceous limestone, which, as on the west flank of the Malverns, is characterized by cross veins of pink and white calcareous spar. The strongest bed (about 10 feet thick) is a tough, impure, earthy limestone, largely extracted for the roads, which again is covered by a band of purer limestone, followed by shale and an upper bastard limestone. Now, although the whole of this subformation, d^1, does not contain limestone courses which have an united thickness exceeding 20 or 30 feet; still the rock, together with its interpolated beds of sandy shale, is seen to occupy a very great surface on the map, from its being so equably spread out at a low angle of inclination. Among the chief fossils of the limestone, which I have recently collected on the spot, are the two trilobites of the following woodcut, viz. Bumastus Barriensis, Foss. 9. f. 2., and Homalonotus delphinocephalus, f. 1. There are also Phacops caudatus (Asaphus, Sil. Syst.); Cornulites serpularius; Orthoceras annulatum; Phragmoceras —— sp. ; Euomphalus sculptus; Spirifer elevatus; Strophomena Phillipsi (imbrex, Davidson); S. pecten ; S. depressa ; Leptæna

transversalis, Dalm.; Orbicula Forbesii; Atrypa reticularis; Rhynconella Wilsoni, &c.; and rarely Orthis elegantula.

FOSSILS, 9.

TRILOBITES OF THE LOWER WENLOCK OR WOOLHOPE LIMESTONE.

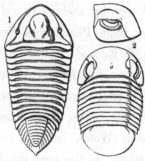

1. Homanolotus delphino-
cephalus, Green. 2. Bu-
mastus Barriensis.

These figures are about one
third the natural size.

Here again, as in Radnorshire, we have a great predominance of
true Wenlock fossils, and one, the Rhynconella Wilsoni, which, as
will presently be seen, is most abundant in the highest Silurian
division, or Ludlow rocks; and thus in every respect, whether as to
position or fossils, the Woolhope and Radnorshire limestones are
identical. The overlying shale, d^2, and the great limestone, d^3, of the
preceding section of Woolhope, p. 106., are precisely the same as the
chief masses of the Wenlock Edge, and these are surmounted by the
Ludlow rocks, e, which dip under the old red sandstone, f.

The Silurian rocks, which range from the Malvern Hills and the
Woolhope district to May and Huntley Hills in Gloucestershire,
exhibit the same general order of succession, accompanied, however,
by modifications of the lithological and zoological distinctions of
the lower member of the Wenlock formation, which it is right to
notice. Instead of being the compact, hard, tough, and strong-bedded
rock of Woolhope, it becomes, on the western flank of May Hill, a
group of nodules and very irregular courses disseminated in shale, and
thus simply forming rather a more calcareous base of the Wenlock
shale than what is seen under the Wenlock Edge and other places

where such nodules occur. Here, indeed, the upward development from a thick nucleus of red-coloured Caradoc sandstone beneath, into a series of interlaminated sandstones and shale, which assume to a great extent the fossil characters of the Lower Wenlock, is well exhibited. At the same time, both the Pentamerus undatus and P. lens, which are Caradoc fossils, occur amongst the shells. It is, in fact, impossible here to draw a rigid line of demarcation between the Lower and Upper Silurian. In one part of this region, as on the west flank of the Malverns, the lower type ascends rather higher, whilst, in another, the upper fauna expands downwards. In no portion of Britain, therefore, are the two formations of Caradoc and Wenlock better linked together than in the Malvern and May Hill region.*

This tract, so elaborately described by Professor Phillips, was also carefully studied by Sir Henry de la Beche himself, when the line of demarcation between the Lower and Upper Silurian rocks fixed upon by the government surveyors was made the same as that which had been originally suggested in the "Silurian System."† On the west flank of the North Malverns, the Caradoc sandstone with characteristic fossils passes upwards into, and is interlaced with, sub-calcareous shales and nodular bands, in which some Lower Silurian species are intermixed, with others which are essentially Wenlock. After lucidly explaining how such transitions are in harmony with well understood operations in nature, Professor Phillips thus writes:—

"Though practically we must draw a line of division, let us not forget,

---

* Professor Sedgwick has recently endeavoured to show, that a sandy portion of the beds on the west flank of May Hill, which occupy this intermediate position, ought not to be classed with the Caradoc, as before alluded to, as they have been, by the government surveyors and myself, but should, from their fossils, be ranged with the Wenlock. If this suggestion be adopted, it will only strengthen the opinions of Professor Phillips and myself, as to the difficulty of every where drawing a decisive fixed line between the older and younger groups of this natural system.

† See Sil. Syst. p. 442. et seq.

that in this locality there is no firm and hard boundary between the
Lower and Upper Silurian periods; for both by mineral and by organic
evidence, the characteristics of these periods are found to overpass each
other; the older characters reappearing within the later deposits, and
the later characters showing themselves amidst the earlier deposits."
"As a general result," he adds, "it is quite evident, that the successive
changes of organic forms, as they are exhibited to us in the successive
groups of strata, are not simply dependent on the lapse of time, nor
explicable as a series developed in proportion to the time, unless we
survey the phenomena over very wide areas, and include in the comparative terms geological periods long enough to neutralize the influence of peculiar physical conditions. These," he truly says, " on
account of their local origin, limited area of effect, and recurrence at
indifferent periods, have at almost every geographical point, at some
epoch or other, broken or mingled the series of organic life." * This is
and has long been my belief, as founded on extensive observation.

A limestone subordinate to shale, and bearing precisely the same
relations to a subjacent sandstone as that of the localities above cited,
occurs on the western and south-eastern flank of the Lower Lickey
Hills in Worcestershire. There, the Silurian rocks consist of Caradoc
sandstone, with its usual fossils, overlapped on its edge by shale, in
which are courses of limestone (at Colmer's End), the whole of the
mass having been thrust up as an irregular dome through the overlying coal measures and red sandstone. Again, in the adjacent tract
near Walsall, and between that town and the Barr Beacon, the same
Lower Wenlock or Woolhope limestone, long worked at the Hay
Head, dips away from a point of Caradoc sandstone†, to pass at a
gentle angle of 8° or 10° under the great body of the shale with its
calcareous nodules and numerous small fossils; the whole being

---

* See Mem. Geol. Surv. of Great Britain, vol. ii. p. 75.

† Mr. Jukes recently indicated to the Geological Society the existence of this
point of Caradoc sandstone. See also his published Survey of this district,
Mem. Geol. Surv. vol. vi.

covered by the thick or chief limestone as exhibited in the great quarries of Walsall.   There also, as described in my original work, this lower limestone, with which the Woolhope limestone is now identified, is of the same small dimensions as in other localities.   In it are nearly all the very same fossils as at Woolhope, and notably the Bumastus Barriensis, Foss. 9. p. 108., named by me from the adjacent village of Barr, and known to fossil collectors as the "Barr Trilobite;" together with many forms common to the whole formation, including orthoceratites or corals.*   In this way, proofs have been obtained, that a limestone whose real place I indicated in several detached districts, as being inferior to the great mass of shale, is truly by its fossils the lowest member of the Wenlock formation.

Wenlock Shale. — This shale, which is infinitely the largest and most persistent member of the formation, is often distinguished by its layers of deposit being marked by a few small elliptical and round concretions of impure and very earthy limestone, which increase in purity as the strata ascend towards the chief or upper limestone. This shale is well exposed on the banks of the Severn, near Coalbrook Dale and the Iron Bridge, where it is called " Die Earth" by the miners, and is thence to be followed all along the escarpment of Wenlock Edge, occupying a broad valley of denudation between it and the ridge of the Caradoc, called Apes Dale.   In the Malvern district it is also a mass of finely levigated argillaceous matter, the lowest part of which is more charged with calcareous matter than in Shropshire, and in which a large spheroidal structure is apparent.   In that district Professor Phillips estimates the thickness of the shale at about 640 feet.   In some parts of Wales the Wenlock shale is as incoherent as in those English counties where it appears; but in Denbighshire, it is a subcrystalline slaty rock. (See map and general section.)

The prevailing fossils of the Wenlock shale, exclusive of trilobites and corals, which are mostly of species unknown in the lower

* See Sil. Syst., p. 488.

deposits, are small brachiopodous shells of the genera Leptæna, Orthis, Strophomena, Atrypa, and Rhynconella. In a general way the fossils of this stratum, the chief of which are given in plates 20. and 22., are the same as those of the overlying limestone. The Orthis biloba of Linnæus (Spirifer sinuatus, Sil. Syst.), O. hybrida, and the large flat Orthis rustica, are characteristic shells in this formation. Leptæna lævigata, and L. transversalis; Pentamerus linguifer, and Athyris tumida, Dalm. (Atrypa tenuistriata, Sil. Syst.); Rhynconella rotundata; R. depressa; R. Stricklandi; R. deflexa, and R. sphærica; and in some parts R. navicula, are the chief brachiopodous shells.

Among them, however, are several common to this deposit and the Lower Silurian rocks, such as Orthis elegantula, so common in the slates of Snowdon; Strophomena pecten and S. (Leptæna) depressa, Atrypa marginalis, and A. reticularis, and the Spirifer plicatellus. Most of these have a great range, and pass upwards through the Wenlock, high into the Ludlow formation. Aviculæ, Nuculæ, and a few other bivalve shells occur frequently; but few of them are characteristic except the Cardiola interrupta. Of spiral shells, Euomphalus funatus and E. alatus, Pileopsis (Nerita) Haliotis, Turbo cirrhosus, Bellerophon Wenlockensis, and B. dilatatus are the most frequent. Theca Forbesii and T. anceps are pteropod molluscs, not uncommon in this deposit. Certain Orthoceratites, such as O. annulatum, O. filosum, and O. angulatum, are rare; but many of the thin-shelled species, O. subundulatum, O. primævum, and others, abound in these muddy sediments. They are almost the only shells in this formation, as it is exhibited in Denbighshire and other parts of N. Wales, and occur there in the greatest abundance. Phragmoceras of one or two species, and certain Lituites, e.g. L. articulatus, L. Biddulphii, and occasionally L. giganteus, are conspicuous forms; but these two last-mentioned genera are not commonly met with.

Of trilobites, Encrinurus punctatus and E. variolaris, Calymene

Blumenbachii and C. tuberculosa, Phacops caudatus and its variety
P. longicaudatus, are characteristic, which, with many other smaller
species, are the same with those of the Wenlock limestone. The
genera Trinucleus, Asaphus, and Ogygia are never detected even
in the lowest portions of the shale, these forms being essentially
characteristic of the Lower Silurian rocks.   Cornulites serpularius,
pl. 16. f. 6., and other annelides, are sometimes found, as well as
stems and portions of encrinites, though the perfect fossils are very
rare in the typical region.*   With respect to corals, it may be stated
generally, that they are the same as those of the Wenlock limestone,
but fewer in number; the cup corals, Cyathophyllum, Omphyma, &c.,
Favosites alveolaris, and Stenopora fibrosa, being the most conspicuous.
We ought not to omit to notice Graptolites priodon (Ludensis, Sil.
Syst.), which is a most abundant and characteristic fossil of the
Wenlock shale, and has been already figured as occurring in the Lower
Silurian rocks, Foss. 6. f. 3., and pl. 12. f. 4.

Wenlock Limestone. — The upper member of the formation is a
limestone usually of lighter grey colours than the lower band already

VIEW OF WENLOCK EDGE, AS SEEN FROM THE HILLS OF LUDLOW ROCK ON THE S. W.
(From Sil. Syst.)

The linear ridge on the left is the Wenlock limestone dipping under the higher ridge of Ludlow
Rocks.   The valley of Apesdale on the extreme left is in the Wenlock shale.

* One of the most elegant of the Silurian encrinites, the Actinocrinus pulcher,
Salter, which is found in strata of this age in North Wales, will be mentioned in
chapter 9.

I

described, and in every respect identical with the well known limestone
of Dudley, but which I named after the sharp rectilinear ridge of
Wenlock (see Map), because its relations to inferior and superior
deposits are there better seen than in any other part of the British
Isles.

It consists of thick masses of grey, subcrystalline limestone, occasionally, but very rarely, of light pink colour, in part very argillaceous, in others more crystalline, which is highly charged with
encrinites and replete with corals.   The rock is essentially of a concretionary nature, and thus differs much from the flat-bedded Woolhope limestone ; being for the most part marked by nodules of small
size.   Occasionally the concretions are very large, and are then
locally termed " ballstones."

This limestone is underlaid and overlaid by shale of pale grey and
greenish tints, copiously charged with small nodules of argillaceous
limestone, of very irregular persistence.   The large concretions or
"ballstones" (some of which near Wenlock have a diameter of

OLD QUARRIES IN THE WENLOCK LIMESTONE.

The dark cavities indicate the places from whence the ballstone or best crystalline limestone
has been extracted.   The fossils chiefly occur in the surrounding layers of impure earthy
limestone and shale.

80 feet) being more crystalline, have been quarried out as the best flux for the smelting of iron, and their extraction has left caverns in the quarries, as expressed in this woodcut.

But though very thick at Wenlock, this limestone thins out so rapidly in its course to the south-west, that at Haven in the interior of the Ludlow promontory (see woodcut in the next chapter) it is represented by thin courses of small concretions only; and on the banks of the river Lugg, west of Aymestry (in both of which localities it was first recognized by the Rev. T. T. Lewis), it is alone represented by a few coralline concretions, varying in size from 2 inches to 2 feet, but still full of the characteristic fossils.   Thinning out entirely in Radnorshire, it is scarcely to be recognized throughout the counties of Brecon, Carmarthen, and Pembroke; its place being only marked in the cliffs of Marloes Bay, to the west of Milford Haven, by some characteristic fossils and a very small quantity of impure limestone immersed in grey and sandy shale. (See lower, vignette, p. 66.)

In the districts of Malvern, Woolhope, May Hill, and Uske, however, the Wenlock limestone is copiously and instructively developed; and exhibits in numberless natural sections and quarries the same characters as in Shropshire. On the west flank of the Malverns, where in the Ridgeway of Eastnor Park it assumes the same linear outline as at the Wenlock Edge, the limestone is estimated, by Professor Phillips, as having a thickness of 280 feet. There, in the flexures between the Ridgeway and Ledbury, it is admirably exposed, together with the overlying Ludlow formation. (See section, p. 94.)

To the north of the town of Dudley, this limestone rises up into various domes called the Castle Hill, Wren's Nest, and other smaller elevations, which have been protruded from beneath the surrounding coal strata.   The Castle Hill is here represented to the left hand of the spectator, who is standing on the slopes of the Wren's Nest, and looking over the town of Dudley, to the hills of basalt near Rowley.

Castle Hill.   Rowley Hills.   Hagley Hills.

DUDLEY, FROM THE WREN'S NEST.
(From a Drawing by Lady Murchison, Sil. Syst., p. 480.)

As the signs of violent igneous action are apparent, both in the subterranean rocks of this rich mining tract, and in the various outbursts of basaltic or trappean rocks which have been extended to the surface (Rowley and Pouk Hills, &c.), it is fair to infer that these domes have been thrown into an inflated and arched form by subterranean expansion.* The Wren's Nest and Castle Hill thus exhibit on each side two courses of limestone, which, from its superior quality, has been alone worked out, first in open quarries and afterwards by deep subterranean galleries. The annexed diagram, which exhibits a section through the Wren's Nest, shows how the two bands of the best limestone, d^3, have been extracted, leaving

---

* See the account of the igneous rocks of this district, Sil. Syst., p. 496. First described by Mr. Keir and others, and subjected to experiments by Mr. Gregory Watt, their subterranean relations to the strata, were ably described by Mr. Blackwall of Dudley, after the publication of the " Silurian System." The descriptions of these and other phenomena connected with the region around Dudley, are given in detail by the Government Surveyors, Records of the School of Mines, vol. i. pt. 2. p. 240.

SECTION ACROSS THE WREN'S NEST.

(From Sil. Syst., p. 484.)

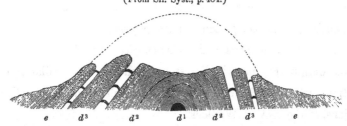

d^1. Lowest Wenlock shale (place of Woolhope limestone?). d^2. Wenlock shale. d^3 Two bands of limestone separated from each other by bavin. They are represented by the white spaces, from which the best limestone has been quarried, leaving only arches of support. e. Overlying bavin and shale passing up into Ludlow rocks.

only a few arches of support; whilst the other beds, consisting of impure and nodular earthy limestone, thus form the framework of the hill. The dotted arch indicates what the dome must have been when entire, whilst the tinted portion of the drawing represents the body of the hill, the upper surface of which, having been hollowed out towards the centre, has obtained the name of Wren's Nest. It is, therefore, an elevated, inflated dome, which was probably truncated at its summit during the same period of disturbance and denudation which gave to the mass its peculiar form.

When viewed from below on its southern face, and where neces-

THE SOUTH END OF THE WREN'S NEST.

(From a Sketch by the Rev. W. Whewell, D.D. Sil. Syst., p. 485.)

The spectator, looking to the north, sees how the limestone strata fold over a central dome of shale.

sarily the excavated outline of the summit cannot be seen, the flank of the dome of shale and "bavin," from whence the limestone has been entirely removed, presents this appearance.

In the exterior nucleus, which also consists of shale, the calcareous matter disappears; but the limestone itself is quite similar in composition, colour, and large ballstones, as well as in its small concretions, to the rock at Wenlock.

At Woolhope in Herefordshire (see p. 106. and woodcut) a similar uprise from the centre has there produced a much grander phenomenon. Strata lower than any seen at Dudley are there, as before explained, brought to view; and the denudation of the valleys which lie between the ridges that encircle the central dome has been so complete, as to render it the finest known example, within the British Isles, of a valley of clean denudation as well as of elevation. Not only have no extraneous loose materials been translated to it from other tracts, but every fragment derived from the mass of rocks which must once have arched over it, has been swept out of the central and encircling hollows; a striking proof of the forcible agency exerted in the denuding operation.*

In making an approximate estimate of the thickness of the Wenlock formation in the original Silurian region, I spoke of 700 feet for the shale, and about 300 for the limestones, including the nodular beds. Subsequently measuring the strata of the same age near the Malvern Hills, Professor Phillips gives them a thickness of about 920 feet, which, with the addition of the Lower Wenlock of Woolhope limestone and shale, which were not incorporated in the formation when I wrote, amounts to about my estimate. If, however, the upper or chief limestones, including all the beds of "bavin" or impure nodular limestone, be measured at Dudley, that portion of the formation must be considered as considerably thicker

* Sil. Syst., p. 436.

than in Salop or Hereford. In North Wales, though void of lime-stone, the formation is often of much greater dimensions, amounting probably to more than 2000 feet.

In the above-mentioned districts of England, the Wenlock lime-stone passes upwards gradually into a thick mass of pale-coloured shale, undistinguishable from that beneath the solid rock. From its physical relations, and from its forming usually a part of the same hills as the mass of the Ludlow rock, it is classed with that formation; though the reader must understand, that in reality it is only an intermediate band intimately connecting the Wenlock and Ludlow in the Upper Silurian group, and is almost equally related to both by its structure and fossil contents.

The most prominent fossils of the limestone, among which corals abound, have been long known to collectors, who have derived their

FOSSILS, 10.

CORALS OF THE WENLOCK LIMESTONE.

1. Favosites cristata, Linn. 2. F. Gothlandica, Linn. 3. Do. variety, and 3*, 3**, magnified portions of two varieties. 4. Favosites alveolaris, Goldf. 5. Alveolites Labechii, Milne Edw. 6. Favosites oculata, Goldf.? 7. Stenopora fibrosa, Goldf. 8. Do. variety incrusting shells (Alveolites fibrosa, Sil. Syst.).

chief supplies from Dudley and its environs. My friends who are searching for Wenlock fossils, may, however, be told that their labour will be better rewarded by a search at the northern end of the

Wenlock Edge, near Coalbrook Dale, than even at Dudley. The corals, encrinites, and shells, along this rich escarpment, are not only more easily detached from the matrix, but are less eagerly collected by dealers. I would particularize Benthall Edge, which overhangs the Severn in so picturesque a form, and Gliddon Hill, as excellent localities.* This rock is indeed distinguishable from all the inferior strata, by the very great abundance of its corals, the profusion of which make it resemble in many places a coral reef. These fossils were admirably described for me in the original " Silurian System " by my valued friend Mr. Lonsdale.

Fossils, 11.

CORALS OF THE WENLOCK LIMESTONE.

Millepore Corals of the genus Heliolites (Porites, Sil. Syst.), most nearly related to the " blue coral " Heliopora cærulea, Blainv., of the Australian coral reefs.

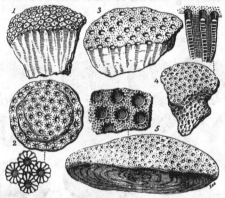

1. Heliolites tubulata, Lonsd. 2. H. petalliformis, Lonsd. 3. H. megastoma, M'Coy. 4, 5. H. interstinctus, Wahl.

Three woodcuts, representing the most common, perhaps, of these corals, will be given in chapter 9.

In addition to the Halysites catenulatus, or universal chain-coral, Foss. 12. f. 6. ; and the Stenopora fibrosa, Foss. 10. f. 7.; which are

* Some of the finest corals originally published in the " Silurian System," were collected by the Rev. T. T. Lewis, in the gorge of the river Lugg, above Aymestry.

Fossils, 12.

CORALS OF THE WENLOCK LIMESTONE.

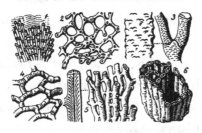

no less common in the Lower Silurian rocks (see chap. 8., p. 176.), the following species of corals are everywhere typical of the Wenlock limestone : —

Heliolites interstinctus (Porites pyriformis of my old work), Foss. 11. f. 4., 5., both the large and small-celled varieties; H. tubulatus, f. 1.; H. petalliformis, f. 2.; Favosites alveolaris, Foss. 10. f. 4.; F. cristata, f. 1.; F. oculata, f. 6.; Cœnites juniperinus, Foss. 12. f. 3.; Syringopora ramulosa, f. 5.; and its young or creeping form, fig. 2. (Aulopora serpens, "Sil. Syst.") ; S. bifurcata, f. 4.; Omphyma turbinatum, chap. 9. Foss. 34. f. 4.; Cyathophyllum truncatum, ib. f. 2., and C. articulatum, f. 1.; Acervularia ananas, f. 6. ; with many others.

Of the crinoids the more perfect of the forms published in my old work are alone reproduced, in the plates of this volume.* Perhaps the large species Periechocrinus (Actinocrinus) moniliformis, pl. 13. f. 1, 2., is the most characteristic; for it covers large surfaces of the limestone at Dudley, and is found in disjointed fragments in many other localities.

Among the mollusks of this formation, which are figured in plates 20. to 33., Orthoceratites are very abundant, O. annulatum and O. Brightii being frequent species.   Bellerophon dilatatus, B. Wenlockensis, and the singularly beautiful Conularia Sowerbyi, pl. 25., are characteristic.   Among the most frequent spiral shells are five species of Euomphalus, viz., E. carinatus, E. sculptus, E. discors, E. funatus, and E. rugosus (pl. 24, 25.).

Ordinary bivalve shells are less common ; but Orthonota cingulata,

* See Plates 13. to 15.

Avicula reticulata, Pterinea retroflexa and P. planulata, chap. 9.
Foss. 42. are abundant species. Of the brachiopods, Strophomena
euglypha; S. filosa, and especially S. depressa; Pentamerus galeatus;
Spirifer plicatellus (S. trapezoidalis, S. radiatus, and S. interlineatus,
Sil. Syst.); S. crispus and S. elevatus; Orthis rustica, O. elegantula,
and O. hybrida, Leptæna transversalis, and L. lævigata*, are all of
them shells described in the " Silurian System;" and of the Terebratulæ, Rhynconella borealis (R. lacunosa of my old work), pl. 22. f. 4.
is by far the most common species; it is generally in company with
R. cuneata, R. Salterii, and R. Wilsoni — the latter sometimes very
abundant.

Trilobites are very common; the most frequent of them is the
so called Dudley Locust or Calymene Blumenbachii, a species which,
as we have seen, also occurs deep in the Lower Silurian rocks,
and, as we shall presently find, ascends also into the Upper Ludlow.
Other forms of these creatures are also prevalent; a few may be
noticed. Encrinurus punctatus, chap. 9. Foss. 46. f. 5. and E.
variolaris, f. 6. are very common in both the limestone and the shale.
Phacops Downingiæ is one of the most characteristic trilobites, particularly in the environs of Dudley. The Phacops Stokesii (Calymene macrophthalma of my former work), Phacops caudatus,
Acidaspis Brightii, and Cheirurus bimucronatus, are frequent fossils,
and, as before stated, belong also to the inferior formations. The
Phacops caudatus is especially abundant in the Malvern Hills;
whilst the Calymene Blumenbachii is the reigning fossil at Dudley.
Bumastus Barriensis and Homalonotus delphinocephalus, which have
been cited from the lower Wenlock or Woolhope limestone, are
also found in this rock, the former very frequently. The annelides
Cornulites serpularius and Tentaculites ornatus occur on almost
every specimen of the limestone at Dudley.

* See Plates 20. to 22.

A further acquaintance with the fossils of the Wenlock formation must be obtained by consulting the 9th chapter, as also the numerous descriptive monographs mentioned in the Preface. The object of this work is not to direct attention to rare and curious species, but chiefly to those forms which are characteristic of the rocks.

CHAPTER VI.

UPPER SILURIAN ROCKS — *continued.*

THE LUDLOW FORMATION, GENERAL CHARACTER OF.   ITS SUBDIVISION IN THE
TYPICAL DISTRICTS, INTO LOWER LUDLOW ROCKS, AYMESTRY LIMESTONE, AND
UPPER LUDLOW ROCKS.

IN a general sense the Ludlow rocks of the Silurian region of
England and Wales must be viewed as a continuation of the
same argillaceous masses which prevail in the underlying Wenlock
formation.   Such, however, is more particularly the case in the
lower beds of this deposit.   The central portion consists in several
tracts of an argillaceous dark-grey limestone.   The upper member
being more sandy and somewhat calcareous, and yet retaining in
parts much of the "mudstone" matrix, is in great measure an imperfect, thin-bedded, grey-coloured, earthy building stone.   Occasionally the highest stratum is composed of light-coloured sandy
freestones and tilestones, through which the formation graduates
lithologically and conformably into the lowest beds of the Old Red
or Devonian rocks.

Such is the general order near the town of Ludlow, which stands
upon the higher strata of the formation.   Its central and inferior
masses are best seen either in the escarpments of the adjacent hills
on the S. W., or in that ridge which for a distance of twenty
miles on the N. E. is interposed between the Wenlock Edge and
the old red sandstone of Corve Dale and the Clee Hills.   The
section and view at pp. 102. 113., will give the reader an adequate
idea of the succession.   (See also Map and coloured section.)

In following the formation from the Ludlow tract on its strike or

direction to the S. W., its included limestone, like that of the subjacent deposit, is also soon seen to thin out and disappear.   Scarcely

LUDLOW CASTLE.
(From a Sketch by Lady Harriet Clive, Sil. Syst. p. 195.)

In this sketch the river Teme is seen to flow in a chasm of the Upper Ludlow Rocks, the strata
on which the spectator is supposed to be standing, being the same as those on which the
castle is built.   The basaltic Titterstone Clee Hill is in the distance.

has the geologist quitted the north-western corner of Herefordshire,
than he finds the central band attenuated to a mere course of calcareous grit, which is entirely lost in Radnorshire.   There the upper
Silurian rocks of the mountain of Radnor Forest, which are laid open
in the ravine of "Water break its Neck" and other gullies, expose a
gradual succession from the Wenlock through the whole of the
Ludlow formation up to the junction beds of the old red sandstone,
and with scarcely a trace of limestone.   As such also the formation
ranges for the most part through Brecon and Carmarthen, the central
part being nowhere a workable limestone, and only here and there
calcareous, except through the occasional presence of the shelly
remains of a few fossils.   In Marloes Bay, Pembrokeshire, where
the Silurian rocks are exposed in the sea-cliffs (see view, p. 66.),
it is difficult to say more than that sandy calcareous shale and very
impure limestone containing Wenlock fossils, are surmounted by
ferruginous and hard sandstone, rarely calcareous, and in parts a
conglomerate, but full of characteristic Ludlow rock fossils.   Though
the uppermost of these beds are not unlike some of the rocks of

Ludlow, yet without this change in mineral character, the same order of superposition prevails in Pembroke as in Salop and Hereford, or in the Clyro and Begwm Hills, on the banks of the Wye, between Hay and Builth, where I first marked this succession. (See Map, and Sil. Syst., pp. 5. 312.).

Let us now consider the nature of the different members of the Ludlow formation, where they are most conspicuously characterized by a diversity of mineral matter and the greatest quantity of fossil remains.

Lower Ludlow Rock.—These strata, which are, I repeat, simply an upward prolongation of the uppermost member of the Wenlock formation, are composed of dark grey shale, very rarely in any degree micaceous, with small concretions of impure limestone. My chief reason for grouping them with the Ludlow rather than with the Wenlock rocks was, that every where in the typical districts of Shropshire and Herefordshire, these shales occupy the base of the escarpments of the same ridges of which the harder masses of Aymestry limestone and Upper Ludlow rocks form the summits and outward slopes.

The clearest and finest examples of this are seen in the deeply excavated interior of the Ludlow promontory, particularly in descending from the Comus Wood of Milton* into the depression of

SECTION ACROSS THE LUDLOW PROMONTORY.

(From Sil. Syst., pl. 31. f. 5.)

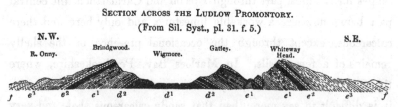

d^1. Wenlock shale. d^2. Wenlock limestone. e^1. Lower Ludlow. e^2. Aymestry limestone. e^3. Upper Ludlow. f. Old Red Sandstone (bottom beds of).

* Milton passed some time in Ludlow Castle, in his day a border Welsh fortress, and his "Mask of Comus" was performed in it. The scene immortalized by the name of Comus Wood is one of the deep depressions which vary the surface of the Ludlow promontory.

Mary Knoll Dingle, and thence into the great valley of denudation of Wigmore, in which the Wenlock formation is exposed as expressed in the preceding section.

The inferior strata, for the most part argillaceous, are often arranged in spheroidal masses, showing a tendency to concretionary structure, which rapidly exfoliate under the atmosphere, and break into shivery fragments. Calcareous nodules, differing only from those of the Wenlock deposit in being usually of a blacker colour, have often been formed around. an Orthoceras, Trilobite, or other fossil, as a nucleus.

One of the most prevalent of these fossils is our old friend Calymene Blumenbachii, whose acquaintance the collector may have first made in the lower rocks. (See also ch. 8.) It is accompanied very frequently by the long-tailed Asaphus, or, as it is now called, Phacops longicaudatus. These two may be considered the characteristic trilobites of the formation, though there are several other species. And, with these, the Graptolites priodon, or Ludensis*, occurs abundantly. Nor are the Cardiola interrupta, pl. 23. f. 12., and Murchisonia Lloydii, pl. 24. f. 5., less characteristic fossils.

In ascending, the strata become somewhat more sandy, constituting thick, earthy, and very slightly calcareous flagstones, the "pendle" of the workmen, in which orthoceratites prevail, the flaglike separation being due to laminæ of sand.

These strata form the support of the Aymestry or Ludlow limestone, from which they are usually separated by soft soapy beds, in parts an imperfect fuller's earth. It is the decomposition of this unctuous fuller's or "Walker's" earth of the natives, beneath heavy masses of the limestone, which rest upon it, which has occasioned numerous landslips both near Ludlow and in other parts of Herefordshire, one of the most striking of which will presently be mentioned.

* Ludensis is the Latin word, signifying "of Ludlow."

The lower shale occupies the escarpments and contiguous valleys of the Ludlow rocks which range from Shropshire by Presteign to Radnor Forest, and also large undulating tracts of the western parts of Shropshire or contiguous parts of Montgomeryshire, such as the Long Mountain, and other tracts around Welshpool and Montgomery (see Map). From the escarpments west of Kington, it extends, together with the upper members of the Ludlow rocks, to the banks of the Wye, and is finely exposed in the noble escarpment at the western end of the Forest of Mynydd Epynt in Brecknockshire. A good idea of the outline of that tract is conveyed by a sketch given in the 10th chapter, where the rounded and soft outline of the escarpment in the middle ground, as seen from the slaty hills on the west, is due to the soft nature of this deposit. The same rock is also well developed in the Malvern tract, where Professor Phillips assigns to it a thickness of about 750 feet. (See also section, p. 94.)

In the Woolhope elevation and the group of Usk, or as lying between the Dudley and Sedgley (Wenlock and Aymestry) limestones in Staffordshire, it is everywhere the same dull, non-micaceous shale, which, from its incoherence, has, for the most part, been denuded; thus giving rise to a deep valley, which separates the harder parts of the Wenlock and Ludlow rocks from each other.

In the environs of Ludlow, and in many parts of the Silurian region, this inferior member of the Ludlow rocks is specially characterized by a profusion of straight or curved chambered shells, Orthoceratites, Lituites, and Phragmoceras, a genus named by my friend Mr. Broderip, and which was unknown before the publication of the "Silurian System." Orthoceratites abound; not less than eleven species having been originally figured as characteristic of this rock. But extended researches have shown that, in this case as in many instances of other fossils alluded to, several of these chambered shells occur in much older as well as much younger members of the system. Among the orthoceratites, the largest, and perhaps the most common, are the O. Ludense, pl. 28. f. 1., and O. filosum, pl. 27. f. 1.; these are generally accompanied by a smaller, thin-shelled species, which appears to be the O. subundulatum of Portlock, Foss. 44. f. 3. The others, though often found, are by no means so abundant or characteristic. Of Lituites, the only common species is a very large one, eight or ten inches in diameter — the L. giganteus of pl. 33. f. 1, 2. The Phragmoceras, just noticed, is a most remarkable shell with the mouth or opening contracted into the shape of a key-hole. Some of its forms are flattish, broad, and shaped like a hatchet-head (pl. 32.); whilst others, known as the pear orthoceratites (pl. 30.), are suddenly swelled into a balloon-like shape above, and end below in a tapering point.

Among the lamellibranchiate shells, Cardiola interrupta, Broderip, is the most common (pl. 23. f. 12.); yet this same species, formerly believed to be peculiar to this zone, has also been found in the Llandeilo formation! Another bivalve equally characteristic, and as yet known only in Upper Silurian rocks, is the Cardiola striata (pl. 23. f. 13.). This, in company with the chambered shells above noticed, is to be found in all the fossil-bearing localities of Shropshire and the neighbouring regions. Orthonota rigida, pl. 23. f. 8., and Pterinea retroflexa, f. 17., are not such common species; and the latter is far more frequent in the Upper Ludlow rock.

Many species of fossils have indeed been found common to this stratum and the Wenlock formation; and this is especially the case with the brachiopodous shells. Among these may be cited the Pentamerus (Atrypa) galeatus, Strophomena depressa and S. euglypha, Leptæna lævigata, Atrypa reticularis, and Rhynconella (Terebratula) Wilsoni. Nor are there any brachiopods strictly peculiar to it, unless it be the small but very characteristic Lingula lata, pl. 20. f. 6. In this view, the Lower Ludlow shale might be classed with the Wenlock formation; but the other forms of mollusks, above noticed, give to it nevertheless a distinct character, and entitle us to class it in the Ludlow formation.

Aymestry or Ludlow Limestone.—The want of persistence over wide areas of any mass of solid limestone in the centre of the Ludlow formation has been already adverted to. In some parts of South Wales, where the calcareous matter is absent, it is even difficult to trace the place of this band in the Ludlow rocks; but in Herefordshire and Shropshire, and again at Sedgeley in Staffordshire (near Dudley), it is a dark-grey limestone, worked pretty extensively for use. And even where the lime is very sparingly distributed, the rock is a highly calcareous flagstone, and may generally be recognized by its predominant fossils and well-defined joints.

This central member of the Ludlow formation was named by me after the beautiful village of Aymestry, where the rock is well laid open, and where its relative position and fossil contents were elaborately worked out by my friend, the Rev. T. T. Lewis.* It is a subcrystalline, earthy limestone, arranged in beds from one to five feet

WHITEWAY HEAD. (Escarpment of Amyestry Limestone. Strata dipping to the S. E.)
(From Sil. Syst., p. 243. Sketched by the late Rev. W. R. Evans.)

* See Sil. Syst., p. 201.

thick; the laminæ of deposit being marked by layers of shells and other fossils. In the escarpment of the south-western limb of the Ludlow promontory, this rock is frequently seen to form bluff cliffs, the inclined strata of which appear, as in the centre of the preceding woodcut, resting upon the Lower Ludlow shale, and plunging under the Upper Ludlow rock.

When quarried into, the rock is of an indigo or bluish grey colour, in parts mottled by the mixture of white calcareous spar. The quarries, like those in all the harder bands of the Ludlow formation, present natural backs or divisions, as in the above diagram, usually coated by a dirty yellow or greenish shale. These are the faces of joints more or less vertical, which, when open, occasion the rock to separate into rhomboidal masses, which are easily detached where the strata are much inclined. The rock is, therefore, subject to slides or subsidences, particularly where the underlying saponaceous " Walker's, or Fuller's Earth " prevails. Examples of these slides may be seen at many spots, but no where more instruc-

THE PALMER'S CAIRN LANDSLIP.

(From Sil. Syst., p. 248.)

The woodcut exhibits the slope of the beds and the vertical joints, which in conjunction with the upper and lower lines of bedding divide the rock into lozenge-shaped detached lumps.

tively than at the Palmer's Cairn, or Churn Bank, S. W. of Ludlow, as represented in the foregoing woodcut. The area there affected exceeds fifty acres.

This limestone often occupies the summit or capping of the escarpment of Ludlow rocks, as in the ridges of Mary Knoll and Brindgwood Chase west of Ludlow; and in the hills extending from the View Edge and Norton Camp, where it forms a conspicuous band parallel to, and loftier than, the Wenlock Edge (see woodcut, p. 113.).

At Aymestry, the limestone occupies both banks in the gorge of the river Lugg; but, as above said, it rapidly thins out to the southwest. On the whole it is much less of a concretionary character than the subjacent Wenlock limestone, and partakes more of the flat-bedded character which is observable in the "pendle" beds beneath, and is indeed characteristic of all parts of the Ludlow formation.

In the Woolhope valley of elevation, the Aymestry rock (see section, p. 106.) assumes precisely the same external or geographical features as on the flanks of the Ludlow promontory, in having from its hardness resisted denudation better than other portions of the deposit. It thus forms the crest of the external and encircling ridge, and is prominent in the hills of Marden, Seager, and Backbury. Although it there differs in being less of a limestone than near Ludlow, it contains many of the same fossils, even the Pentamerus Knightii, its characteristic shell (pl. 21. f. 10.), having been found in the Woolhope district since the "Silurian System" was published.

On the outermost western slopes of the Malvern Hills, and on the sides and summits of their northern prolongation the Abberley Hills, the Aymestry rock, though containing less calcareous matter than near Ludlow, is still the well-defined central portion of the formation. In some of these tracts (as near the Hundred House) it might be used for lime, if the Wenlock limestone of such superior quality were not in juxtaposition. The same may be said of the districts of May Hill, Usk, &c., where this limestone is represented by the harder and somewhat calcareous central part of the Ludlow rocks.

At Sedgeley, in Staffordshire, the rock again becomes once more a complete limestone, in which the predominant Aymestry fossil, the Pentamerus Knightii, abounds. There it is known as the "black limestone," in contradistinction to that of Dudley, on which, with an intermediate shale or "bavin," it reposes.*

The fossils which pervade the Aymestry limestone, in addition to the Pentamerus Knightii, are the Rhynconella (Terebratula) Wilsoni, Lingula Lewisii, Bellerophon dilatatus, Avicula reticulata, &c., and many of the same shells, corals, and trilobites, which are common in the subjacent Wenlock limestone. Indeed, except in the less number of species and the occurrence of some of the shells more characteristic of the next member (the Upper Ludlow rock), this limestone is scarcely to be distinguished by its fossil contents from the Wenlock limestone.

In some tracts the place of this rock is marked only by the shelly courses, which near Ludlow form its immediate cover; and wherever that stratum occurs it is replete with the Rhynconella (Terebratula) navicula (pl. 22. f. 12) and the small Leptæna lævigata. It is this band which forms the base of the Upper Ludlow rock.

Upper Ludlow Rock.—This is the most diversified in structure and contents of the three subdivisions of the highest Silurian formation, and is remarkably interesting in exhibiting a transition from its highest members into the next overlying system, the Old Red, or Devonian. Its lowest stratum may be considered the calcareous shelly band, charged with Rhynconella (Terebratula) navicula, which has just been mentioned as forming the roof of the Aymestry limestone, and which occasionally attains a thickness of 30 or 40 feet. This is surmounted by grey, argillaceous masses so common throughout the Silurian rocks, and which from their incoherent nature easily de-

* See Silurian System, sections, p. 481. *et seq.*

compose to mud. Like other sediments of higher antiquity in the same system of rocks and fossils, it has a tendency to run into large spheroids, and occasionally contains small concretions of sandy clay, which, being more destructible than the pure argillaceous matrix, have often weathered out in the faces of the escarpments, and exhibit the lines of stratification by small elliptic cavities like swallow-holes.

The chief and distinguishing portion of the Upper Ludlow contains more calcareous matter and sand than the beds immediately beneath. It is, on the whole, a slightly micaceous thin-bedded stone of bluish-grey colour within, but weathering externally to a brown and rusty grey, and remarkable for its transverse, symmetrical joints, as exhibited in this view.

UPPER LUDLOW ROCKS AT THE BONE WELL.*
(From Sil. Syst., p. 250.)

* This well is so named because bones of mice, frogs, and other small animals, are from time to time washed out from the open joints of the impending rock, and offer what was formerly considered a great marvel. Old Drayton, in his Polyolbian, makes the bones of frogs into those of fishes : —

> " With strange and sundry tales
> Of all their wondrous things; and not the least in Wales ;
> Of that prodigious spring (him neighbouring as he past)
> That *little fishes'* bones continually doth cast."

Though quarried extensively, these stones, either when not well selected or not placed in the wall in the direction of the layers, are very prone to decomposition. Some strata of this character appear in the foreground of the vignette, p. 125., which represents the Castle of Ludlow standing on the rock out of which it was built.

I have elsewhere compared this member of the Upper Ludlow with the " Macigno " of Italy, and particularly with those dark greyish portions of it which occur between Perugia and Florence. This close lithological resemblance of so old a rock to so young a deposit (for the Italian Macigno is no older than the London clay), is cited to show how similar sandy, muddy, and calcareous materials, collected at the bottom of a sea, often necessarily resemble each other when formed into stone, though originally deposited at such very widely separated periods. In this comparison it is also to be specially noted, that the " Macigno," or young Italian rock, is infinitely more hard, compact, and durable, than the ancient stone of Ludlow.*

The surfaces of these Upper Ludlow rocks are occasionally covered by small wavy ridges and furrows, here and there crossed by little, tortuous, raised bands, the first of which are supposed to have resulted from the rippling action of waves, when the sediment was accumulating under the sea, and the last to have been the traces left by worm-like animals frequenting a sandy and muddy shore as the tide receded.

It is chiefly in this portion of the formation, that the best defined organic remains are found, often preserving the sharpness of their forms, and the remains of their original shelly coverings. Here we meet with a profusion of the following fossils: Chonetes (Leptæna) lata, pl. 20. f. 8.; Orthonota amygdalina, pl. 23. f. 7.; Goniophora cymbæformis, pl. 23. f. 3.; Avicula lineata, f. 16.; Pterinea retroflexa, f. 17.; Orbicula rugata, pl. 20. f. 1, 2.; Orthis

---

* See my description of the Macigno of Italy, Alps, Apennines, &c. Quart. Journ. Geol. Soc., vol. v. p. 280., where I showed that it was of tertiary age.

elegantula (var. orbicularis), pl. 20. f. 9, 10.; O. lunata, f. 11.; Rhynconella (Terebratula) nucula, pl. 22. f. 1.; Turbo corallii, pl. 24. f. 1.;
T. octavia, f. 4. &c.; and the coiled shelly serpuline body, Serpulites
longissimus, pl. 16. f. 1.   The Cornulites serpularius, and a small
Tentaculite are far from uncommon.

Corals are but scarce; yet Stenopora (Favosites) fibrosa is found
frequently encrusting particular species of shells,—Turbo corallii and
Murchisonia corallii, as their names imply, being its favourite
habitats; it is figured in a previous woodcut, p. 119.   Foss. 10. f. 8.

Orthoceratites, occasionally of large size, occur, and trilobites.
Of the latter, Phacops caudatus and the Encrinurus punctatus, with
a rare sample of the Calymene Blumenbachii, reach the top, but
they are not common; nor indeed are trilobites at all abundant,
except a fine species of the genus Homalonotus, H. Knightii,
pl. 19. f. 7., which may be found throughout the whole range of this
formation, sometimes of very large size.

In the cliffs at Ludlow, the chief rocks are surmounted by what
I termed the fucoid bed.   This is a greenish-grey argillaceous sandstone, almost entirely made up of a multitude of small, wavy,
rounded, stem-like forms, which resemble entangled sea-weeds.   In
this mass is found, and always in a vertical position, the singular
body (pl. 15. f. 4.) named Cophinus dubius.   It is generally of an
inversely pyramidal shape, and its sides are scored with elegant
transverse grooves.   I am assured by Messrs. Sowerby and Salter,
who have studied it attentively, that it is the impression made by
the stems of encrinites, which, rooted and half buried in the micaceous
mud, produced, by their wavy and somewhat rotatory motion, the
beautiful pattern, every line of which answers to one of the projecting bosses or rings of the jointed stem.   The best proof that such
has been the cause, is that these stems are always present, and lying
contiguous to the markings.   The slow trailings of these stems have
probably left their traces in virtue of the micaceous character of the

mud, which may have impeded the perfect fusion of the separated portions of the semifluid mass after the stem had passed through them.

The Upper Ludlow rock is, in this respect, the most interesting of the divisions of the Upper Silurian group, that it is the oldest band in which any remains of fishes and land plants have been discovered. The upper and lower parts of this stratum, as seen at Ludlow, are finely laminated, earthy, greenish-grey sandstones, containing a few remains of ichthyolites, with several shelly remains characteristic of the Upper Ludlow. It was the central part only of this stratum, or a gingerbread-coloured layer of a thickness of three or four inches only, and dwindling away to a quarter of an inch, which exhibited, when first my attention was directed to it*, a matted mass of these bony fragments, for the most part of very small size and of very peculiar character. These, with a very few remains of shells, and a rare crustacean, the Pterygotus problematicus, occur in a cement in which varying proportions of carbonate of lime, iron, phosphate of lime, and bitumen are disseminated. Some of the fragments of fish were of a mahogany hue, but others of so brilliant a black, that, when discovered, the bed conveyed the impression of being a heap of broken beetles.

The supposed fishes of this bed, as exhibited in plate 4 of my original work, must now, however, be reduced in number. At all events, besides the remarkable Pterygotus, which is here figured in pl. 19. f. 4, 5., and which was removed by Agassiz himself to the class of crustaceans, Professor M'Coy has diminished the list of fish remains by proving, that some of the supposed fish *defences* should also be removed to that group. One of these, to which he has applied

---

* This course was discovered by my friends and excellent Ludlow coadjutors, the Rev. F. T. Lewis and Dr. Lloyd, the latter, now alas! removed by death. By their assistance, and that of the late Rev. J. Evans, I traced this fish-bed in several other parts of the Ludlow promontory. See Sil. Syst. p. 198. 605.

the name Leptocheles Murchisoni (see pl. 19. f. 1, 2.), is figured
in my original work as an Onchus, or fish defence.*

This bed, enclosing minute fragments of fish-bones and skin or
shagreen, has since been detected in tracts very distant from Ludlow.
The upper portion of the Ludlow formation, or capping of the
bone-bed, is composed of light-coloured, thin-bedded, slightly micaceous sandstones, in which quarries are opened out near Downton
Castle on the Teme. (The Downton Castle stone, Sil. Syst. p. 197.)
The uppermost layers of the whole system, and which form a transition into the Old Red sandstone, consist of tilestones and sandstones, occasionally reddish, in which, besides other fossils found in
the Ludlow rock, the Lingula cornea, pl. 34. f. 2., with crustaceans
(pl. 19. f. 1. 3.), and defences of fishes, often occur.

Being compelled in my researches to draw a line of demarcation
between the upper part of the Ludlow formation and the bottom
of the overlying Old Red sandstone, I formerly included the
tilestones in the latter; particularly as in most parts of the region
they decompose into a red soil, and thus they afford a clear physical
line of demarcation between them and the inferior rocks, which
facilitated the construction of the geological map.†

Even then, however, the fossils which were figured as characteristic of such tilestones exhibited little else, as I showed, than

* For further illustration of this point, see Quart. Journ. Geol. Soc. Lond.
vol. ix. p. 12.

† The reader, who may refer to my original map, must recollect, that it was
constructed between the years 1831 and 1836, when even the geographical outlines
of a large portion of the country had not been defined by the publication of the
sheets of the Ordnance Survey. Geologists who have had to labour with little
topographical assistance in a region like this, which had been wholly unexplored by
miners, are not those who will criticise those errors of detail, which have been
remedied by the government surveyors who followed me. The classification and
chief outlines of my map and sections, as far as they relate to my own Silurian
rocks, have indeed been completely sanctioned by Sir Henry de la Beche and his
associates; and this is the highest reward which a pioneer like myself could expect
to receive.

species common to the Ludlow rock itself. This zoological fact, and subsequent researches in other parts of England, above all those of Professor Sedgwick in Westmoreland, where the Upper Ludlow strata are much developed, have, for eleven years, led me to classify these tilestones with the Silurian rocks, of which they form the natural summit. For, even in their range from Shropshire through Hereford and Radnorshire, into Brecon and Carmarthenshire, whether they are of red or yellow colours, they are charged with Orthoceras bullatum, Chonetes (Leptæna) lata, Spirifer elevatus, Orthis lunata, Rhynconella nucula, Cucullella ovata, Bellerophon trilobatus, B. expansus, Trochus helicites, Holopella (Turritella) obsoleta, and the minute bivalved crustacean, Beyrichia tuberculata. (See pl. 34. and its figures). All of these are the most common fossils of the Upper Ludlow rock; although a few of them descend as low as the Caradoc sandstone.

Including the Downton Castle building stones, this band of transition contains the oldest traces of terrestrial vegetation yet known in the British Isles. The specimens which I found were usually small, and little more than carbonized stains. At the bottom of the detached basin of Old Red sandstone of Clun Forest in Shropshire, I detected thin layers of matted and broken vegetables (frequently carbonized) in the tilestone or "firestone" beds of that tract (Sil. Syst. p. 191.). Since then, the government collectors have enlarged our acquaintance with them, and in the museum in Jermyn Street there are to be seen many more specimens of this stratum, among which are the minute globular bodies, pl. 35. f. 30., called "bufonites" in the "Silurian System," but which, as will presently be explained, are now shown to be vegetables.

Though not every where divisible into the portions above described, the Upper Ludlow rock maintains, on the whole, a decisive aspect and position in its range from Shropshire through Hereford and Radnorshire into Brecon and Carmarthenshire, until last seen

in the cliffs of Marloes Bay in Pembrokeshire. Frequently somewhat calcareous, the deposit is for the most part a harder and
sandier stone, partaking much more of the character of the Italian
"Macigno," as before said, than any other rock of the Silurian system.

On the eastward slopes of Bradnor and Hergest Hills, near Kington, and particularly in the ridges extending thence by Gladestry
and Pains Castle to the Trewerne Hills on the banks of the Wye,
this sub-group is admirably exposed in slightly inclined masses replete with fossils. In the escarpment of the Trewerne and Begwm
Hills *, and in many other places, this grey, shelly, thin-bedded mass
of rock is strikingly exhibited, dipping under the bottom beds of
the Old Red sandstone, as in the long sectional woodcut in the
10th chapter.

Along the outer or western edge of the Malvern Hills, the Upper
Silurian, reposing on the Lower Silurian as before described, p. 94.
exhibits, in like manner, Upper Ludlow rocks dipping under the Old
Red marls, descriptions of which, as occurring at many places, from
Ledbury northwards, may be consulted in the " Silurian System,"
and in the monograph of Professor Phillips.†

Like passages upwards, from an inferior grey-coloured series, to a
superior red system, are observable at Usk, and in tracts around
that town, where these beds repose on Wenlock limestone.‡

In Brecknockshire, to the south of Builth, the Ludlow rocks, surmounting a noble escarpment of the other members of the Upper
Silurian division on the right bank of the Wye, but in which no
limestones occur, exhibit a fine upward development, as they pass
under the expanse of the Old Red sandstone in the wilds of Mynydd
Epynt (see the long vignette, chapter 10.). There the Upper Ludlow
rises from beneath the Old Red, in a rapid anticlinal flexure at Alt-

* This was my first Silurian section, 1831, in passing from the known Old Red to
the then unknown Ludlow rocks, Sil. Syst. p. 5, 312..

† Mem. Geol. Surv. Gt. Brit., vol. ii. part 1.   ‡ Sil. Syst. p. 408.

fawr and Corn-y-fan, as here represented; the central and lower
members of the formation forming the underlying strata.

BRECON ANTICLINAL OF LUDLOW ROCKS, THROWING OFF OLD RED SANDSTONE.
(From Sil. Syst., p. 211.)

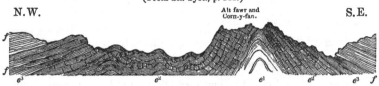

e^1. Lower Ludlow. e^2. Middle Ludlow with a calcareous band representing the Aymestry limestone. e^3. Upper Ludlow. f. Old Red junction beds.

Thence into Carmarthenshire, the junction of the Ludlow rocks
with the Old Red sandstone is well laid open in numerous places,
especially in the narrow valley of Cwm Dwr, between Trecastle and
Llandovery, where the tilestones on which Horeb Chapel stands,
are full of the casts of shells, among which are characteristic forms,
such as the Trochus helicites, Turbo Williamsi, Bellerophon trilobatus, and many others.

The banks of the river Sowdde, in Carmarthenshire, east of Llangadock, also expose a good junction of these highly micaceous
Upper Silurian flagstones with overlying Old Red marl; the
whole at very high angles of inclination. Thence, in their range
to the mouth of the Towy, the Upper Ludlow becomes a compact
hard sandstone, every where surmounted by Old Red.*

In Pembrokeshire, similar junctions with the Old Red sandstone
are seen near Tavern Spite, Narberth, Freshwater East, Freshwater
West, and in Marloes Bay. In all these places, strata of dull
greenish-grey argillaceous sandstone, minutely micaceous, differing
chiefly from the type of the Upper Ludlow of Shropshire in being
harder and thicker bedded, and which repose on rocks with Upper
Silurian fossils, plunge under red and green strata (the red rab
of Pembroke), or bottom beds of the Old Red sandstone. (See p. 66.)

* See woodcuts, pp. 72, 73.

In the valley of Woolhope, the same succession is very apparent
all round the external rim of that remarkable elliptical elevation (see
diagram, p. 106). In nearly all parts of that circumference, the Upper
Ludlow is well exposed, and occasionally contains, in one or two
spots, the rare remains of fishes.

At Hagley Park, distant only two miles from the north-western
end of the Woolhope ellipse, and four miles east of Hereford, the
uppermost beds of the Ludlow formation have been recently unco-
vered from beneath their covering of red clay and marl; and there
the thin bed, containing fish-bones and the crustacean Pterygotus,
has been found by Mr. H. Strickland to be just in the same relative
position as at Ludlow.* This spot marks a minor undulation, or
dome of the Ludlow rocks, the surface of which only is visible;
whilst a much greater mass of the formation protrudes in the adjacent hill of Shucknall, as formerly described (Sil. Syst.). At
Hagley Park, the fish bed, scarcely exceeding an inch in thickness,
lies between the beds of brownish and yellowish sandstone (the
Downton Castle stone) and a grey micaceous shaly rock full of Upper
Ludlow fossils. The fish remains are chiefly those of the minute
shagreen scales (pl. 35. f. 18.), the fish defences Thelodus parvidens and
Onchus tenuistriatus (fig. 15 and 17.), with Coprolites (fig. 21-27.);
whilst in the sandy beds above these there are the remains of carbonized vegetables, and among them Mr. Strickland detected some
of the minute globular or spherical bodies mentioned above, which
Dr. Hooker has ascertained to be a fossil of the natural order of
Lycopodiaceæ. Similar vegetable traces have been observed, again,
at various places around the external rim of the Woolhope ellipse.

Again, in the southernmost prolongation of the Silurian group
of May Hill and Huntley Hill, where I formerly described the

* I have visited the spot since the discovery, in company with Mr. Strickland.
See his descriptions of these beds in the S. of Herefordshire and in Gloucestershire. Quart. Journ. Geol Soc. Lond. vol. viii. p. 381., and ib. vol. ix. p. 8.

whole Silurian series as reduced to one thin mass of Ludlow rock, having the Old Red sandstone on one side, and the New Red on the other, my friend, Mr. Strickland, has very recently detected the existence of the very same thin fish-bone layer in precisely the same place as at Ludlow.

Lastly, in the arched, or dome-shaped masses of Upper Silurian rocks, which rise out from beneath the Old Red sandstone at Pyrton Passage on the Severn, Professor Phillips has noted the remains of small fish-bones in the Upper Ludlow rocks.*

The recognition of this thin band of the uppermost Ludlow rock, in which the oldest recognizable land plants and fishes are associated with certain shells for a distance of some forty-five miles from the tract in which I first described it, is truly remarkable. For whilst it shows the value of close and minute researches, it is a strong proof that, after many years of hard labour, the Ludlow rock still remains, as formerly described by me, the lowest stage in the crust of the globe in which any ichthyolites have been detected, a fact which will be reverted to in ensuing chapters.

* Mem. Geol. Surv. of Great Brit., vol. ii. part 1. The succession of the Silurian rocks in the dislocated tract of Tortworth, to the S. E. of Pyrton Passage, is described in detail in the Silurian System, p. 99.

CHAPTER VII.

SILURIAN ROCKS OF BRITAIN BEYOND THE TYPICAL REGION OF ENGLAND
AND WALES, AS SEEN IN CORNWALL, THE NORTH-WEST OF ENGLAND,
SCOTLAND, AND IRELAND.

THOUGH occupying large spaces in other parts of the British Isles, the Silurian rocks, as separated into formations and characterized by fossils, are no where so clearly defined as in that typical region of which we have taken leave. In no other tract of the United Kingdom have geologists been able to show so clearly the relations of the lowest fossiliferous or protozoic rocks to others beneath them which exhibit no signs of former life; nor is there any where else so clear an ascending order from the Silurian into the next overlying deposit, the Old Red Sandstone or Devonian rocks.

In Cornwall, for example, the discovery within the few last years only of certain fossils, has proved that some of the quartzose sandstones forming its southern headlands are Lower Silurian. The fossiliferous sandstone in question passing to the south of the Dodman, and coming out to view in Gorran Haven, contains several species of Orthidæ as well as trilobites most characteristic of that division.*

There, however, no one can show an unbroken descending sequence beneath those beds, nor an ascending order from them to the contiguous and younger deposits of Devonian age; and as large

* These fossils were collected by Mr. Peach, and, from their inspection and a visit to their chief localities in 1846, I pronounced the rocks in which they occur to be Lower Silurian rocks (Trans. Roy. Geol. Soc. Cornwall, 1846, p. 317.), and, as such, they were inserted in a new edition of my Geological Map of England and Wales, published by the Society for the Diffusion of Useful Knowledge.

portions of the strata of Cornwall have been highly altered, and mineralized, so is this southern tract much dislocated. In such a region, therefore, we cannot expect to meet with proofs of succession. It is sufficient to state, that the band of siliceous grits and quartzites in the south of Cornwall, which I had termed Silurian in 1846, presents much of the character and aspect of the opposite rocks of France, which the French geologists have mapped and described as Lower Silurian, as will hereafter be noticed. This is precisely one of those broken and insulated tracts of older sedimentary rocks, where the geologist has no other test by which he can recognize their age than by their imbedded organic remains.

Professor Sedgwick, who has visited the localities since I described them, shows, indeed, that these strata are inverted; the Lower Silurian (which he now calls Cambrian) *overlying* the Devonian or Old Red rocks. At the same time, the chief fossils defined by Sowerby and M·Coy *, consist of the simple plaited Orthidæ so very common in the Lower Silurian rocks, viz., Orthis calligramma, O. flabellulum, with O. elegantula, and O. testudinaria; Strophomena grandis; and the trilobites Homalonotus bisulcatus; Calymene brevicapitata of Portlock, and Phacops apiculatus, Salter. That these are true and original Lower Silurian types cannot be doubted.

In the north-western and mountainous part of England the Silurian rocks appear in great force in the counties of Westmoreland and Cumberland and the adjacent tracts of Lancashire and Yorkshire. Though some of their members there assume a different lithological aspect to what they maintain in the Silurian and Welsh region, they have been clearly paralleled by several geologists with those original types. Professor Sedgwick, who has most studied the lake region (which I have traversed on three occasions

---

* See Quart. Journ. Geol. Soc. Lond. vol. viii. p. 13.

L

for purposes of general comparison), and who has described it in
a series of valuable memoirs, has recently grouped the lowest fossiliferous limestones, or those of Coniston, in his Cambrian rocks,
though during several years he identified them by means of the
published fossils with the Lower Silurian and even with its upper
portion.

In truth, the region of Siluria, as geologists now admit, afforded
the key by which the fossiliferous strata in the north-western tracts
of England were brought into order and had their proper places
assigned to them. For, in Cumberland, where the lowest members
of this primeval series rise up into the lofty mountains of Skiddaw
and Saddleback, the inferior masses of crystallized schists which pass
downwards into fine, glossy, chiastolite slates, have been too much
metamorphosed by eruptive, granitic rocks, to allow us to hope for
the detection in them of any regular order or symmetry, still less
of fossil contents. These beds are followed by green slates and
porphyries, in which also no fossils have been found, owing, as it
is believed, to the great abundance of igneous matter. In vain,
therefore, in such a country, does the geologist seek for the
equivalents of the bottom rocks, described in the earlier chapters as
overlaid conformably by strata with Lingulæ and trilobites. No
where in the lake region have the repeated labours of geologists or
fossil collectors, including that close searcher Mr. John Ruthven, detected the lower fossil band or the Lingula flags of North Wales,
unless it be represented, as Professor Sedgwick thinks probable,
by the few graptolites and fucoids of the Skiddaw slate. With the
exception of rare sea-weeds and zoophytes, not a trace of any thing
has been detected in the thousands of feet of strata, which, with
interpolated igneous matter, intervene between the slates of Skiddaw
and the Coniston limestone, with its overlying flagstones, &c. At
that zone only do we begin to find any thing like a true fauna;
and judging from its remains it may be admitted to be a repre-

sentative of the Llandeilo formation of the Lower Silurian rocks. For, among the fossils, the same species of simple plaited Orthidæ, with trilobites and corals, are again met with which characterize the limestones of Llandeilo or Bala.

From physical data, and the great thickness of the strata, we may indeed hypothetically admit, that Professor Sedgwick is right in assigning to the schists of Skiddaw (in which one or two graptolites alone have been found) as high an antiquity as the lowest *fossil* beds of North Wales and Shropshire. But all we can safely say is, that, reasoning from the graptolites, the whole is a Lower Silurian series, which is metamorphosed and obscured by igneous eruptions in its lower parts, and exhibits beneath it no clear representative whatever of the Longmynd or bottom rocks.

In proceeding, however, to the south and south-east from the more crystalline and mountainous part of this north-western tract of England, the Lower Silurian rocks, whether they be ranged with the Llandeilo or with the Caradoc subdivision, are seen to be succeeded by younger deposits, which, though of very different mineral characters, unquestionably represent by position and fossils the Wenlock and Ludlow formations. I cannot more tersely and clearly express these relations than in the language which Professor Sedgwick himself applied to this region in 1845, *i. e.* six years after the publication of the "Silurian System." "Thus, the fossiliferous slates," says he, "present, first, *the Lower Silurian rocks in a very degenerate form, and secondly, the Upper Silurian in a noble series.*"*

Nor can I convey a better idea of the succession of rocks in this district than by referring the reader to sections and descriptions by the same author, published in 1846, when combined with his last memoirs of 1852. It is enough for me, on this occasion, to state, that these show a conformable succession of deposits from the crystalline chiastolite schists, resting on a granitic nucleus or centre,

---

* See Quart. Journ. Geol. Soc. Lond. vol. i. p. 443.

up through Skiddaw slates, and a vast thickness of green slates and
porphyry, to a thin band of limestone (Llandeilo and Bala), and a
considerable thickness of overlying flagstone, with Lower Silurian
fossils, followed, as the Professor has recently discovered, by the
equivalent of the Caradoc sandstone (the " Coniston Grits"). These
are surmounted by a copious Upper Silurian series, in which the
Wenlock and Ludlow formations are recognized by their position
and fossils, though in their mineral aspect they differ much from the
original types.

This Upper Silurian series, forming the rocks on both sides of
Windermere, and thence extending far to the south, consists in the
first instance of the Ireleth slates. These rocks have a slaty
cleavage oblique to the beds, are in parts calcareous, and have
proved to be the equivalents of the Wenlock formation, or more
particularly of its hard slaty form which occurs in Denbighshire.
The overlying masses are sandy and pebbly rocks. Then follow
other coarse slates, grits, and flags, called by Mr. Sharpe " Windermere rocks," in which Lower Ludlow fossils occur, and then a very
distinct representative of the Upper Ludlow, as proved both by
position and fossils, but differing in being a much harder sandy
flagstone than the rocks of like age in Siluria. The series terminates upwards in true tilestones, which, on the banks of the
Lune and near Kendal, are the exact counterparts of the uppermost
Silurian zone, and are charged with numerous fossils, some of the
most striking of which will be treated of in chap. 9.*).

* The reader who desires to study the data by which our present knowledge of
the geology of the Lake district has been acquired must read the various memoirs
of Professor Sedgwick in the Proceedings of the Geological Society, vol. ii. p. 675.,
and the Quart. Journ. Geol. Soc., vol. i. p. 442., vol. ii. p. 106., vol. iii. p. 133.,
vol. iv. p. 216., vol. viii. pp. 35. 136.; the Memoirs of Mr. Sharpe, Proc. Geol.
Soc., vol. iv. pp. 23. 70.; and those by Professor Phillips, Geol. Trans. N. S.,
vol. iii. p. 1., and vol. iv. p. 95.

In this region, however, the ascending series is interrupted, as Professor Sedgwick has shown.* Instead of a complete series of the Old Red Sandstone, as it has been described in Shropshire, Herefordshire, or South Wales, and a conformable gradation and passage upwards into its overlying sandstones and marls, the Silurian rocks of all ages are at once unconformably overlapped by red masses, chiefly coarse conglomerates, which alone represent the great and complex group to be considered next in order.

Silurian Rocks of Scotland. — In the early days of Scottish geology, its illustrious founders Hutton and Playfair considered the schistose mountains in the south of Scotland to be void of all traces of life, until their able associate, Sir James Hall, detected a few fossil shells in a limestone at Wrae Hill, in Peebleshire, which had been considered to be primary. The merit of transferring these strata from the primary to the transition class is also in great measure due to our contemporary Professor Jameson; for though he made little account, as was usual when he wrote, of organic remains, he gave a clear general view of the rocks.

" On the one discovery of Sir James Hall at Wrae Hill, our Scottish geologists," says Hugh Miller in a recent sketch, " seem to have hybernated for more than forty years."†

It was not until some years after the publication of the " Silurian System" that the researches of Professor Nicol first really indicated the relations of the Wrae fossils and the associated schistose

---

* Besides this general transgression and unconformity, it has been pointed out by Professor Sedgwick, that this " old red" contains many fragments of the Silurian tilestones, which must have been solid before the conglomerate was formed. — Quart. Journ. Geol. Soc. vol. i. p. 449.

† For a complete historical sketch, see a notice by Mr. Hugh Miller (Witness newspaper, November 24th and 27th, 1852). A single fossil was found, but not described, by Laidlaw, the friend of Walter Scott; and orthoceratites were afterwards discovered by Mr. Charles Maclaren in the Pentland Hills, and noticed in his excellent work, entitled " the Geology of Fife and the Lothians," 1839.

masses to the known members of the series; and, since then, other researches of that author, as well as of Moore, Sedgwick, Cunningham, Stevenson, Harkness, and myself have brought the Scottish masses into a distinct comparison with their true types.

Silurian rocks, and particularly their lower members, are now known to occupy a very large region of the south of Scotland. Ranging on the whole from E.N.E. to W.S.W., they appear in considerable masses in Berwickshire and Roxburghshire, and thence spread out in still larger areas over the counties of Selkirk, Peebles, Dumfries, Kirkcudbright, Galloway, Wigton, and Ayr. In short, they constitute, on the whole, the undulating moory hills, which, from their prevalent wildness of aspect, have been called the South Highlands. Subjected, as they have been, to numerous eruptions of granite, syenite, porphyry, greenstone, and other igneous rocks, some of these greywacke schists have long been known to geologists for the remarkable curvatures they exhibit in the sea-cliffs of Berwickshire. A drawing is here annexed, which represents such

VIEW OF THE CLIFFS NEAR ST. ABB'S HEAD.
(From a sketch by Sir A. Alison, Bart.)

flexures and breaks, as seen in the cove, called Petticur Wick, near St. Abb's Head.*

As numerous similar contortions are also seen to have taken place in the western prolongation of the same great series to the coast of Wigton and Ayr, we ought to be cautious in drawing determinate sections across the interior portions of a region where the surface of the round and undulating hills is obscured by moss, heath, and bog, and where exposures of the bare rock are rarely to be met with.

Difficult, however, as it has been to determine the oldest portion of these inland masses of the Scottish greywacke, we have now reached that starting-point. When traversing the tract between Dumfries and Moffat, in 1850, it occurred to me that the dull reddish or purple hard greywacke to the north of the former town, which so resembled the bottom rocks of the Longmynd, Llanberis, and St. David's†, would prove to be of the same age, and, therefore, the true South Scottish axis. For, this rock (in great part similar to the hard greywacke of St. Abb's Head) seemed evidently to throw off anthracitic schists, with graptolites, both to the south and to the north. Mr. Harkness established this order, after my visit, by detailed sections in that parallel, ranging by the Dryfe Water.‡

Professor Nicol, who led the way in opening out the proofs, both physical and fossiliferous, that the South Scottish Hills were really Silurian, has now satisfied himself that this real axis of the old unfossiliferous greywacke ranges by Teviotdale.§   Proceeding from

* The woodcut is taken from one of several rapid but very clever sketches made in my note-book, in the autumn of 1833, by the historian Sir Archibald Alison, from a boat in which Sir John Hall of Dunglass and Professor Sedgwick were also my companions. Some of the lower buttresses are porphyries and other igneous rocks, which form the adjacent headland of St. Abb's. The convoluted masses form a portion of the Longmynd or *bottom rocks* of the region.

† Quart. Journ. Geol. Soc. Lond., vol. vii. p. 162.

‡ Quart. Journ. Geol. Soc., vol. vii. p. 52., and vol. viii. p. 393.

§ Meeting of British Association, Belfast, 1852.

SILURIAN ROCKS OF THE SOUTH OF SCOTLAND.

Length of section fifty miles.

S.

Cheviot Hills. Old Red conglomerate, f. Mountain limestone and Coal formation, g. Trap and Porphyry, c. Enton Beds. Anthracite. δ¹

Stobbs. ROXBURGHSHIRE. Teviot. a SELKIRKSHIRE. Ettrick. 1600 ft. Ale. δ¹ 1600 ft. Yarrow. 2000 ft. Thornielee slates. δ² Graptolites. Annelida. Alum shale. Anthracite. Tweed. δ* Griestou slates. Felspar Porphyrites. Graptolites. Annelida. Trilobites. Anthracite. 2200 ft. PEEBLES-SHIRE. Trilobites. Wrae limestone. δ³ Lituites. Orthida. Trilobites. Encrinites. 1800 ft. Old Red sandstone, f. Carboniferous limestone and coal-fields, g. Trap and Porphyry. LOTHIANS. Pentland Hills 1700 ft. Orthoceratites. c

N.

that centre, he has prepared the annexed general section, which (minor flexures being omitted) exhibits a short ascending section to the Cheviot Hills on the south, and a very long one to the Pentland Hills, near Edinburgh, on the north.

This traverse shows in the clearest manner an upward development of the lowest divisions of the series; so that after placing other transverse sections in parallel with it, whether along the east coast or from Dumfries and Moffat, we will then pass to the west coast, where we meet with higher and much more fossiliferous masses of the Silurian system.

Casting his eye over this diagram*, the reader will readily understand the comparison which is drawn between these Scottish strata and those of the typical region of Siluria and Wales, described in the second, third, and fourth chapters. Beginning at the geological axis in Teviotdale, he

* Reduced from a large coloured section exhibited at the Belfast Meeting of the British Association for the Advancement of Science, 1852.

perceives that the older and unfossiliferous greywacke, *a*, which is the Longmynd or bottom rock, throws off schists, *b*[1], both to the south and to the north; the former passing under the old red sandstone and carboniferous strata associated with porphyries of the Cheviot Hills. To the north of the Teviot axis, the schists and greywacke, *b*[1], which are in parts alum slates (Etterick), may be well supposed to represent in time, as suggested by Professor Nicol, the schists of North Wales with Lingulæ, though as yet the Scotch strata have afforded no such fossils. These are surmounted by anthracitic schists, containing many coiled annelids or sea-worms, *b*[2], and form the base of a vast thickness of graptolite schists, which, after a synclinal flexure in the Yarrow Hills, show another anticlinal in the valley of the Tweed, *b**. Whilst the Thornielee slates, with their graptolites, annelides (Nereites, Crossopodia), &c., are inclined southward, the Grieston slates, on the north bank of the anticlinal, plunge to the north, and then, for the first time in ascending order, we find trilobites in addition to the graptolites and annelids.

Associated with felspar porphyries (the whole representing pretty fairly the trappean and graptolitic group of North and South Wales) these rocks rise into mountains 2200 feet above the sea, and from them the sources of the Clyde, as well as of the Tweed, take their rise. Thence the northern dip being continued, the whole of the preceding masses are seen to be overlaid by the Wrae limestone of Peebleshire, *b*[3], which, from its organic remains of trilobites and shells, has been paralleled with the Llandeilo limestone.†

† Among the fossils of the Wrae limestone found by Professor Nicol are the Illænus Bowmanni, Salter; Harpes parvulus, M'Coy; Asaphus tyrannus, Sil. Syst.?; Phacops Odini?; Cheirurus; Orthoceras arcuo-liratum, Hall; Orthis calligramma, Dalman; O. biforata, Schlotheim; Strophomena tenuistriata, Sil. Syst. Salter, in Quart. Journ. Geol. Soc. Lond., vol. iv. p. 205., and Professor M'Coy in the " Pal. Foss. Woodw. Mus." *passim.*

All the associated schists and greywacke were, therefore, termed by Professor Nicol, Silurian (1846); the more so as the fossiliferous limestone reposed on schists containing graptolites. In pursuing this traverse northwards, other schists and greywacke rocks are seen to overlie the Llandeilo limestone of Wrae, though these are soon lost under and covered over by the old red sandstone, *f*, and the carboniferous formation of the Lothians, *g*. Beneath those deposits, *f, g*, we know not what flexures may have occurred in the subjacent or older rocks of the great intervening trough; but on reaching the Pentland Hills, hard greywacke strata are again met with, nearly parallel to those we have left, and, like them, inclined to the N. N. W., at very high angles. Under these circumstances it is impossible to assign to this greywacke of the Pentland Hills any very definite place in the Silurian system. Penetrated as it is by igneous rocks, and often in a much altered condition, this rock has still afforded some casts of orthoceratites; so that with this fact and an apparent superposition, it may be suggested that these strata are not far removed from the age of the Caradoc sandstone.

A second traverse of the chain near the eastern coast or towards the sea-board of the Lammermuir Hills, exposes the lower members of the series just described. There the Teviotdale anticlinal axis, *a*, is marked by the course of the Lower Tweed. Thus, the purple greywacke seen to the north of Berwick, and of part of which the outlines have been sketched (p. 150.), appears to belong to the same bottom rocks of the series. Further northwards, higher beds, the equivalents of the Peebleshire groups, occur, and in these, Mr. Stevenson of Dunse, who has published some valuable observations on this district*, has found, on the Dye near Byrecleugh, a graptolite, an obscure coralline, and slates with annelid impressions.

---

* See Proc. Geol. Soc. Lond., vol. iv. pp. 29. and 79.; and Quart. Journ. Geol. Soc. Lond., vol. vi. p. 418. I long ago examined the purple greywacke or bottom rock, near Dunse, in company with Mr. William Stevenson.

As these beds lie in the line of strike of the Grieston and Thornielee slates, they probably coincide with them in age. The predominating dip of the strata is to the north-west, showing a general ascending section, notwithstanding the numerous convolutions, which are beautifully seen on the coast, as already adverted to.

A third traverse of these South Scottish hills, made along the line of the Caledonian railroad *, exhibits the same general features, and many of the same details, as the preceding parallel sections. The Dumfries axis of the older and unfossiliferous greywacke, before alluded to, throws off anthracitic and graptolitic schists to the south at Lockerby, and at Moffat to the north. The details of these must be sought for in the memoirs of Professor Harkness, who has shown to what a great extent some of the schists are charged with graptolites. Among them may be cited the wide-spread forms of the Llandeilo flags—Diplograpsus pristis, D. folium, D. teretiusculus, and Graptolites sagittarius—all found in the alum slates of Sweden; Graptolites lobiferus, a Bohemian fossil, with other species found in Lower Silurian strata elsewhere in the British Isles, or peculiar to these deposits. About twenty species have already been enumerated from this district alone.† (See Foss. 6. p. 46.)

All these rocks, anthracitic and aluminiferous, which are charged with graptolites and annelids, dip northwards from Moffat to Abington, and thus pass under the only other masses of Silurian age which are visible in this direction. These strata, which occupy the tracts of the Lead Hills and other lofty summits, are metalliferous schists and greywacke, in parts much altered, which, penetrated by felspar porphyries and other igneous rocks, were celebrated for yielding gold ore in the reigns of the Fourth and Fifth James of Scotland.

---

* Professor Nicol and myself examined this section together in 1850. See Quart. Journ. Geol. Soc., vol. vii. p. 137.

† See Harkness, Quart. Journ. Geol. Soc. Lond., vol. vii. p. 48 ; M'Coy, Palæozoic Fossils, Cambridge Museum.

With them, or rather overlying them, is a rude breccia, in parts calcareous, which may possibly represent the Wrae limestone of the preceding diagram. But, further northward, other and still thicker strata of Silurian age are visible; the edges of the older strata being covered by the coal-fields of Lanarkshire. Though satisfactory in exhibiting a good ascending series from an unfossiliferous base of the sedimentary series, this section is less perfect in carrying out the order to beds on or above the horizon of the Llandeilo limestone, than that which is figured above. (p. 153.)

The fourth, or west coast traverse, from Luce Bay across Wigtonshire and the north part of Ayr, does not develope the order of the lower masses, but adds much to our acquaintance with higher and more fossiliferous strata of the Silurian system, in exhibiting beds more copiously charged with fossils than any rocks of this age in Scotland.*

It is to the northern portion of this region that I have paid most attention. So contorted and fractured are the strata of the southern portion of this series, that it is indeed no easy matter to place the component parts in their exact relative places; not even after much labour bestowed on the coast sections of his native county by my friend Mr. John Carrick Moore. The black glossy slates of Cairn Ryan, and certain red schists containing graptolites, might lead us, from analogy, to suppose that they represented the older graptolitic portion of the previous sections, which lie above the Longmynd or bottom rocks. On the other hand, they are associated with a coarse conglomerate, containing pebbles of granite and porphyry, with here and there blocks two feet in diameter, and this conglomerate

* Accompanied by Professor Nicol, I examined this tract in 1850. Our fossil collector was Mr. Alexander M'Callum, of Girvan, who has searched every locality with great assiduity, and is specially recommended to geologists visiting those parts. See details in my Memoir on the Silurian Rocks of the South of Scotland, Quart. Journ. Geol. Soc., vol. vii. p. 137.

so resembles a rock of the Ayrshire district about to be described, which clearly dips under certain fossiliferous schists (Kennedy's Pass), that the pebble-beds of Wigton and Ayr may be eventually placed on the same parallel. They are, however, separated by so much intrusive and eruptive trap, which occupies the coast from Correrie Burn to the Stinchar River, that it becomes impossible, without long and continuous labour, accurately to coordinate the disrupted masses. From the mouth of the Stinchar to Kennedy's Pass, south of Girvan, the eruptions of porphyry, greenstone, and syenite are, indeed, on a still grander scale (Knockdolian, Bennan Head, &c.). There, owing to these great extrusions from beneath, accompanied by much serpentine and some metamorphosed schists, the Silurian strata have been thrown into at least three flexures in the horizontal space of a few miles, on each of which, or along the Stinchar, Assell, and Girvan rivers, limestones of the age of the Llandeilo flags are brought to the surface. These limestones are overlaid in the contiguous troughs by other rocks, some of which, consisting of coarse conglomerates and shelly sandstone, may represent either the Upper Llandeilo or the Caradoc sandstones, whilst certain flagstones and schists might seem, by some of their fossils (though other forms oppose it), to indicate a passage into the lower member of the Upper Silurian rocks.

Referring for details to the memoir above cited, I will here very briefly describe some of the salient features of these Ayrshire deposits. The limestones in the Stinchar and the Girvan rivers contain the following fossils: — Orthis calligramma, Dalm.; O confinis, Salter; O. elegantula, Dalm.; Leptæna sericea; Cheirurus gelasinosus, Portl.; Pleurorhynchus dipterus, Salt.; and some terebratuloid shells, with corals of the genera Heliolites, Favosites, Omphyma, Strephodes, &c.

All these are Lower Silurian types, and with them is also found a species of the genus Maclurea of Hall, which is peculiar to the Trenton limestone, one of the marked Lower Silurian deposits of North America.*

These limestones, usually of dark grey colours, are supported by schists, and overlaid by shelly, fine micaceous dark-grey sandstones, separated by thin courses of shale, in which many fossils occur, particularly in the Mulloch Hill, on the right bank, and in the Saugh Hill, on the left bank of the Girvan water.

Among the fossils are: — a variety of Atrypa hemisphærica, Sil. Syst.; Orthis reversa, Salt.; O. biforata, Schloth.; O. elegantula (canalis, Sil. Syst.); Stropho-mena pecten, Dalm.; Bellerophon dilatatus, Sil. Syst.; Murchisonia cancellatula, M'Coy, a beautiful fossil; M. simplex, a Welsh species; Trochus Moorei, M'Coy; together with the well-known Silurian corals, Heliolites (Porites) inter-stinctus and H. tubulatus; Petraia subduplicata, Favosites alveolaris; Ptilodictya (Stictopora) acuta.

Almost all these forms occur in the Llandeilo flags, though some of them, such as the Atrypa hemisphærica, which is by much the most abundant of the shells, are also characteristic of the Upper Caradoc sandstone. Again, among the trilobites, we have here the Lichas laxatus, a common species in the limestone at Llandeilo in South Wales, and, in the very same matrix, the Calymene Blumen-bachii. Now, although this last-mentioned crustacean has been found throughout nearly the whole system elsewhere, it is by far more common in the Upper than the Lower Silurian, and is here associated with the Phacops Stokesii, and Encrinurus punctatus, both of them fossils which usually pertain to the Wenlock forma-tion. So that here again in Scotland, as in England and Wales, and particularly as we ascend in the series, we meet with rocks in which the upper and lower types are mixed together.

It may also be remarked, that in the greywacke sandstones of Saugh Hill, wherein the Atrypa hemisphærica abounds, we

* This species, which, though differing in some particulars, closely resembles the M. magna, Hall, has been so named by Professor M'Coy, who has described for Professor Sedgwick the fossils collected in this tract by Mr. John Ruthven.

have also, for the first time in the Scottish ascending series, Pentamerus oblongus*, Petraia subduplicata, and Tentaculites, which fossils, though occasionally found in lower strata, are peculiarly abundant in the ' Upper Caradoc,' properly so called, of Siluria; and in North America, as in England, these fossils mark the division between Lower and Upper Silurian.

Although the moory, mossy, and covered nature of the hilly grounds, which intervene between the upcasts of the lower limestone and the shelly greywacke which they throw off, prevent very clear sections, I am disposed to consider, as stated elsewhere, that the conglomerates, orthoceratite schists, and shelly sandstones of the coast section south of Girvan, are among the youngest masses of these environs. They contain a large orthoceratite, which Professor M'Coy has recently published as O. politum, and which M. Barrande has identified with a species occurring in the lower part of his Upper Silurian group of Bohemia; whereas in Ayrshire it is associated with Lower Silurian graptolites. Here, also, we have the Orthoceras angulatum (virgatum), which I published as an Upper Silurian form. On the other hand, there is a large orthoceratite, not to be distinguished from the O. vaginatum, Schloth., so common in the Lower Silurian of Scandinavia and Russia, and also O. bilineatum, a Lower Silurian fossil of America.

The Cyrtoceras, a genus which is also here, was, a few years ago, known only in the upper, but it is now found also in the lower rocks; and with this fossil occurs a thin, finely striated Orbicula, the O. crassa, Hall, closely resembling O. striata of the Ludlow rocks.

* Several of the species, particularly the Atrypa hemisphærica, and Pentamerus oblongus, occur in the Clinton group which forms the base of the Upper Silurian of the United States. In addition to the Upper Silurian fossils mentioned in the text, Professor Wyville Thomson has recently informed me that he has detected, near Girvan, the Bumastus Barriensis in the same beds with the Calymene Blumenbachii.

There are, too, both double and single graptolites, — the latter belonging to a Lower Silurian species.

Whether, therefore, I judged from the clear proofs exhibited on the coast near Kennedy's Pass, where these fossiliferous schists repose upon a great mass of coarse conglomerate and sandstone, or from the specific character of many of their imbedded fossils, I came to the conclusion that they indicate an ascending series in Ayrshire, although they chiefly contain Lower Silurian forms.

Although this is no place for details, a few words must be said of the very striking powerful conglomerate which underlies the above-mentioned sandy schists. Among its rounded and waterworn pebbles, Professor Nicol and myself distinguished upwards of twenty varieties of rock, varying in size from musket-bullets to blocks of two and three feet diameter. They consist of small specimens of earthy greywacke, and larger blocks of hard silicious greywacke, Lydian stone, hornstone, felspar porphyries of various colours, greenstone, syenite, and granite. If my view be correct, this conglomerate, on which the above fossil strata rest, is of about the same age as the pebble-beds of the typical region which occur towards the higher part of the Lower Silurian.

Such conglomerates, which may be indicative of the powerful and long-continued action of waves on a coast, are, however, only to be viewed as local phenomena, and may therefore be looked for in various parts of the primeval series of Scotland; just as they have been shown to occur at various levels in the Silurian rocks of England and Wales. They occasionally appear, indeed, in the coarse grits of the Longmynd or bottom rocks; and, even whilst I write, I am reminded by Professor Ramsay that coarse conglomerates occur in the same unfossiliferous subjacent greywacke of Carnarvonshire.

. All the shelly Silurians of Ayrshire being covered towards the north by the old red and carboniferous deposits, there is no evidence of a younger order of things beyond what may be viewed as the transition from the Lower to the Upper Silurian.

On the opposite side of the South Scottish axis, however, it would appear, judging from their fossil contents, that the highly convoluted strata which form the southern headlands of Kirkcudbright Bay (Balmae Head and Little Ross), and the rocks on the east side of Kirkcudbright Bay, are of the age of the Wenlock shale. Hard and intractable as the Ireleth slates of Cumberland, which have been placed in the same parallel, and containing only very rarely nodules slightly calcareous, these argillaceous and siliceous schists have yielded at one or two spots a good many fossils. In them we find the lower forms, it is true, of Orthoceras tenuicinctum (p. 195.), and the Leptæna sericea, Sil. Syst.; but from the same beds Mr. Salter has catalogued the following Upper Silurian types, viz., Phacops caudatus, Beyrichia tuberculata, Orthoceras annulatum, Chonetes sarcinulata (Leptæna lata of the " Sil. Syst."), Rhynconella nucula, Avicular lineata, Grammysia (Orthonota) cingulata. With these are associated other fossils which pervade the whole system, such as Atrypa reticularis, Halysites catenulatus, Graptolites priodon, and Bellerophon trilobatus. The last, which in Siluria occurs in the uppermost Ludlow rocks, is, in Ireland, associated with lower types, and may therefore belong to either division.

In this way we have no means of defining, with greater precision, the age of the fossil promontories of Kirkcudbright, than by saying that, being unequivocally Silurian, they seem to overlie the great mass of the older Silurians; and, containing a fauna of younger date in that era than any other part of Scotland has afforded, they are more referable to the Wenlock than to any other formation. It may also be surmised, that the time was, when there existed a large and visible upward development of still younger Silurian rocks, which ranged from the south of Scotland to meet the equivalents of the Wenlock and Ludlow rocks of the Cumbrian or

M

Lake districts of England — all which, if such existed, are now concealed beneath the sea.

In Scotland, however, still more than in the north-west of England, great links are wanting in the upward continuation of the palæozoic succession, which is so well exhibited in the Silurian region of England and Wales; for along the shores of Kirkcudbright these Upper Silurians, some of which indicate that they have been formed on a shore, are at once unconformably surmounted by strata of the carboniferous age.*

In this sketch of the older sedimentary strata of the South of Scotland, I have made but few allusions to the great masses of igneous rocks which have been intruded among the Silurians. In numberless cases, whether around the granite of Criffel and Cairnsmuir or the porphyry of Tongueland immediately to the north of Kirkcudbright, the schists which are in contact with such eruptive masses are so highly metamorphosed, that no one who looks at such effects can doubt as to the cause. So far we deal with facts only. At the same time, desirous as I am of avoiding speculative geology in a work like this, I am bound to state, that the view of much more extended metamorphism, to which I have long inclined, and which I suggested

---

* The phenomenon of the unconformable superposition of the carboniferous strata (with mountain limestone, fossils, &c.) to the Silurian rocks, along the coast of the parish of Kerrick, in Kirkcudbright, has very recently been described in great detail by Mr. Harkness, in a memoir read before the Geological Society, April 1853. This author has also explained, by elaborate sections, the extreme curvatures of all the rocks which he considers to be Upper Silurian, and parts of which he shows were shore deposits. He further indicates the points or dykes of porphyry which protrude among them, and has endeavoured to mark the line of separation between them and the Lower Silurian, which on the whole rise up to the N.N.W., their upper beds being conglomerates and coarse sandstones. Professor Nicol and myself examined Balmae Head in 1851. Mr. Stevenson described the nodular shale of Little Ross Head, Ed. Phil. Mag., vol. xxxv. The organic remains above alluded to were collected by the Earl of Selkirk and Mr. Fleming.

publicly in 1851*, has recently received a vigorous support from my friend Professor Nicol, who has for some time also entertained similar ideas. I then endeavoured to show, that certain bands of clayslate, chloritic and micaceous schist of the southern zone of the Highlands, with their interstratified limestones, were probably nothing more than metamorphosed Lower Silurian rocks, similar in age to the natural strata we have been considering, and which, ranging parallel to them, come out again in broad undulations, associated with many granitic and other eruptive rocks, to the north of the great Caledonian trough of old red sandstone, and its overlying coal-fields. †

One of these great undulations, which occur in the rugged mountains of the west called Argyll's Bowling Green, exhibits an axis of mica-schist throwing off to the north and south great thicknesses of chloritic schist and altered sandstone with included and regularly stratified limestones; all trending from W. S. W. to E. N. E., or parallel to the Silurian rocks of the south of Scotland.

In a subsequent memoir, Professor Nicol has supported this idea by striking facts and good reasoning. Describing the geological structure of Cantyre, and stating that the so-called mica-slate of that tract is only a partially altered micaceous sandstone, which seems almost to pass upwards into the old red sandstone, he shows how all the other crystalline strata comform and bend round the granitic nucleus of Goatfell in Arran. In confirmation of the probable identity of some of the crystalline strata of the north with the Silurian rocks of the south, he mentions the illustrative fact, that calcareous matter prevails most extensively towards *the western extremity of both groups of rocks.* In the South Scottish or Silurian strata, no calcareous beds are known at their eastern termination;

* See Quart. Journ. Geol. Soc. Lond., vol. vii. p. 168.   † Ib. p. 160.

limestone only begins to appear in feeble courses in Peebleshire, near the centre of the chain, and becomes very abundant (as above described) in Ayrshire. It is the same with the crystalline strata of the north. In the coast section from Stonehaven to Aberdeen, no calcareous bands occur; in Forfarshire and Perthshire several are known; they become still more abundant in the north of Argyleshire; and in Cantyre (as in Ayr) we have a great group of limestone rocks.*

The probability of the change of large portions of the Scottish Lower Silurians into crystalline rocks will presently be rendered more evident by an example derived from the west of Ireland.

Silurian Rocks of Ireland. — Rocks which from their included organic remains must be classed as Silurian, occupy a large portion of Ireland. In a retrospect, however hasty, of the progressive steps in the palæozoic classification of Irish rocks, the first allusion is due to the veteran geologist Weaver.† The first publication, however, in which any of the fossils of this ancient date were illustrated, was the detailed survey of parts of the North of Ireland by Colonel Portlock, in which that author described, from a single small district of schistose greywacke, near Pomeroy, in Tyrone, a multitude of interesting types, which enabled him distinctly to class these strata with the Lower Silurian group.‡

When the first edition of the geological map of Ireland, by Griffith, appeared, the existence of rocks containing fossils of this age had been ascertained only in a few localities, and the term Silurian was restricted by him to the precise spots from which those organic remains were described, expressly for that important work,

---

* Quart. Journ. Geol. Soc. Lond., vol. vii. p. 23.

† See Trans. Geol. Soc. Lond., vol. v. old ser. p. 117.

‡ See Report on Londonderry and parts of Tyrone and Fermanagh. Dublin, 1843.

and under that name, by Mr. M'Coy.* But the progress of research, of late years, on the part of Mr. Griffith and his assistants, has
much extended the areas in Ireland in which Silurian fossils occurred,
and has prodigiously increased their lists of fossils.

The labours of the government surveyors, first under Professor
Oldham, and now under Mr. Jukes, have since led to a still more
methodical elucidation of these rocks. In this way it has been ascertained that under the one term of greywacke, the following masses
were grouped : — First, the old dull red and purplish schist and grey
quartzite, to the south of Dublin, or bottom rock of the Longmynd,
in which no other fossils, as before stated, have ever been found, except the small zoophyte Oldhamia (p. 32.) Occupying the headlands of Bray, the mountains of the Sugarloaf, and the gorge of the
Dargle, to the south side of Dublin Bay (and probably many other
parts of the kingdom, as yet undetermined), these elder or bottom
rocks are succeeded by schists, which contain undoubted Lower
Silurian remains at intervals, and range in large but broken bands to
the south, through Wicklow and Wexford, reappearing from beneath
the old red sandstone and carboniferous limestone in the cliffs of
Waterford to the south of the bay of that name. Flanked, as these
Lower Silurians are, on their western side by dominant ridges of
eruptive granite that form the Wicklow mountains, they are there
either in a crystalline or subcrystalline state, and thus exhibit a third
condition of the greywacke of old authors, which, in the new maps of
the geological surveyors, is distinguished by a peculiar tint.

The silvery sheets of the well-known waterfall of Powerscourt are
precipitated over highly inclined crystalline chloritic and micaceous
schists of this class, which in Ireland are the equivalents of the crystalline rocks of Anglesea (see p. 23.). In the central parts of the south
of Ireland, or on the western side of the granite of the Wicklow

---

* Griffith's Silurian Fossils of Ireland, described by M'Coy. 1846.

Hills, the limestones and schists of the Chair of Kildare, which
are so replete with fossils, or the schists on the flanks of the
Slieve Bloom Mountains, are classed as Lower Silurian.* So also are
several masses of these rocks to the north of Dublin, as at Portrane,
and especially near Pomeroy, in Tyrone, the locality so rich in fossils,
which, as above stated, was the first clear Silurian type described in
this island. It is truly remarkable that so many of the characteristic fossils of the Llandeilo formation should have been detected by Colonel Portlock in this small tract, and one of no larger
dimensions in Fermanagh. Several of the Trinuclei and other
trilobites, such as Phacops, Calymene, and Illænus, are identical in
species with those from Shropshire and Wales; and so is it with
the single plaited Orthidæ, Leptænæ, and Strophomenæ, some spiral
shells, and many orthoceratites. Commingled, however, with these
are peculiar forms, first made known from this district, and which
are very rare indeed in any other tract in our islands. Such, for
example, among the trilobites are the Remopleurides, Foss. 31.
f. 5. in chap. 8. Harpes, ib. f. 4., Amphion, Bronteus, and, lastly,
the smooth forms of Asaphus, called Isotelus by American authors,
and which in Ireland, as on the other side of the Atlantic, are
very abundant, whilst they are exceedingly rare in Wales or
England, and not known at all on the Continent.

In Wexford and Waterford, as at the Chair of Kildare, the older
strata have been so much disturbed and insulated, that it is difficult
to detect anything like a good sequence of order. But still the
fossils, together with the physical characters of the rocks, clearly
mark the Lower Silurian era. In the cliffs on the bank of the river
Suir, below Waterford, and in the maritime headlands of Newtown
near Tramore, and of Bon Mahon, the schists, with feeble impure con-

* It is believed that the schists of Down are of the same age as the graptolite
schists of Wigton and Galloway.

cretionary limestones, are charged with the simple plaited Orthidæ, trilobites of the genera Trinucleus, Asaphus, and Phacops, double graptolites (Diplograpsus), and the coral Stenopora fibrosa, var. Lycopodites, in as great abundance as in the strata of the same age in Scandinavia.*

In the western districts of Cork and Kerry the extent has not yet been well ascertained to which strata, as distinguished by fossils * or infraposition, are separated from the large masses of the old red sandstone or Devonian system, which there rise into the loftiest mountains of Ireland (including Macgillicuddy's Reeks and Mangerton), around the Lakes of Killarney, and whose flanks are extended to the west coast opposite to the Island of Valentia.

In the district of Dingle, however, or the western part of Kerry, which I have personally examined, a vast number of fossils (trilobites, mollusks, and corals) have been of late years found, which I have recently inspected, either in the cabinets of Mr. Griffith, or those of the government surveyors.

Some of these fossils, particularly from the eastern parts of this promontory, may belong to the same Lower Silurian type which prevails through Wicklow, Kildare, Wexford, and Waterford. Others, however, including most of the species found in Ferriter's Cove and at Dunquin, are as certainly Upper Silurian fossils, which are well known in England and Wales. Among the commonest of these may be mentioned, Euomphalus funatus, Cardiola interrupta,

---

* Most of the fossils described by Mr. Weaver in his memoir on the south of Ireland, Trans. Geol. Soc. Lond., vol. iv. *old series*, are now known to pertain to the carboniferous system ; but certain localities which he notices, such as Smerwick Harbour or Ferriter's Cove, and Bon Mahon, near Waterford, were long ago recognized by Mr. Griffith, and his able assistant Mr. Kelly, to be Silurian. In addition to the public collections in Stephen's Green, Dublin, and the rich collection of Mr. Griffith, the geologist will find an instructive assemblage of these Lower Silurian fossils of Waterford in the possession of Mr. Nevin, of that city.

Murchisonia Lloydii, Pterinea (Avicula) retroflexa, Orthis lunata, O. elegantula, Rhynconella (Terebratula) Wilsoni, R. navicula, Phacops (Asaphus) caudatus, Encrinurus punctatus, and a multitude of common Upper Silurian corals.

It is, therefore, to be hoped, that the exact order of that promontory between Dingle and Tralee Bays may soon be unravelled. I have, indeed, little doubt, from what I have been told by Sir Henry de la Beche, Mr. Jukes, and Professor E. Forbes, that the public will soon have presented to them a complete ascending order in this region of the south of Ireland, from Lower through Upper Silurian rocks into a full development of the old red sandstone surmounted by the carboniferous limestone.

In passing along the west coast to the Connemara district in the county of Galway, the geologist again meets with tracts lying to the north of the picturesque mountains, known as the Bins of Connemara, which are charged with Silurian fossils. These fossil bands occupy the bold and precipitous sides of the deep bay of Killery, and range over a considerable space eastwards, by Leenane, Maam, and Oughterard, to the shores of Loughs Corrib and Mask.

Whilst on the eastern side of the island the eruption of the granite of the Wicklow mountains has metamorphosed, as above said, the contiguous Lower Silurians, so in Connemara on the west, we have a still more striking example of the metamorphism of a very large mass of strata which, in my opinion, must be also considered inferior members of the same series. The annexed diagram is reduced from one of Mr. Griffith's transverse coloured sections across this remarkable district, which, after a personal survey of most of the Silurian tracts of Ireland, has appeared to me best adapted to explain how certain crystalline rocks which, from their mineral aspect, have hitherto been supposed to be of higher antiquity, are, in all probability, nothing more than the altered lower members of the sedimentary deposits under consideration.

SECTION ACROSS THE BINS OF CONNEMARA, AND KILLERY HARBOUR.
(From a coloured section by Mr. Griffith, prepared by Mr. Ryland Byron).

N.

Knockaskeheen.

Glenkeen, 1974 ft.

Bundorragha, 794 ft.

Killery Harbour.

Benbaun, 2395 ft.

T'ievebreen.

Ben Glenisky, 1710 ft.

Ballynahinch Lake.

Derryadd West.

Roundstone Bay. S.

a. Metamorphosed lower slaty rocks with white, green, and grey crystalline limestone (Connemara marble). b. Hard quartz rocks passing into flagstones and courses of grey limestone, subordinate to mica-schist in one tract not traversed in this section. This limestone completely resembles that of Inverary and Loch Fyne in Scotland (see p. 163.). c. Silurian rocks composed of conglomerate, dark schist, and reddish sandstone slightly unconformable in this traverse to the underlying metamorphic rocks, but in other places completely discordant. Fossils numerous in several localities. * Igneous rocks, Granite; † porphyry and greenstone.

In proceeding from Roundstone Bay on the south, low isles and headlands of granite, *, followed by syenitic and hornblendic rocks, †, throw off highly altered micaceous and quartzose schists, with courses of white and grey crystalline limestone, a, which, as you ascend the mountain of Ben Glenisky, one of the twelve Bins of Connemara, are seen to be followed by strong masses granular, white, quartz rock (manifestly an altered sandstone) which alternate with mica-slate, b.

Interlaced in the lower part of this series, and lying in depressions between Ben Glenisky and Ben Bawn (2300 feet above the sea), and thence ranging to the west, are certain crystalline limestones, one course of which is grey and white, and another is the beautiful green serpentinous marble, or serpentine of Connemara. These are again overlaid by the granular quartz rocks and mica-slates, b, which, occupying the greater portion of these mountains, are, in some places, surmounted on their northern face (with a slight

unconformity* only) by conglomerate sandstone and schists, c, containing fossils, as seen at Lettershanbally, Blackwater Bridge,
Leenane, Maam, and other places. Now, these fossils, all unequivocally Silurian, are by no means types of the lowest members
of the system. On the contrary, though some of them are found
in Lower Silurian strata, their prevalent species refer the beds
rather to the upper part of the Llandeilo, or even to the Caradoc
formation. In fact, the strata resemble, both zoologically and
lithologically, the fossiliferous greywacke on the banks of the
Girvan water, in Ayrshire; many of the same species occurring in
the two districts. (See p. 159.). Several of the Irish fossils are,
indeed, much more common in the Upper Silurian of England and
Wales.

For instance, in the beds near Maam, Professor Nicol and myself
collected fossils, some of which would be considered Lower and
others Upper Silurian; there being among the latter the Bellerophon trilobatus, first published from the tilestones, or summit of
the Ludlow rocks. With the widely spread fossils Strophomena
depressa and Atrypa reticularis, which occur in nearly all the
Silurian rocks, we here met with such shells as Pentamerus oblongus,
Orthis calligramma (virgata), O. reversa, Atrypa hemisphærica,
which elsewhere are only known in the lower division, associated
with Rhynconella cuneata and R. Wilsoni, Bellerophon trilobatus,

---

* In a traverse of this tract, which I made from Clifden to Killery Bay,
in the company of Professor Nicol, the green marble seemed to me to be enclosed
in mica-schist and quartzose rocks. We also, however, met with a grey limestone,
subordinate to the upper portion of the mica-schist, which, as in Argyleshire on
both sides of Loch Fyne, exhibits many of the appearances of ordinary stratified limestone, in respect to bedding, joints, and wayboards; but we could detect no fossils
in it. Though the above section exhibits the strata, c, containing Silurian fossils,
reposing with a slight unconformity only on these micaceous rocks, there are
many other such junctions (as around Maam) where the relations are quite
discordant.

and some lamellibranchiate shells, such as the Pterinea or Avicula retroflexa, and Cucullella antiqua, which were described as true Upper Silurian fossils. Here, also, in beds of junction with the mica-slates, are the chambered shells, Cyrtoceras approximatum, Orthoceras bullatum, O. ibex and O. angulatum,— species formerly supposed to mark the Upper Silurian only, but since also found in some of the lower strata. Among the trilobites we have even our constant friend Calymene Blumenbachii, and the Encrinurus (Amphion) punctatus, together with a genus only yet known in Lower Silurian strata, the Asaphus, apparently the same species, found in Tyrone, and Cyphaspis megalops, a Dudley fossil. Again, the corals from one locality (Kilbride) present us with such species as Favosites polymorpha, F. Gothlandica, F. multiporata, &c., which are usually upper types, commingled with shells, some of which are Upper, whilst perhaps still more are Lower Silurian.

This statement is not made to weaken the value of the separation on a general scale of the Silurian system into two groups; but simply to show that there are many tracts where the appropriation of certain strata absolutely to one or the other becomes very difficult, if not impossible.

In addition to the clear types described by Portlock from the centre of the north of Ireland, we have indeed the best proofs, in a multitude of trilobites, orthidæ, and other remains, that the lower fossil-bearing schists of Wicklow, Kildare, Wexford, and Waterford, as well as the Tyrone beds, are unquestionably Lower Silurian; whilst in Connemara and the adjacent tracts the indications are as clear, that we have reached much higher members of the series, while around Dingle Bay the fossils of the upper division preponderate. Where, then, is the true lowest Silurian in the Connemara tract? My belief is that it is represented by the great underlying crystalline masses of the Bins of Connemara, and that the limestones included in those mica-schists are nothing more than the altered Lower

Silurians; some of them being of the age of Llandeilo and Bala in Wales, or of Wrae and Girvan in Scotland.

In this Galway region we cannot, indeed, pursue a clear ascending order, any more than in Scotland. We can only observe that a coarse, chloritic, quartzose conglomerate overlies the fossiliferous beds in question — a conglomerate which, according to Mr. Griffith, rises into mountainous masses on the north side of the Killery Bay. Alternating with green and purple slate and interstratified felspar porphyries, †, these rocks, after an undulation, repose upon a vast mass of greenish, grey, and reddish slate, *a**. The base of the fossiliferous series to the north of the Killery harbour is composed of a dark and reddish brown sandstone, under which is a conglomerate of mica-schist and other rocks; the whole resting, as on the other side of the trough, on mica-schist. (See section.) Then follows a boss of granite,* (Knockaskeheen), which throws off crystalline schists on both flanks, — the masses in the north being subtended by the unconformable and horizontal carboniferous limestone of Clew Bay in Mayo.

The reddish-brown sandstone of this tract, occasionally fossiliferous, which is well seen a little to the north of Maam, has, indeed, a mineral aspect which so much resembles the old red sandstone of other countries, that if the same rock should accidentally have been misnamed in some parts of Ireland where fossils do not occur, the mistake is one which, in a region so broken, and which is to so great an extent obscured by drifted materials (the " escar" of Ireland), must almost necessarily have been made before the classification of rocks through their organic remains was established. Considerable masses of the brown and purple sandstone of Ireland will be found to belong, I apprehend, to the Silurian age.

The progress of research will also, I trust, bring out other evidences to support the views here entertained respecting metamorphism; and if the proofs of a still more perfect Silurian series be obtained in the Dingle and Killarney districts, then the above interpretation of the Connemara tract, with its crystalline and highly altered bottom rocks and fossiliferous upper strata, may be applied to the still more altered and crystalline rocks of Donegal. The latter may, indeed, be brought into coordination and connection with those true and shelly Silurian rocks in Fermanagh and Tyrone which have been described by Colonel Portlock. This is a task worthy of the government surveyors, and which, I have no doubt, they will ably accomplish. No one who has examined the small patch near Pomeroy * could imagine, from the aspect of the schists, whether argillaceous or quartzose, that it could have proved so highly fossiliferous; and hence we need not despair of finding many more such former burying-places in various parts of the large area of Ireland, over which Silurian rocks are believed to extend; but which are now to so great an extent metamorphosed, and so sterile in organic remains.

Besides many fossils which are common in Great Britain, Ireland presents us with other species of crustaceans and shells, a full acquaintance with which must be sought for in the works of Portlock and M'Coy, in the tables of Griffith, and in the illustrations of Irish geology which are preparing for publication by the government surveyors.

Some of the most characteristic of these fossils will be noticed in the two ensuing chapters, which treat of the chief organic remains of the Silurian system.

Vertical dimensions of the Silurian rocks of the British Isles. We have as yet no means of accurately estimating the thickness of the

* I recently examined the Tyrone tract under the guidance of Mr. John Kelly, the assistant of Mr. Griffith, an excellent and exact field geologist, to whom I am indebted for valuable assistance in examining other localities of Silurian rocks (Dublin district, Chair of Kildare, Dingle, &c.) We were accompanied in the Pomeroy district by M. de Verneuil, M. Pierre de Tchihatcheff, and Professor Nicol.

older deposits of Scotland and Ireland which have been treated of in
this chapter; but I find, on consulting with Professor James Nicol,
that the Scottish section given at p. 152. can hardly represent less
than 50,000 feet, although we have no indication that the bottom
of the sedimentary series is reached, nor have we any thing like a
completion of the upper Silurian rocks. With the extension of the
geological survey to Ireland (a benefit which it is hoped Scotland
may also soon enjoy), we may ere long be furnished with the requisite data respecting the sister isle.

In the meantime, reverting to the typical region of Wales and
the adjacent English counties, as described in the earlier pages to
Chapter V. inclusive, we can appeal to the admeasurements of the
Government surveyors. In Shropshire, the Longmynd or unfossiliferous bottom rocks (the Cambrian of the Survey) are said to have
the thickness of 26,000 feet, or about three times that of the same
strata in North Wales; whilst my original Lower Silurian strata of
Shropshire to the west of the Longmynd exhibit a width of 14,000.
On the other hand, in the region between the Menai Straits and the
Berwyn Mountains, where the bottom rocks are so much less copious than in Shropshire, the fossiliferous Lower Silurian, from the
base of the Lingula flags to the top of the Llandeilo or Bala formation (including the stratified igneous rocks), swells out to about
19,000 feet, and the Caradoc sandstone, on the borders of Radnor
and Montgomery, has a thickness of from 4,000 to 5,000 feet.
Taking the greatest dimensions, we are, therefore, presented with the
prodigious measurement of about 50,000 feet of sedimentary strata,
in the lower half of which no fossils have been found, the upper part,
as above described, bearing a group of fossils to which allusion has
already been made, and whose chief characters will be specially considered in the sequel. Although of such vast volume in parts of
the region described, it must be observed that the Lower Silurian
rocks of other tracts, though precisely of the same age, as proved

by their imbedded organic remains, are often comparatively of very small dimensions.

Though more replete with fossils than the inferior group, the Upper Silurian rocks attain nowhere* a greater thickness than from 5,000 to 6,000 feet, the Ludlow rocks being for the most part more developed than the Wenlock formation. In this way the whole of the fossiliferous Silurians of England and Wales, measured from the Lingula beds to the Ludlow rocks inclusive, have the enormous maximum dimensions of about 30,000 feet; and if we add the conformable underlying sedimentary masses of pretty similar mineral aspect, but in which no fossils have been found, we have before us a pile of subaqueous deposits reaching to the stupendous thickness of 56,000 feet, or upwards of ten miles!

* This observation applies to the original Silurian region and Wales. The Upper Silurians of Cumberland and the lake country may be of larger dimensions ; but I am not aware that their thickness has been estimated.

CHAPTER VIII.

ORGANIC REMAINS OF THE LOWER SILURIAN ROCKS.

A FULL acquaintance with the Silurian fossils of the British Isles can only be gained by a study of the various works in which they have been successively described.   The first of these in order of time, as also the first in which the organic remains were classified and placed in their true geological position, is the " Silurian System ;"* the next is a report on the Silurian fossils of Tyrone, by Colonel Portlock; a third is on the Silurian fossils of Ireland, by Professor M'Coy.   Then follow various publications of the government geological survey, particularly the volume on the Malvern and Abberley Hills, by Professor Phillips and certain monographs, descriptive of both Lower and Upper Silurian forms, in the Decades of the geological survey, by Professor E. Forbes and Mr. Salter.   Valuable notices of these fossils have also been published by Mr. Sharpe and other writers in the volumes of the Quarterly Journal of the Geological Society of London; whilst an excellent description of the Upper Silurian brachiopods by Mr. Davidson, has appeared in the Bulletin of the Geological Society of France, in which nearly all the British species of that age are figured and described.   Even since the period of this last publication, and whilst these pages are printing, the same author has published, in the volumes of the Palæontographical Society, a highly valuable monograph, detailing the generic characters of all the Silurian brachiopods.

* The shells of the Silurian System were described by James de C. Sowerby, Mr. Salter, then a very young man, assisting, and drawing many of the forms. The corals were described by Mr. Lonsdale, and the trilobites by myself.

In all these British works, and in the volumes of my contemporaries which relate to the continents of Europe and America, the fossils described have been invariably referred to as Lower and Upper Silurian. In one work only, — published last year by Professor Sedgwick, — on the palæozoic fossils of the Woodwardian Museum of the University of Cambridge, has this nomenclature been changed; the term Lower Silurian being disused by him, and that of Cambrian substituted.

The reader, who will consult the various works alluded to, will find, that whilst a marked division was at first particularly insisted on, as existing between the Lower and Upper Silurian, subsequent researches, extended over large areas, have shown that the two groups are really much more knit together in one natural series than when my classification was proposed; it being now well ascertained, as explained in the preceding pages, that a very great number of the most characteristic fossils are common to the inferior and superior members of the system. In the course of the observations which follow, this generalization will be placed in a clearer point of view.

Graptolites have been already alluded to, as being an *exclusive* and highly characteristic Silurian form of polypes, which inhabited a sea with a muddy bottom (p. 46.). Of those species which specially mark the Lower Silurian rocks, the double graptolite, Diplograpsus pristis, p. 46., Foss. 6. f. 4., ranges from some of the inferior strata to the upper beds of the Caradoc sandstone; whilst the Graptolites priodon, Foss. 6. f. 3., first named G. Ludensis from being found in the superior member of the whole system, or the Ludlow rocks, has been also abundantly found in the older strata of Scotland, and also in those of North Wales.

But besides these animals of ancient muddy seas, which are, as before said, exclusively Silurian in all quarters of the globe, other forms of zoophytes are found in the lower portion of these deposits, which multiply very much as we ascend in the series. Among these certain species of corals of the genera Halysites, Heliolites or Palæopora (Porites, Sil. Syst.), Stenopora, and Petraia, are the most prevalent.

Fossils, 13.

Lower Silurian Zoophytes.

1. Stenopora fibrosa, Goldf. 2. The same species, variety Lycoperdon, Hall. 3. Petraia subduplicata, M'Coy.

4. Halysites catenulatus, Linn. 5. Ptilodictya dichotoma, Portlock. 6, Retepora Hisingeri. M'Coy

One of the most striking of these bodies, and one of the most widely spread in various countries, is the Halysites catenulatus of Linnæus, f. 4., the well known " chain coral" (Catenipora escharoides of the Silurian System). Another, still more abundant in the lower rocks, is the Stenopora (Favosites) fibrosa, whether branched, f. 1., or hemispherical, f. 2.; a third, the Petraia subduplicata, f. 3. Others, less common, are the Stromatopora striatella, chap. 9., Foss. 32.; Heliolites (Porites) interstinctus, p. 120. Foss. 11.; H. in-

Fossils, 14.

Lower Silurian Zoophytes.

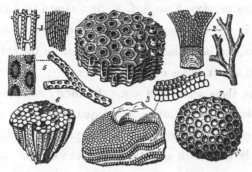

1. Fenestella sub antiqua, D'Orb. 2. Ptilodictya acuta, Hall. 3. Nidulites favus, Salter. 4. Sarcinula organum, Linn.

5. Heliolites inordinatus, Lonsd. 6. Favosites Gothlandica, Goldfuss. 7. Heliolites megastoma, M'Coy.

ordinatus, Foss. 14. f. 5.; H. megastoma, f. 7.; Favosites Gothlandica, f. 6. ; F. alveolaris, p. 119., Foss. 10.; and the Sarcinula organum, Linn., Foss. 14. f. 4., found in the slates of Westmoreland, and in the Upper Silurian limestone of Gothland in Sweden.

The reader has now before him in these woodcuts, some of the corals which most frequently occur in Lower Silurian strata, but if he wishes to study many other forms of these fossils, particularly those in the Upper Silurian rocks, where corals are much more abundant, he must consult the clear and faithful descriptions of them given by Mr. Lonsdale in the " Silurian System," in which work 62 species are accurately figured.   By reference to the authority of that valued friend, who assisted me so materially in preparing my original volumes, and to the works above cited, it will be seen that, of the ten species here mentioned as frequent in the Lower Silurian, five of them, *i. e.*, the chain coral, the Favosites alveolaris and F. Gothlandica, the Heliolites interstinctus, and H. megastoma, are still more abundant in the upper division of the system*, as is also the case with H. pe-talliformis, ch. 5. Foss. 11. f. 2., and H. tubulatus, ib. f. 1.

Heliolites favosus, M'Coy, and H. subtilis, of the same author, are rare Lower Silurian species of the last named genus.   Nebulipora, a new genus of Professor M'Coy, contains two or three species of minute-celled corals, resembling Favosites.   One, the Nebulipora favulosa of Phillips, is figured in the woodcut, p. 86., Foss. 7. f. 13. It is very common in the Llandeilo flags.

These are not, of course, all the corals of the Lower Silurian rocks. Two or three other species of Petraia have also been described from

---

* The palæontologist will also naturally consult the writings of Milne Edwards and Jules Haime, Archives de Museum d'Histoire Naturelle, vol. v., who have shown that several of the corals which were supposed to be common to the Silurian and Devonian rocks are not so.   I must also specially refer the naturalist to Danas' magnificent and profound work on zoophytes, recently published as part of the results of the "United States Exploring Expedition ;" and to the descriptions and figures of Professor M'Coy, in Professor Sedgwick's work above quoted.

the Llandeilo flags, and the common P. bina, abundant in Wenlock strata, is even more common in Caradoc sandstone. (Chap. 9. Foss. 33. f. 7, 8.)

Others of these zoophytes, although possibly allied to mollusca, so resemble corals in the eye of the field geologist as to pass for such. These are certain forms of Bryozoa or Ciliobrachiata. Among them the Ptilodictya dichotoma, p. 176., Foss. 13. f. 5., and the P. acuta, ib. Foss. 14. f. 2., with some new species, are not unfrequent fossils of this age. A coral of a net-like form, the Retepora Hisingeri, Foss. 13. f. 6., is common in the slates of Wales. Glauconome disticha, of the Wenlock limestone, ch. 9., Foss. 35. f. 5., is often found in calcareous bands of the Llandeilo flags, as are also the Fenestella Milleri, ib. f. 4. and Fenestella subantiqua, D'Orb. (F. antiqua, Sil. Syst.), Foss. 14. f. 1. Of these the three last are also common Upper Silurian species.

Of Crinoids, or the lily-shaped tenants of the deep, most of which were attached by their root or base to submarine rocks, the inferior strata of the Silurian epoch have in general afforded fragments only. Referring the reader therefore to the plates 13 to 15. of this work, as containing the forms best known in the Silurian rocks when my work was published, it is enough now to state, that, for the most

FOSSILS, 15.

Glyptocrinus basalis, M'Coy.   A Lower Silurian crinoid.

part, broken portions of the stems only of Encrinites (pl. 12. f. 3.) have been detected in the earlier portion of these deposits in Britain.

One brilliant exception, however, is the Glyptocrinus basalis,

M'Coy, a crinoid found in the Lower Silurian slate rocks of North Wales; fine specimens of it are to be seen in the cabinets of the Museum of Geology in Jermyn Street, and in the Woodwardian collection of Cambridge.

Next in the natural ascending order come those remarkable animals, whose globular forms and strong external plates have often enabled them to resist destruction better than the delicately constructed and slender stemmed Encrinites, viz. the Sphæronites of old authors, the Cystideæ of Von Buch. Abounding in the Lower Silurian of Scandinavia and Russia, they were long ago described by Linnæus and Gyllenberg, whilst they have recently had much new light thrown

Fossils, 16.
Lower Silurian Cystideans.

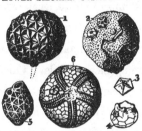

1. Echino-sphærites Balticus, Eichw. 2. E. punctatus, Forbes. 3. Ovarian pyramid of E. granulatus, M'Coy.

4. Caryocystites munitus, Forbes. 5. C. granatum, Wahl. 6. Agelacrinites Buchianus, Forbes.

on their natural affinities by Von Buch, Volborth, H. I. H. the Duc de Leuchtenberg, and Edward Forbes. They constitute, in the primeval era, the representatives of the sea-urchins or Echinidæ of the secondary and tertiary periods and of the present day. They have affinity with Crinoids on the one hand (some possessing very perfectly formed arms and tentacles), and with sea-urchins on the other *, and have been divided into genera called Echino-sphærites, Echino-encrinites, Sycocystites, Hemicosmites, Cryptocrinites, &c., as here figured.†

* The views expressed in the text are those of Edward Forbes.

† Other forms of these Cystideæ are given in the work, "Russia and the Ural Mountains," vol. ii. See also the monograph of the eminent Leopold Von Buch, "über Cystideen," Trans. Berlin. Acad., 1845; Volborth, "über die Echino-Encrinen," Bull. Sc. Acad. St. Petersburg, T. x., and Bull. Phys. Mat., ib. T. iii. No. 6.; Beschr. Thier. Silur. Kalk. v. Max. Herz. v. Leuchtenberg. St. Petersb. 1843; and E. Forbes, Mem. Geol. Surv. vol. ii. pl. 2. 1848.

They usually occur in clusters, on the shaly surface of beds of
limestone, resembling bunches of enormous grapes; and in Sweden
and Russia, where thus developed, they are associated with Orthidæ
and other shells of the Lower Silurian rocks.   In Britain they had
not been found when the Silurian classification was published; but
by the researches of the government surveyors they were detected,
and pretty plentifully, in strata at Sholeshook, Pembrokeshire, laid
down by me as of Llandeilo age (see p. 68.).   The usual forms in
our country are species of Echino-sphærites and Caryocystites.   The
Echino-sphærites aurantium, the commonest species in the north of
Europe, is rare in Britain; but E. Balticus, Foss. 16. f. 1., and Caryocystites granatum, f. 5., both common in the Lower Silurian of Scandinavia,
are found profusely in Pembrokeshire, as above, in one of my old fossil-
bearing localities; and thus our British strata are well paralleled with
their foreign analogues.   E. punctatus, f. 2., and Caryocystites munitus, f. 4., are common species at Bala.   One exceedingly rare and
curious form, possessing arms, which are affixed like the spokes of a
wheel on the upper surface, has been found in North Wales, and
figured in the Memoirs of the Geological Survey: it has received from
Professor E. Forbes the name of Agelacrinites Buchianus, in honour
of Leopold v. Buch, and is represented above, f. 6.

Nor must we omit two species of star-fish, first discovered by
Professor Sedgwick and Mr. Salter at Bala.   One of these, described
by Professor Forbes, belongs apparently to the modern genus Uraster,

Fossils, 17.

Uraster obtusus, Forbes.   A Lower Silurian star-fish.

so common on our own shores: the other was unfortunately lost
when it had been ascertained to belong to the true Ophiuroid group,

and named provisionally O. Salteri, by Professor Sedgwick. The Uraster obtusus, from the slates of Wales and Ireland, is represented in the preceding woodcut.

Among the shells of the Llandeilo strata, above the lowest or primordial zone described in the last chapter, Lingula again occurs, but of different species. Lingula attenuata, pl. 5. f. 1., is indeed one of the most characteristic shells of the Llandeilo formation, occurring plentifully at Builth and Llandeilo, in South Wales, and at Shelve, in Shropshire. Other species will be mentioned afterwards.

But by far the most abundant bivalve shells of these rocks belong to the group of brachiopods represented more particularly by the genera Orthis, Leptæna, and Strophomena. With these are associated several forms of Terebratula, or rather Rhynconella of modern authors, the species in the old rocks being chiefly those which have a sharp beak, with the perforation beneath it; while in true Terebratulæ the hole is in the beak itself. The genus Atrypa is also common, as well as Pentamerus; the various species of the last genus being almost exclusively Silurian.

Although all the well established forms of these genera which were known at the period of the publication of the " Silurian System" are now re-produced from the original etchings in the annexed plates (5 to 8.), attention will here only be specially directed to the figures of those which are found in abundance, and are therefore useful indicators, either as characterizing one zone or band of strata, or which, being found to ascend through a number of beds, have therefore lived through a long period of primeval time.

The genus Orthis is one of the most prolific forms of brachiopod of this period. The species have generally one valve convex, the other flat, and are of a roundish outline. Those which most characterize the Lower Silurian division are arranged, as before said, in plates 5, 6, and 7. Some are figured in the two following woodcuts.

Out of ten species originally published as Lower Silurian types, at
least six are now well ascertained (see p. 87.), to range from the Llandeilo formation, as exhibited among the rocks of Snowdon, to the
uppermost beds of the Caradoc sandstone of Shropshire inclusive,
where these are succeeded by the Wenlock shale. These are Orthis
elegantula, pl. 6. f. 5.; O. testudinaria, pl. 6. f. 1, 2.; O. vespertilio,
pl. 5. f. 16—20.; and the simple plaited species figured in the following woodcut viz., O. flabellulum, f. 1.; O. Actoniæ, f. 2.; and the
variable O. calligramma (virgata, Sil. Syst.), f. 3, 4.

Among the other species most frequent in the lower division, and, as
far as I know, peculiar to it, are the O. insularis, Pander; O. alternata,
pl. 7. f. 4.; O. confinis, Salter; O. crispa, M'Coy; O. reversa, Salter, the
last being very common in the Lower Silurian rocks of Scotland and the
West of Ireland, and not rare in England. O. protensa (including O. lata

FOSSILS, 18.

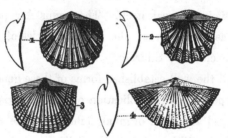

1. Orthis flabellu-
lum. 2. O. Actoniæ.
3, 4. O. calligram-
ma, Dalman; two
varieties.

Characteristic
Lower Silurian
forms of simple-
plaited Orthi-
des.

of my old work), pl. 5. f. 22—24.; O. turgida, M'Coy; O. sagit-
tifera, M'Coy; and O. Hirnantensis, M'Coy, are other species of the
Lower Silurian, found in Wales.

Wherever several of these forms of Orthis are found, the geologist
may be pretty sure that he is working in Lower Silurian rocks;
though even among these there are species, such as O. elegantula and
O. calligramma, which occur in the Upper Silurian; — the former
most abundantly.

Three of the above species, the O. insularis, O. testudinaria,
and O. calligramma, are found in Scandinavia and Russia; while

another Lower Silurian species, O. striatula, Emmons, Foss. 19. f. 3., not published in the " Silurian System," but still a very common British species, is the most abundant Orthis of this epoch in North America. The Orthis biforata or lynx, Foss. 19. f. 4., formerly called Spirifer lynx, is common in the Lower Silurian of North America, Russia, and Sweden; and is abundant in those British localities originally called by me Lower Silurian, as well as in North Wales and Ireland. It also occurs, though rarely, in the Wenlock limestone and well-defined Upper Silurian of Britain and Gothland. On the other hand, the Orthis biloba of Linnæus (Spirifer sinuatus Sil. Syst.), pl. 20. f. 14., which is most abundant in the Wenlock shale, has recently been detected in the Llandeilo flags of Builth (Radnorshire), in Pembrokeshire, and in Denbighshire.

FOSSILS, 19.
LOWER SILURIAN BRACHIOPODS.

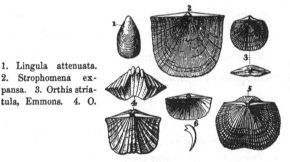

1. Lingula attenuata.
2. Strophomena ex-
pansa. 3. Orthis stria-
tula, Emmons. 4. O.

biforata, Schlotheim.
5. O. porcata, M'Coy.
6. Leptæna sericea.

The closely allied genus of Strophomena, consisting chiefly of the flatter forms of the so called genus Orthis of the " Silurian System," exhibits some very marked and frequent species which are eminently typical of the Lower Silurian. These are the Stropho-mena expansa, pl. 7. f. 1. and Foss. 19. f. 2.; Strophomena tenui-striata, p. 86. Foss. 7. f. 8.; S. grandis, pl. 7. f. 6, 7., which abound in the Llandeilo formation, and have been formerly given in the plates of my work, as species of Orthis and Leptæna; they not only prevail in the Llandeilo and Bala deposits of Wales, but are common in the strata of the same age in Westmoreland, and in the Caradoc sandstone of Shropshire. Strophomena corrugata, Portlock, is a highly ornamental species of the genus, with radiating lines and transverse undulations : it is found in Wales and Ireland. Another shell of this genus frequent at Bala, the S. spiriferoides, M'Coy, Foss, 20. f. 2., is so like Spirifer radiatus of the Upper Silurian rocks, as to be easily mistaken for it. It is a common Caradoc species.

The universally spread Leptæna depressa (Sil. Syst.), now more correctly referred to the genus Strophomena, extends upwards throughout the whole series from the very oldest beds of Llandeilo, to the Upper Ludlow rock.

Of the two species of Leptæna which are prevalent in the lower division, the most frequent is L. sericea, pl. 6. f. 13., and Foss. 19. f. 6.; which, occurring in swarms among the slates of Snowdon, is also frequent in the Caradoc sandstone of Shropshire, and of the Malvern Hills; whilst the L. transversalis, pl. 20. f. 17., published originally as a fossil of the Wenlock shale, is now found in the Llandeilo formation of Wales and Westmoreland. The former of the two last-mentioned species has indeed an universal range; being known in Russia, Scandinavia, Central Germany, the British Isles,

FOSSILS, 20.
LOWER SILURIAN BRACHIOPODS.

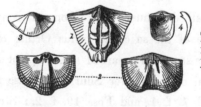

1. Orthis biforata, Schloth. (internal cast). 2. Strophomena spiriferoides, M'Coy. 3. Leptæna quinque- costata, M'Coy. 4. Leptæna tenuicincta, M'Coy; and a lateral view of the same.

and America. Leptæna tenuicincta, M'Coy, Foss. 20. f. 4., and L. quinque-costata of the same author, f. 3., are not unfrequent species of the Llandeilo flags, both in England and Ireland.

The Orbiculoid and Cranioid groups, though not common in the lower division, yet afford some characteristic species. The remarkable Crania (Pseudocrania) divaricata of M'Coy, Foss. 21. f. 2., like its congener the Crania antiquissima of Russia, is a Lower Silurian species. Orbicula punctata of the Caradoc sandstone, f. 1., can with difficulty be distinguished from the beautiful O. cancellata of the Trenton limestone, America; and Siphonotreta micula, M'Coy, f. 3.,* a small species of a genus abundant in the Lower Silurian beds of Russia, is found plentifully in the trilobite flags of Builth.

Lingula, on the other hand, offers some prevailing forms. Such is L. attenuata, the characteristic Llandeilo fossil before alluded to, in p. 183. Its associates are L. granulata, Phil.; L. ovata, and L. tenuigranulata, M'Coy, as here figured. Obolus, a genus closely

FOSSILS, 21.

LOWER SILURIAN BRACHIOPODS.

1. Orbicula punctata. 2. Crania divaricata, M'Coy. 3. Siphonotreta micula, M'Coy; young and full grown, natu- ral size. 4. Lingula granulata, Phill. 5. L. tenuigranulata, M'Coy; the largest fossil species known in Britain.

* As an example of the confusion which might be introduced by substituting the word Cambrian for Lower Silurian (see page 175.), it may be mentioned, that this new species of brachiopod, though alone collected in the old and well known Lower Silurian locality of Llandeilo flags at Builth, has been just published as a Cambrian fossil in the last fasciculus of the palæozoic fossils of the Cambridge Museum. The reader who is unacquainted with my original work, must therefore understand, that very many of the fossils to which the words Bala are prefixed, in the work descriptive of the Cambridge Museum, published only last year, have been alone found in my own original *Lower Silurian localities*, described by me as such between the years 1833 and 1839. Such are all those fossils from the Carmarthenshire localities of Llandeilo, Mandinam, Goleugoed, Llangadoc, Noeth Grüg, &c.; all the Brecknock and Radnorshire localities, near Builth and Llandrindod; the Montgomeryshire localities, near Welshpool and Meifod, such as Allt-yr-Anker, &c.

allied to Lingula, is common in Russia; and though not found
in our Lower, is now known to be present in our Upper Silurian.

Of the genus Atrypa, which is still retained by the naturalists of
the Government Survey, the most characteristic Lower Silurian form
is perhaps the A. hemisphærica, pl. 5. f. 3., figured in the " Silurian
System" from the Caradoc sandstone; it has since been found in
the Lower Silurian strata of Ireland, and of Ayrshire in Scotland,·
and even in Russia and North America.

Atrypa marginalis (Ter. imbricata, Sil. Syst.), which was described
from the Wenlock shale, has since been observed abundantly in some
of my old localities of the Lower Silurian of Carmarthenshire and
Montgomeryshire, and is also very plentiful in Ireland.   (See pl. 22.
f. 19.)   Another and the most typical species of the genus, Atrypa
reticularis (Anomia reticularis, Linn.), pl. 5. f. 4, 5., and Foss. 8. f. 5.,
p. 88., which is one of the most abundant of the Silurian fossils, is unfortunately of little value in fixing the precise age of the rock; for it
passes not only through the whole system, but also into the next
overlying, the Devonian, provided its representative in the later
period (which is usually larger) should not prove to be a separate
species.   This fossil has, indeed, an enormous geographical range,
and in such cases observation has taught us that the animal has
always had a very prolonged existence.*

Spirifer is a very rare Lower Silurian genus, one good species only,
the S. radiatus, Sil. Syst., being known in this division (see pl. 5.
f. 9.).   Atrypa crassa of the " Silurian System" (see pl. 9. f. 6, 7, 8.),
may, however, possibly belong to this genus.   It is not very common.

The genus Terebratula, properly so called, and as represented by
the animal of that name which still lives in the seas of all parts of
the globe, did not exist in the Silurian epoch.   The primeval forms
allied to it are now, as before stated (p. 107.), referred to the genus

* See D'Archiac and De Verneuil, Trans. Geol. Soc., vol. vi. p. 335., and Prof.
Edward Forbes, Trans. Brit. Assoc. Cork Meeting, 1842.

Rhynconella of Fischer. This genus is much less frequent in the lower than in the upper division of the system. Some rather doubtful forms were described in the " Silurian System," under the names Terebratula neglecta, T. tripartita, and T. pusilla, which may probably belong to species found in the Upper Silurian. Several species, however, are known in the lower division. One, a smooth form, the R. (Hemithyris*) angustifrons, is plentiful in the Lower Silurian rocks of Ayrshire, and is figured in the following wood-cut, f. 4. R. furcata, pl. 5. f. 12. is an Upper Caradoc species.

Fossils, 22.

LOWER SILURIAN BRACHIOPODS.

1. Rhynconella serrata, M'Coy. 2. R. nasuta, ib. 3. R. nucula. 4. R. an-

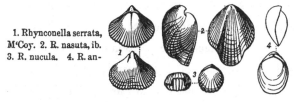

gustifrons, M'Coy. All these are from the S. of Scotland.

Rhynconella nasuta, Foss. 22. f. 2., is another fine Scottish species, and R. serrata, f. 1. is found in Scotland, Ireland, and North Wales. There are some forms, indeed, of this genus, which, appearing in the Llandeilo formation, have been continued for a great length of time. Such, for example, are the Rhynconella (Terebratula) borealis, pl. 22. f. 4., and R. rotunda, f. 18., originally described in my work from the Upper Silurian of Wenlock only. R. Lewisii, Davidson, which has been figured in chap. 9., Foss. 40. f .2., is said by Professor M'Coy to be found in the lower division, as also is R. depressa, pl. 22. f. 17. R. decemplicata, pl. 5. f. 15., and R. nucula, pl. 22. f. 1, 2., occur at Bala, but are much more common in the Caradoc. The latter, indeed, is one of the most characteristic and abundant

---

* Most of the brachiopods here termed Rhynconella in compliance with the views of Davidson, Morris, and other palæontologists, are classed as Hemithyris by D'Orbigny and M'Coy. (See " British Lower Palæozoic Fossils.")

shells of the Upper Ludlow rocks, and is another example of a species
formerly known to me only in the Upper Silurian, which is now found
in the Llandeilo and Caradoc deposits.

A common and most characteristic Silurian form of the Terebratuloid group is the Pentamerus, Sow., some species of which are
figured on the woodcut, Foss. 8.; and of these the species which
are common in the lowest strata are Pent. lens, pl. 8. f. 9, 10.;
P. undatus, pl. 8. f. 5, 6.; and P. oblongus, f. 1—4., including
P. lævis, formerly considered distinct. When the " Silurian System"
was published, the Pentamerus oblongus was cited as specially characteristic of the uppermost member of the Caradoc sandstone only,
and particularly of that band, here and there a limestone, which has
already been alluded to as intervening between it and the Wenlock
formation. This is still the characteristic position of the species, but
now the same shell is detected not unfrequently, both in Wales
and Scotland, in beds which are considered part of the Llandeilo
formation. It is thus shown to be common to all the Lower Silurian
group, except the Lingula flags. There is a rarer species, the P.
globosa of the Llandeilo flags, pl. 8. f. 8. (Atrypa globosa " Sil.
Syst."), to be added to the list.

Pentamerus liratus, also (Atrypa lirata, Sil. Syst.), is abundant
in the Caradoc sandstone of May Hill, and this species is also
found in very similar strata in Pembrokeshire (p. 67.). It must be
recollected that this genus, Pentamerus, is eminently Silurian, though
not exclusively so; distinguishing by its species the lower from the
upper division perhaps as well as any other genus of brachiopods.

Of the next great class, the Lamellibranchiata, or ordinary bivalve
shells, but few had been discovered in the lower group of strata when
my large work was published. Yet they are now ascertained to
be far from rare in the sandy and argillaceous parts of the Llandeilo
flags in England; and they also occur in rocks of the same age in
America. The genus Pterinea (Avicula, Sil. Syst.), so common, as

will hereafter appear, in the upper division, has some representatives in the lower.   P. Triton, Foss. 23. f. 8., for example, is found at Llandeilo, and the common P. retroflexa of Hisinger, pl. 23. f. 16., ranges from the Caradoc sandstone of Malvern to the tilestones of the Ludlow rock.   P. pleuroptera Conrad, and P. tenuistriata, ch. 9., Foss. 42., are two other species of this formation, quoted by M'Coy as ranging upwards to the Ludlow rock.   It is probable that some other species of the genus have an equal range.

FOSSILS, 23.

LAMELLIBRANCHIATA.

1. Modiolopsis postlineata, M'Coy. 2. M. expansa, Portlock.   3. M. modiolaris, Conrad. 4. Nucula varicosa, Salter. 5. Lyrodesma plana, Conrad.   6. Nucula obliqua, Portlock.   7. Pleurorhynchus dipterus, Salter.   8. Avicula Triton, Salter.

Modiolopsis, Hall, is a still more common genus of the Lower Silurian.   M. orbicularis, pl. 9. f. 1., one of the few species published in my former work, is at present known only in the Caradoc sandstone of Shropshire; it was formerly described as an Avicula.   M. modiolaris, Foss. 23. f. 3., and M. postlineata, f. 1., are frequent in the rocks of North Wales.   M. expansa, f. 2., is a flattish species found in strata of the same age in Ireland.   M. inflata, M'Coy, and some other species, are also occasionally met with in Wales; and other forms of the genus are not uncommon.   This genus is closely allied to the common Mytilus or muscle of our coasts.

Nuculæ and shells of similar genera with toothed hinges are not scarce, but few of them have been yet described.   It is well, however, to note that an undescribed species, N. varicosa, Salt. Mss. Foss. 23. f. 4., and N. obliqua, f. 6., are frequent in the lower division.   N. deltoidea, rhomboidea, and lingualis, are three species described by Professor Phillips from the Caradoc sandstone of Malvern. Lyrodesma plana, Foss. 19. f. 5., is nearly allied.to the same group of shells: it is found in the limestone at Bala.

The Orthonota (Cypricardia of the Sil. Syst.) is a genus more frequently found in Upper Silurian rocks, but some are Lower Silurian species. O. nasuta occurs in the Caradoc of Horderley, and is represented in our woodcut, Foss. 12. f. 7. In the illustration on the preceding page are figured a few other forms of Lamellibranchiata, and particularly the curious fossil, Pleurorhynchus dipterus, f. 7., which, though unknown in England and Wales, has been found by Mr. J. Carrick Moore in the Lower Silurian rocks of Ayrshire, in Scotland, and by the government geologists at the Chair of Kildare in Ireland.

The Gasteropods, or univalves of the most ancient natural system of life, are now declared by naturalists to belong mainly to genera which are extinct. Such are Murchisonia of D'Archiac and De Verneuil, formerly included in the genus Pleurotomaria; Holopea and Scalites of my American contemporary Hall; Loxonema of Phillips; and Euomphalus. A few species of Turritella, indeed, are quoted; but these, in all probability, belong to quite a different genus, which has been named Holopella by M'Coy.

It is important to remark, that shells which abound in the younger secondary and tertiary deposits, and at the present day, such as true species of Trochus, Turbo, Littorina, and Buccinum, have not yet been found in these ancient rocks.* The genus Patella, or an allied form, however, does appear to occur in them, and an Irish species, the P. Saturni, Goldfuss? is here figured, Foss. 24. f. 9.

* The reader may be surprised to find numerous species of the four last-mentioned genera spoken of in the former edition of the "Silurian System;" but it is only of late years that the distinction has been drawn, by which such genera have been excluded from the Silurian rocks, and the accuracy of the exclusion is yet doubted by some good naturalists, so much have these shells the external appearance of the above-mentioned genera. Euomphalus is probably a very near ally of the modern

Among these old genera, the genus Euomphalus is of very early creation, four or five species being known in the inferior strata (one or two are given in pl. 6.). Like other genera we find that Euomphalus pervades the whole Silurian system ; but it is far less characteristic of the Lower than of the Upper Silurian, and it is not certain that any species extends upwards from the Llandeilo flags into Wenlock strata.

FOSSILS, 24.

GASTEROPODA.

1. Maclurea Logani, Salter? ; operculum of the genus, f. 1a.
2. Scalites — sp. undescribed.
3. Murchisonia obscura, Portl.
4. Turbo rupestris, Eichw. 5. Holopea concinna, M'Coy. 6. Murchisonia gyrogonia, M'Coy. 7. M. semirotundata, Portl. 8. Helminthochiton Griffithii, Salter. 9. Patella Saturni, Goldfuss?

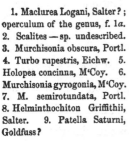
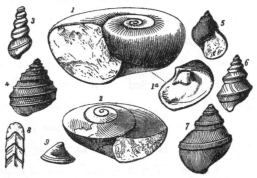

One of the most emblematic univalves of this age is the Murchisonia, more complete specimens of which are now given, in the annexed woodcut, Foss. 24. f. 6, 7., than any figured in my former work, where they were called Pleurotomariæ. In Portlock's report several species of this genus are described from the Lower Silurian rocks of Tyrone: one of these, M. turrita, is very elegantly sculptured ; another, M. semirotundata, conspicuous for its broad rough band, is also found there, and is here given, Foss. 24. f. 7. M. gyrogonia, f. 6., and M. scalaris, Salter, are common in Wales, and there

genus Delphinula. These changes of the names of fossils made by palæontologists, although troublesome, may, in some instances, have an important bearing on the philosophy of the science, in enabling us to understand the conditions under which certain animals lived and died; whilst the field geologist has to note carefully the new forms discovered; his chief occupation being to mark the extent to which the old and characteristic species are found to extend their vertical range.

are many other similar species. The Turritella-like shells of
the Lower Silurian rocks are either smooth forms of the genus last
mentioned, in which case Mr. Salter informs me that they show the
characteristic band (M. obscura, Foss. 24. f. 3.), or belong to the
newly proposed genus Holopella, in which the striæ of growth run
straight across the whorls, and are not bent into a notched or
angular form. H. (Turritella) cancellata of my old work, Foss. 8. f. 8.,
is an example of this latter genus from Caradoc sandstone; and the
common Upper Ludlow rock species, H. obsoleta and H. gregaria,
are both found in the upper beds of the same stratum. Others have
been described from the Llandeilo flags, but all are of the same
general elongated and beaded shape, resembling the Murchisonia
obscura, above quoted. Trochus lenticularis of the " Sil. Syst.,"
see pl. 10. f. 2., and several similar discoid and angular univalves, are now referred to the genus Scalites, Hall, (or its section
Raphistoma). This genus is exceedingly common in the Lower
Silurian rocks of America : it also occurs in Scandinavia and Russia;
and a species of it found at Llandeilo is here given, Foss. 24. f. 2.

Holopea, of Hall, though rather an obscure genus of smooth,
rounded univalves, must now include many Lower Silurian species.
One is figured in pl. 10. f. 1. as H. striatella (Littorina, Sil.
Syst.); another, H. concinna, is represented here, Foss. 24. f. 5.

Other shells, still provisionally called Turbo, e. g. T. rupestris,
f. 4., and T. sulcifer, Eichw., T. crebristria, M'Coy, &c., may probably
be ascertained to belong to extinct genera allied to Holopea. Turbo
rupestris occurs in the Lower Silurian limestones of Ireland, and often
still retains the coloured bands that decorated the shell when alive.

Though not yet detected in England or Wales, a characteristic
genus of the Lower Silurian rocks of North America, Maclurea,
Hall, has been found in strata of the same age in Scotland. The
shell (Foss. 24. f. 1.) appears to be sinistral or reversed, the top flattish,
and the base umbilicate. In all probability, however, it should be

viewed the other way upwards, and so appear with a sunken or concealed spire. Its mouth is closed by a remarkable operculum, of which f. 1 *a* shows the inner side. I have myself collected this genus in the Lower Silurian strata of Ayrshire, associated with Orthidæ, trilobites of the genus Asaphus, and other Lower Silurian fossils. A Canadian species, approaching to M. magna of the United States of North America, and named by Mr. Salter, Maclurea Logani, after the able geologist who discovered it in Canada, is the form to which our British fossil is most allied. The operculum is figured from that species.

Bellerophon, a palæozoic form of the mollusks termed by some naturalists Heteropoda, is one of those genera which specially link together the Lower and Upper Silurian divisions in one system of life. Thus, Bell. trilobatus, Foss. 8. f. 9., and B. dilatatus, Foss. 25. f. 8., are common to both divisions, being found in the Lower Silurian of Ireland and Scotland, and ranging from the Caradoc sandstone to the Ludlow rocks in England and Wales. B. carinatus, Sil. Syst., appears also to occur in both divisions, if f. 7, which represents the Caradoc form (B. acutus), be really referable to the same species. On the other hand, there are several which as yet are known only in the lower division, and are very characteristic of it. B. bilobatus, figured in chap. IV., Foss. 7. f. 10., and B. ornatus, f. 11., are among the common fossils of the Llandeilo flags, and the lower part of the Caradoc sandstone. Both these species are equally characteristic of the same formations in North America, and the former is also abundant in Bohemia. B. perturbatus, Foss. 25. f. 6. (Euomphalus, Sil. Syst.), is one of the common fossils of the black slates of Wales. There are other less known species, and some are yet unpublished.

Certain naturalists regard the genus Bellerophon as the shell of a cephalopodous animal, differing from the ordinary forms of that class by the want of septa or partitions in the shell. Although there is some ground for the supposition, these shells are usually believed to be Nucleobranchiata (orHeteropoda), allied to the floating Carinaria, or glass shell, which they much resemble both in form and sculpture.

Mr. Salter suggests that the Irish Lower Silurian fossil, Ecculiomphalus Bucklandi, Foss. 25. f. 5., may be compared with the modern genus Atalanta ; if its thin shell, triangular section, and finely triated surface do not betray affinity with the group next mentioned.

Nor must a notice be omitted of the genera now clearly ascertained to be forms of Pteropoda — one of the inferior groups of mollusca. Conularia, for example, a most beautifully ornate form, has two or three species in the Lower Silurian.   C. elongata, the one figured in the next woodcut, Foss. 25. f. 3., is frequent in Ireland; and C. Sowerbii, pl. 25. f. 10., ranges upwards from the Llandeilo flags to the Ludlow rocks. M. Barrande has figured many curious forms of Conularia from the Silurian rocks of Bohemia, one of which is spirally curved: Then again Theca, a genus exceedingly like the modern Clio, has two species at least in the lower division; viz., T. triangularis (Orthoceras of Portlock), Foss. 25. f. 2., and T. Forbesi, f. 1.; the latter apparently being distributed through the whole Silurian system.

FOSSILS, 25.

HETEROPOD AND PTEROPOD MOLLUSCA.

1. Theca Forbesi, Sharpe. 2. T. triangularis, Portl. 3. Conularia elongata, Portl. 4. Pterotheca transversa, Portl.

5.Ecculiomphalus Bucklandi, Portl. 6. Bellerophon perturbatus. 7. B. acutus. 8. B. dilatatus.

Figs. 5. and 8. are species often 3 inches wide — our figures are much reduced.

Pterotheca is a new genus proposed by Mr. Salter (Proc. Brit. Assoc., 1852) for a wide shell like Cleodora, — P. transversa is here given, f. 4.

It is to be remarked, that these ancient specimens of their order were gigantic in comparison with their modern representatives.

Cephalopodous or chambered shells of the genera Orthoceras, Cyrtoceras, and Lituites, occur in the Lower Silurian rocks. In the original Silurian region they are, indeed, less abundant in the lower than in the upper divisions of the system; but on the Continent of Europe, particularly in Scandinavia, where they occur under different conditions, in beds full of calcareous matter, they are more frequent in the Lower than the Upper Silurian. Among the earlier developed British species are to be noted Orthoceras politum, M'Coy, a large smooth shell seen by myself *in situ* in Ayrshire, Scotland (see Quart. Journal Geol. Soc., vol. vii. pl. 10.); O. vagans, Foss. 26. f. 1.; O. bilineatum, f. 2.; O. tenuicinctum, f. 3.:

FOSSILS, 26.

CEPHALOPODA.

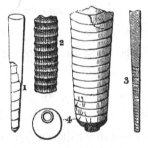

1. Orthoceras vagans, Salter. 2. O. bilineatum, Hall. 3. O. tenuicinctum, Portl. 4. O. Brongniarti, Portl.

(Some forms of Orthoceras most common in the Lower Silurian rocks.)

the two latter are Irish fossils. Another species characteristic of the Lower Silurian strata, and having a large lateral siphuncle, is O. Brongniarti of Portlock, f. 4.; it is supposed to be identical with one from North America, where similar forms are common in the Lower Silurian rocks. The great Orthoceratite of the same strata of Scandinavia and Russia, O. vaginatus, Schlotheim, is occasionally three feet long; it has been found in Scotland (see Quart. Geol. Journ., as above). The smooth Orthoceratites, on the contrary, with slender central siphons, are not confined to either the upper or lower division.

It must here be specially noted, that in the progress of research no less than nine species of Orthoceratites have been found in England, Wales, and Ireland to range from strata of the age of the Llandeilo flags to the Upper Silurian*; several of them even into the Upper Ludlow rock: *e.g.* O. angulatum, Wahl. (O. virgatum, Sil. Syst.); O. ibex, Sil. Syst.; O. subannulatum, Munst.; O. annulatum, Sil. Syst. (O. undulatum of foreign authors); O. filosum, Sil. Syst.; O. subundulatum, Portl. (Creseis Sedgwickii, Forbes); O. tenuicinctum, Portl.; and O. primævum (Creseis primævus, Forbes). There are probably others which have as great a range (see Plates 26 to 29, and Foss. 44. in Chap. IX.).

The singular O. bisiphonatum of the Sil. Syst., only found in one locality, near Llandovery in South Wales (pl. 11˙ f. 5.), appears to have two siphuncles. But this is only in appearance; for the species does not really belong to Orthoceras, and should form, in the opinion of Mr. Salter, a new genus in some respects like the curious Gonioceras of North America, which occurs in rocks of the same age. He proposes for this fossil the name of Diploceras, in reference to its apparent structure.

In Lituites undosus, pl. 11. f. 3., we have another example of an extinct genus, which so much resembles the rare Nautilus of the present day, that, when the " Silurian System " was written, Mr. Sowerby very naturally so termed it. There are several other Lituites in these old strata. L. cornu-arietis, Foss. 27. f. 2., and L. Hibernicus, f. 3., are good examples of our British species. The former is also found in Scandinavia. The Cyrtoceras, which is distinguished by being more slightly curved than Lituites, includes several species in Britain; one, C. inæquiseptum, from Ireland, is here represented, Foss. 27. f. 1.; another, C. approximatum, is figured pl. 11. f. 4.; and

* See British Lower Palæozoic Fossils, Sedgwick and M'Coy, pp. 313, &c. The Museum of Practical Geology in Jermyn Street exhibits several examples of these widely spread fossils.

one of the Ayrshire cephalopods is probably the Cyrtoceras multi-cameratum, Hall, of the United States.

FOSSILS, 27.

CEPHALOPODS.

1. Cyrtoceras inæquisep-tum, Portl. 2. Lituites cornu-arietis.   3. L. (Trocholites) hibernicus, Salter.

Curved forms of Cepha-lopoda, rare in British Lower Silurian strata.

Annelids, or sea-worms, prevailed in some of the earlier sediments in which traces of organic remains are found. Specimens of these were figured, as previously stated, from Llampeter, in South Wales where they occur in strata which the government surveyors have shown to be equivalents of the Llandeilo rocks; e. g. Nereites Cambrensis, Foss. 28. f. 3.; N. Sedgwicki, f. 2.; and Myrianites M'Leayi, f. 1.

FOSSILS, 28.

ANNELIDS; OR MARINE WORM-TRACKS.*

1. Myrianites M'Leayi, Sil. Syst. 2. Nereites Sedg-wicki, ib. 3. N. Cambrensis, ib. 4. Crossopodia scotica, M'Coy.

Tracks of sea-worms in fine muddy sediments, now altered into slates. S. Scot-land and Wales. (The spe-cies figured in the "Silurian System" were drawn on stone, and these figures could not therefore be re-peated in the plates at the end of the volume.)

* See Sil. Syst. pp. 363. 699, &c.

Another form of these creatures, Crossopodia Scotica, f. 4., is given from specimens collected by Professor Sedgwick, in the Lower Silurian and Graptolite schists near Moffat, in Dumfriesshire, a locality which I have also examined, and where I found other species of these long sea-worms. The length of some of these creatures must have been prodigious, probably many yards, judging from the frequent parallel coils which are exhibited on the surface of the schists. (The above figures are very much reduced). It should, however, be borne in mind, that in many cases we see only the track of the worm and not the impression of its soft body, which could rarely be preserved.

The Tentaculites and Cornulites of the earlier primeval strata must also be mentioned as remains of animals of this order. They were worms with shelly tubes like those of Serpula, but distinguished easily by their annulated form and cellular structure* (see pl. 16. f. 7, 8. for the form and magnified sections). Tentaculites annulatus, pl. 1. f. 1., even when found alone, is an excellent index of the relative age of the rock in which it occurs; for in Britain it is a character- istic Lower Silurian fossil; (T. scalaris of old authors, pl. 1. f. 2, is the interior cast of it). Cornulites serpularius, pl. 16., is also a cosmopolite Silurian fossil, ranging from Sweden to North America; but, unlike the Tentaculite just mentioned, it ascends from a low position in the Llandeilo formation to the very summit of the Ludlow rocks. It is therefore, like the Graptolite, a clear manifestation that the rock in which it occurs is Silurian, and is also an example of how a hardy species may endure through variations of condition, and changes of the surface, during which more delicate organisms passed away.

Crustaceans. — The most highly developed articulata which have yet been detected in the Lower Silurian rocks are those of the

---

* These fossils, Tentaculites and Cornulites, have been assigned to various groups of animals, the notion that they were parts of crinoidal creatures being the most generally accepted. They were, however, shelly tubes, not jointed tentacles or stems, and could, by no means, be parts of such animals. — I. W. S.

extinct family of entomostracous crustaceans called Trilobites. The Silurian era was evidently one in which these animals flourished most; for they became infinitely less abundant in the Devonian, and expired before the close of the carboniferous era, during which, as will hereafter be seen, very few genera prevailed. As already indicated, we find some forms of the Trilobite in the very earliest accumulations in which any animal remains occur; and it is now to be remarked, that whilst, on the whole, certain genera and species of these creatures are more exact indicators of the successive strata than the other classes of animals, and that several genera are absolutely peculiar to the Lower Silurian rocks, there are some species that are widely diffused in strata of this era over the world, which are common to both divisions. Mr. Salter reminds me, indeed, that there is not one genus of the upper division which has not previously existed in the earlier period.

Fossils, 29.

TRILOBITES TYPICAL OF THE LOWER SILURIAN ROCKS.

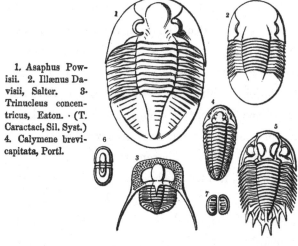

1. Asaphus Powisii. 2. Illænus Davisii, Salter.   3. Trinucleus concentricus, Eaton. · (T. Caractaci, Sil. Syst.) 4. Calymene brevicapitata, Portl.

5. Lichas laxatus. M'Coy. 6. Agnostus trinodus, Salter.   7. Beyrichia complicata, Salter; (the last is not a trilobite, but a bivalved crustacean).

The genera peculiar to the Lower Silurian, not only in this country, but on the Continent and in America, are: — Olenus,

Agnostus, Paradoxides, Cybele, Trinucleus, Ogygia, Asaphus, Remopleurides, and Illænus. On the other hand, the genera Ampyx, Calymene, Lichas, Proetus, Homalonotus, Cheirurus, Encrinurus, Bronteus, and that division of Illænus which I still believe to be a true genus* (Bumastus, Murch.), are found both in the Lower and Upper Silurian.

Of these there is perhaps no one species which has a greater vertical range, or is more widely diffused in geographical space, than the long-known Dudley fossil, Calymene Blumenbachii, Brongn. When this trilobite was described in the " Silurian System," it was considered to be typical of the upper rocks only, whether of Wenlock or of Ludlow age; but the geologists of the Government Survey have since found it in abundance in the lower strata of the Llandeilo formation, even near Snowdon; and I have myself procured it (at least a common variety of it) from beds of Caradoc sandstone in Shropshire. Again, I have seen the true species in the Lower Silurian rocks of Ayrshire in Scotland; and thus, like many examples of the other classes of marine animals which lived in primeval days, this crustacean links together the Lower and Upper Silurian in one system of life. (See pl. 18. f. 10. The head is represented, Foss. 7. f. 1.) It has also been found in Sweden, Bohemia, North America, and Eastern Australia, and always in company with its usual associates.

Again, Encrinurus punctatus, Foss. 8. f. 10., and chap. 9., Foss. 46. f. 5., is another species which has lived throughout the older and younger periods; and Phacops (Asaphus, Sil. Syst.) caudatus, pl. 18. f. 1., has an equal range. There are also several others, such as Acidaspis Brightii, pl. 18. f. 7.; Phacops Stokesii, f. 6.; Cheirurus bimucronatus, pl. 19. f. 11., which occur in both divisions. The Lower Silurian trilobite, pl. 2. f. 3, 4., called Asaphus Vulcani in the " Silurian System," is probably a species of Homalonotus, a genus formerly thought to be peculiar to the Upper Silurian, but of which several examples are now known in the lower division.

* The pygidium is simple, *not tri-lobed* as in most species of Illænus.

The principal forms, however, which have proved unerring indices of the Lower Silurian of Britain are Olenus, Agnostus, and the true species of Asaphus, such as A. Powisii, Foss. 29. f. 1.; A. tyrannus, pl. 1. f. 3, 4., and pl. 2. f. 1., with five species of Trinucleus, pl. 4., all of which were formerly published as true Lower Silurian types.

Since the period of the original publication, various species of trilobites, which were then supposed to be peculiar to the Caradoc sandstone or upper portion of the lower division, have been found to range downwards all through the lower division, excepting only the 'Lingula slates,' where Olenus, and perhaps Paradoxides, are the only trilobites yet known in Britain.

Thus, for example, Asaphus Powisii, pl. 2. f. 2., which was first collected by the Rev. T. T. Lewis from the upper beds of the Caradoc formation, has also been found very low down in the slate rocks of Snowdon and in many other parts of N. Wales, and is, indeed, one of the most characteristic trilobites of that slaty region. Again, Trinucleus concentricus (Caractaci, Sil. Syst.), pl. 4. f. 2—5., and Foss. 29. f. 3., has exactly as extensive a Lower Silurian range, both in Europe and America. This species varies greatly in

Fossils, 30.

Trinucleus concentricus, three Specimens, distorted by Slaty Cleavage.

In illustration of the various forms which a single species may assume under the influence of that crystalline change operated in the strata which is called slaty cleavage, (see p. 35.). The woodcut represents a group of forms, apparently very dissimilar, but all of the same species, here given in their natural position on a fragment of slate. (Phillips, Proc. Brit. Assoc. v. xii. pp. 60, 61.)

form and markings, and, being a very common species, it is met with in every variety of distortion and compression, so that it might often be mistaken for a different species. T. fimbriatus, pl. 4. f. 7.;

T. Lloydii, f. 6.; and T. seticornis, Hisinger, a species also found in
Sweden, are good indications of the Llandeilo flags, and occur in
them only.

Better specimens of some species of trilobites having been found
than those which were collected at the time of the publication of the
" Silurian System," figures of them are given in the following woodcut, Foss. 31., viz. Phacops conophthalmus, f. 3.; and Ampyx (formerly
Trinucleus) nudus, f. 7. Several species, not there described, are
added, viz. Cybele verrucosa, f. 2.; Harpes Flanagani, f. 4.
Remopleurides dorsospinifer, f. 5.; Acidaspis bispinosus, f. 6.; Cyphoniscus socialis, f. 8.; Cheirurus clavifrons, f. 1. And in the
former woodcut (Foss. 29.) Lichas laxatus, f. 5.; Illænus Davisii,
f. 2.; Agnostus trinodus, f. 6.; and a minute bivalved crustacean of
the phyllopod tribe, Beyrichia complicata, f. 7. Of this last genus
several other forms are known.

FOSSILS, 31.

OTHER LOWER SILURIAN TRILOBITES.

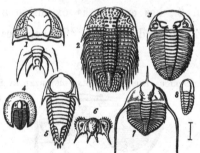

1. Cheirurus clavi-
frons, Dalm.   2. Cy-
bele verrucosa, Dalm.
3. Phacops conophthal-
mus, Bœck.   4. Harpes
Flanagani, Portl. 5. Re-
mopleurides dorso-spi-
nifer, Portl. 6. Acidas-
pis bispinosus, M'Coy.
7. Ampyx nudus. 8.
Cyphoniscus socialis,
Salter.

The comparatively large animal of the phyllopod tribe, Hymenocaris
vermicauda, mentioned in chap. 2, as occurring in the very lowest
fossiliferous beds, must not be forgotten (see Foss. 3. p. 42.). No
other tribes of crustaceans than the Trilobites, and the single phyllopod just mentioned, have been found in the Lower Silurian rocks of
Britain. In the sequel it will be seen, that an order of crustaceans,
apparently of higher organization, is found in the highest band of the
Upper Silurian strata of England.

In grouping the various habitats of the British Silurian trilobites, I would direct the attention of the reader to the description of the lower palæozoic fossils in the Cambridge Woodwardian Museum, by Professor M'Coy, in order that he may see how many species, even in that one rich collection, are common to what have been so long and so generally called the Lower and Upper Silurian rocks.

Again, he will see that out of fifty-one species of trilobites there enumerated, although double the list given in the " Silurian System " when first published, nearly all the new forms (at least twenty-eight) have been found, many of them exclusively, in localities of Montgomery, Radnor, Brecon, Caermarthen, and Pembroke, which were laid down by me as Lower Silurian on the map; a pretty strong indication that the original Silurian region still affords the best fossiliferous types.

In concluding this chapter it is important to remark, that the impressions of footmarks, in the Potsdam sandstone or lowest Silurian rock of North America, which had been supposed to have been made by tortoises, have, in consequence of the discovery of better specimens, been referred by Professor Owen himself to crustaceans.* Hence this last class of animals may still be considered the highest type of life of the earliest primeval division. It is, indeed, a remarkable fact, that the most sedulous research in many parts of the world has failed to discover the trace of any vertebrated animal in the lower division of the Silurian system. All the marine animals, from zoophytes to crustaceans, and which probably amount to more than 1000 species already known, belong to the invertebrated classes, and no true fish has yet been observed. In the sequel it will be seen that this

---

* The zeal of that acute and laborious geologist, M. Logan, in procuring casts of a very great number of these large and curious impressions, cannot be too much commended. In default of these labours, very erroneous ideas might have been propagated respecting the fauna of the Lower Silurian rocks. (See Journ. Geol. Soc. Lond., Vol. VIII. p. 199.)

observation applies also to the Upper Silurian rocks, with the exception of their highest stratum, as formerly expressed in the "Silurian System."

Having thus referred to the figures first published in my original work (and now reproduced at the end of this volume in plates 1 to 11), as well as to other types of Lower Silurian age which have been subsequently discovered, let us now pass on to consider the nature of the Upper Silurian fossils.

CHAPTER IX.

FOSSILS OF THE UPPER SILURIAN ROCKS.

THE reader who has perused the preceding chapters, will have perceived that many species of fossils, which were once supposed to be
peculiar to the Lower or to the Upper Silurian rocks respectively,
are now ascertained to be common to both.

This fact is the result of the researches of many geologists and
palæontologists, both in the region which I first explored, and in
tracts of far greater extent which have been since paralleled with it.
Similar results have, in truth, invariably accompanied the full and
broad development of all the natural geological groups of the secondary
and tertiary strata, which were described and classified before the
older rocks, of which we now treat, had been brought into order,
or even into notice. Thus, for example, when the different members
of the lias, or of the superjacent oolitic formation, were studied
in one tract only, as on the eastern coast of Yorkshire, they were
seen to be there composed of a series of zones, each of which is
for the most part sharply separated from contiguous deposits by
fossils peculiar to it; but, in following the same strata to remote distances, the remains, which were so typical of one member only, are
seen to become common to several formations, and thus to combine
the whole in one natural system — the oolitic or jurassic.

This is just what has happened in my own case, now that the original
Siluria, its inferior and superior masses, and its chief formations and
their subdivisions, have been ascertained to occupy nearly all North
and South Wales, large tracts of Cumberland, Westmoreland, and
Lancashire, a great region in the south of Scotland, and very extensive tracts in Ireland, France and many distant regions. It has

therefore followed, that the two chief members, which from a general similarity of characters were grouped into one system, are now infinitely more knit together or united than when the original classification was proposed. For, although at that time I recognized a general resemblance throughout the system, I was acquainted with but few species of fossils which were common to its lower and upper divisions; whilst it is now ascertained that *not less than a hundred forms of creatures,* which have lived on from the earlier to the later days of the Silurian era, have been already collected in the British Isles.

To some of the most striking of these remains which pervade the whole system, allusion has already been made (Chap. VIII.), and notice will now be taken of some other forms, of like duration in time, which are more abundant in the upper division. The attention, however, of the reader will chiefly be directed to the types which are peculiarly and exclusively Upper Silurian. (See Plates 12. to 35.)

The Graptolites, which, as before stated, are so very abundant in the shaly and schistose portions of the Lower Silurian of Wales and Scotland, become infinitely more scarce as to species in the Upper Silurian; and though there are many tracts where the Wenlock shale and Lower Ludlow rocks are perfectly crowded with them, it is only with one* species, and that the wide-spread Graptolites priodon (Ludensis), that we are acquainted in these upper divisions of the system. When muddy sediments abound in these rocks, the Graptolite is rarely absent; but with the cessation of such peculiar conditions, this zoophyte, which grew on the fine mud at the bottom of the sea, disappeared, and it is but rarely to be detected in the upper or more sandy division of the Ludlow rocks. Thence upwards in the

---

* One other species has lately been detected (but only in one locality, where it is plentiful), in strata believed to be of Wenlock age. It is Grapt. Flemingii, Salter, found in Kirkcudbright Bay, where it was first collected by the Earl of Selkirk.

whole series of palæozoic strata, no true Graptolite, as before observed, has ever been found; and hence, I repeat, it is a marked Silurian type.

It should be specially observed, that in Britain neither the double or foliaceous Graptolites (Diplograpsus), nor the twin Graptolites (Didymograpsus) are ever met with above the horizon of the Caradoc sandstone; and they are most excellent indices of the lower division. But in Bohemia, the foliaceous types are mixed together with the single-sided forms (Graptolites) in the lower part of the upper division.

The true corals (Zoantharia of naturalists) are far more characteristic of the upper deposits than of the lower members of the Silurian rocks, and they are more abundant there, both as to species and individuals; the nodular limestone bands of the Wenlock and Aymestry rocks being frequently little else than corals, or concretions having coralline bodies as their nucleus or on their surfaces; the shells and encrinites and trilobites play but a subordinate part in the formation. When Mr. Lonsdale undertook for me his admirable description of the corals of the Silurian region, and carefully superintended the drawing of their forms as lithographs, sixty-two species only, including Bryozoa (not now usually regarded as true corals), were recognizable in our collections. But the number has been greatly increased of late years, chiefly, as regards Britain, by the researches of Professor Milne Edwards and of Professor M'Coy. Several new genera have been formed of those already known, which as they are founded on peculiarities of structure, growth, and reproduction, are likely to prove of permanent value, both in a zoological and geological point of view.

One of the most important of these discoveries, resulting from the labours of Professor Milne Edwards, and his coadjutor M. Jules Haime *, appears to be, that the bulk, if not all the corals of the Silurian era, and probably of the whole palæozoic series, belong to

* Archives du Mus. d'Hist. Nat. vol. v.

P

great divisions of the coral tribe unknown in modern seas, which, with rare exceptions, became extinct at the close of the palæozoic epoch. If this be established, and the large cup and star corals (Zoanth. rugosa), and the massive millepores (Z. tabulata), be quite distinct in structure from the star corals and madrepores of our own coral reefs, we gain a new fact in the history of animal life upon the globe, which is in harmony with results obtained by the study of the crustacea, mollusca, and fish of the older epochs.

We have already enumerated, chap. 5., a number of corals which pass upwards from the Lower to the Upper Silurian rocks, and as these are generally the most abundant species in the upper division, we may here give a brief list of them.

Stromatopora striatella (concentrica, Sil. Syst.), Foss. 32.; Halysites catenulatus, Foss. 12. f. 6.; Favosites gothlandicus, Foss. 10. f. 2, 3.; Favosites alveolaris, Foss. 10. f. 4.; Stenopora (Favosites) fibrosa, f. 7. 8.; Heliolites interstinctus (Porites pyriformis, Sil. Syst.), Foss. 11. f. 4, 5.; Heliolites tubulatus, f. 1.; Heliolites megastoma, f. 3.; Heliolites petalliformis, f. 2.

FOSSILS 32.

A section of Stromato-
pora striatella, D'Orb.

A very common Wenlock
limestone coral.

In addition to those just quoted, we have enumerated in Chap. 6., p. 119—121., several of the conspicuous corals of the Upper Silurian strata. Of all these, it may be truly said, that they swarm in every quarry of Wenlock limestone, and may be found, in diminished numbers, in the Aymestry rock. But they are not so plentiful in the less calcareous and muddy sediments of the Lower Ludlow rocks and Wenlock shale.

Of all the conspicuous species, perhaps Favosites alveolaris and F. Gothlandicus are the most general, occurring in masses varying from the size of a hazel nut to two and three feet diameter.   The former is easily distinguished (p. 119., Foss. 10. f. 4.) on breaking the mass, by the edges of the tubes being toothed, while those of the other (f. 2, 3.) are smooth regular prisms, on which the double row of pores is very conspicuous.   F. cristata, f. 1., (polymorpha of the Sil. Syst.) appears to range up to the Devonian period, and these three species are widely distributed over the Northern hemisphere.   The various forms of the Stenopora (Favosites) fibrosa are not less common; but this latter fossil does not appear to be so abundant in the Upper as the Lower Silurian, nor to assume so many variations in form.   It is globular, lobed or branched; and on the slabs of limestone at Dudley there very commonly occurs a narrow branched variety, with numerous small pores intermixed with larger ones.   This form is very curious, and is figured in my larger work (pl. 15. f. 9.)* as a variety of Favosites spongites, Goldf.   Professor Milne Edwards believes it to be a different species, and has proposed for it, in his large work, the name Chætetes Fletcheri, in honour of a gentleman who has long been a judicious collector of the beautiful fossils of Dudley.

The species we have called Favosites oculata, Foss. 10. f. 6., is very abundant in the Wenlock limestone.   But it is doubtful if it belongs to the genus, and is named Alveolites repens in the work of Milne Edwards above alluded to.

Next in importance to the large species of Favosites are those of the genus Heliolites.   Of these H. interstinctus is the most common, occurring in globular and pear-shaped masses, of a flat or discoid form, or thin and crusting over shells and other corals.

Cœnites juniperinus, Foss. 12. f. 3., may be easily recognized by the curious linear mouths of the cells: this species abounds in the Wool-

---

* For a further acquaintance with the corals of the Wenlock and Ludlow rocks, I must refer the reader to the " Silurian System." The plates of corals in that work, being lithographs, could not be reproduced here, as the shells and trilobites, &c. have been. But the woodcuts above referred to will, it is hoped, supply the deficiency, in reference to the more characteristic corals.

hope limestone of Presteign, and with C. intertextus (Limaria fruti-
cosa, Sil. Syst.), and occasionally a fine large species, C. labrosus,
Milne Edwards, occurs commonly at Dudley.

There are several kinds of tube-corals, such as the Halysites or
chain-coral, and certain species of Syringopora, which are abundant. The latter genus is no less curious in its growth than in its
structure. When quite young, says Mr. Salter, its divaricating,
trumpet-shaped tubes, creep over the surface of the large corals and
shells; and a form of that age was figured in my former work as
Aulopora serpens (see Foss. 12. f. 2.). It next begins to grow upwards, and each open mouth of the tubes lengthens and becomes a
flexuous stem, occasionally throwing out a lateral buttress in concert
with its neighbour. These buttresses or buds are hollow, and where
they touch each other coalesce, and form a connecting tube; and the
mass thus increases in size upwards, the tubes often branching, and
connecting with those nearest them, till the coral attains its full
size and becomes a Syringopora. S. bifurcata, of which Foss. 12. f. 4.
is the lower branched portion, S. ramulosa, f. 5., and S. fasciculata

FOSSILS 33.

CORALS OF THE UPPER SILURIAN ROCKS.

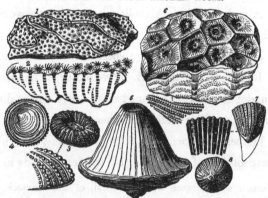

Fig. 1. Thecia Swindernana, Goldf. (Porites expatiata, Sil. Syst.)   2. A section magnified.
3. Palæocyclus porpita, Linn., (Cyclolites lenticulata, Sil. Syst.), and a magnified portion.
4. Under side of ditto.   5. Ptychophyllum patellatum, Schloth, (Strombodes plicatum,
Sil. Syst.)   6. Arachnophyllum typus, M'Coy; and a few of its lamellæ magnified.
7. Petraia bina, Lonsdale, the interior cast, and a portion magnified.   8. End view of do.

Linn. (S. filiformis, Sil. Syst.), are common kinds.   Aulopora serpens,
A. tubæformis, and A. conglomerata of the Sil. Syst. are the young or
basal portion of one or other of these species.   There are one or two
rarer forms, — Thecia Swindernana, Foss. 33. f. 1., and Labechia
Monticularia) conferta (Sil. Syst.) — not so obviously belonging to
the group we have just noticed.

Palæocyclus, Milne Edwards, is considered by that naturalist to be
allied to the modern mushroom-corals or Fungiæ, and is the only
Silurian, or indeed Palæozoic coral of that group.   P. porpita, Linn.
(Cyclolites lenticulata, Sil. Syst.) is the one figured Foss. 33. f. 3, 4.,
and there are other species, P. præacutus and P. rugosus, at Dudley.

All the other cup-shaped corals of the palæozoic rocks, of which
about fifty species occur in Silurian rocks, seem to belong to the
division Zoantharia rugosa; distinguished in general, like the other
great group, by the development of the transverse plates or septa in
the body of the coral (see Foss. 34. f. 5. for example), and including both simple cup-like forms, such as Omphyma (Cyatho-

FOSSILS 34.

CUP CORALS OF THE WENLOCK LIMESTONE.

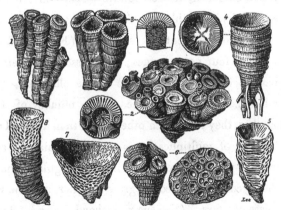

1. Cyathophyllum articulatum, Wahl. (cæspitosum, Sil. Syst.)   2 C. truncatum, Linn.
(Cyathophyll. dianthus, Sil. Syst.); and a single cup, with its marginal disk-buds attached.
3. Strephodes vermiculoides, M'Coy.   4. Omphyma (Cyathoph.) turbinata, Linn., and a
view of the cup with its four basal pits.   5. A longitudinal section.   6. Acervularia ananas,
Linn., and a cup, natural size, with four young buds.   7. Cystiphyllum Siluriense, Lonsdale.   8. C. cylindricum, Lonsdale.

phyllum) turbinata, Foss. 34. f. 4., Cystiphyllum Siluriense, f. 7.
Petraia bina, Foss. 33. f. 7. and 8., branched or composite ones, Cyathophyllum articulatum (cæspitosum) Foss. 34. f. 1. and C. truncatum,
f. 2.; and when such compound corals have grown closely together, so
that the separate corallites or cups press one another into an angular
form, such masses are produced as Acervularia ananas, f. 6., and
Arachnophyllum typus, Foss. 33. f. 6. The latter is now called
Strombodes Labechii by Milne Edwards and Haime. Among these
common species, two are remarkable for the mode of propagation of
their buds or young corallites.

In Cyathophyllum truncatum, Linn. (C. dianthus, Sil. Syst.) just
quoted, the young buds take their origin all round the inner edge of
the parent polype-cup (Foss. 34. f. 2.), and the eight or ten young
corals thus produced quite overtop, and at length cover over the
parent, till in their turn they too produce their young clusters, and
this process is frequently repeated in the upward growth of the
coral. (See the large figure.)

Another species, so like the former that in some conditions it might
easily be mistaken for it, is Acervularia ananas, also described by
Linnæus from rocks in Gothland which are now known to be Upper
Silurian. We have represented a single cup (Foss. 34. f. 6.) which
has given birth to four young cones; but these took their rise, not
from the margin, but from the centre of the old coral, and in all
probability killed the parent at once. As a number of these buds
grow up together, they stunt each other's growth in a lateral direction,
and as the process of multiplication is repeated again and again, the
corals by mutual pressure are forced into an angular form like the
cells of an honeycomb; their edges grow together, and the result is a
compound mass of stars. (See the right-hand figure.) The growth
and reproduction of corals has of late years been much studied, and
has been beautifully illustrated by Dana, Milne Edwards, and other
naturalists.

There are several other cup-shaped corals, which are frequent in the Upper division, and especially in the Wenlock limestone; but they are not so characteristic as those just mentioned. Omphyma (Caninia) lata, M'Coy; O. Murchisoni, Milne Edw.; Cystiphyllum cylindricum, Lonsdale; C. Grayii, Milne Edw.; and C. brevilamellatum, M'Coy; Cyathophyllum angustum, Sil. Syst. pl. 16. f. 9.; Strephodes trochiformis, M'Coy; S. vermiculoides, Foss. 34. f. 3.; Arachnophyllum (Strombodes) diffluens, Milne Edw.; Ptychophyllum patellatum, Foss. 33. f. 5.; Clisiophyllum vortex, M'Coy; Goniophyllum Fletcheri, M. Edw.; a four-sided coral, very like one found in the Isle of Gothland, Sweden. Some cup-corals also are common to Gothland and Britain, viz. Aulacophyllum mitratum and Cyathophyllum Lovèni, besides the two species of the latter genus before mentioned *; and there are some which extend their range to America.

The Ludlow rocks, being for the most part mudstones, do not often contain many corals; but in their calcareous portions, one or

FOSSILS 35.

UPPER SILURIAN BRYOZOA.

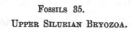
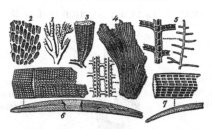

1. Polypora? crassa, (Hornera, Sil. Syst.). 2. Fenestella assimilis (Gorgornia, Lonsd.), a common Wenlock species. 3. F. Lonsdalei, D'Orb. (F. prisca, Sil. Syst.) 4. F. Milleri, Lonsd. 5. Glauconome disticha, Goldf. 6. Ptilodictya lanceolata, Goldf., a reduced figure of a full grown specimen; a part is figured above it, natural size. 7. is the young state; the cells are longer in proportion.

* So many of the Wenlock corals are found in the limestone of Gothland in Sweden, that the stratum might have been identified by them alone — Heliolites interstinctus, Favosites alveolaris, F. Gothlandicus, and F. cristatus; Cœnites juniperinus, Halysites catenulatus, Syringopora fasciculata, Favosites oculata (Alveolites repens) Caninia turbinata, Ptychophyllum patellatum, Acervularia ananas, &c. &c.

other of the species above described are sure to occur. Professor
M'Coy has lately added Cyathaxonia Siluriensis from Kendal to the
scanty list. Stenopora fibrosa, indeed, and Favosites alveolaris, with
some few others, seem to have been indifferent to the nature of the
sediment they lived upon, for they are found throughout the system.

The Bryozoa, which are the highest type of corals, or, as some
naturalists think, distinct from them, are numerous, but not of many
species, in the upper division. We have already quoted the Ptilodictya, or feather coral, in the Lower Silurian. A large species, P
lanceolata Foss. 35. f. 6., is abundant in Wenlock strata (f. 7. is the
young state of it); and P. scalpellum (Eschara, Sil. Syst.) Foss. 36.,
accompanies it: the former occurs also in the Ludlow rocks. Glauconome disticha, f. 5.; Fenestella Milleri, f. 4.; and F. subantiqua,
p. 178. Foss. 10. f. 1., are common Dudley fossils. We have also
figured, Foss. 35. f. 2., another common species, F. assimilis (Gorgo-
nia, Sil. Syst.), and there is a beautiful little cup-shaped Fenestella
in the Wenlock limestone, F. Lonsdalei, D'Orb., Foss. 35. f. 3.,
formerly referred by Mr. Lonsdale to F. prisca of Goldfuss. Polypora (Hornera) crassa, f. 1., is not so often met with.

FOSSILS 36.

Ptilodictya scalpellum,
Lonsdale, (Eschara, Sil.
Syst.), natural size. Also a

portion magnified. It is a
common species in the
Wenlock limestone.

Of Cystideæ the forms are rare, but remarkable and characteristic:
the species in the Upper Silurian being all furnished with what Professor Forbes calls "pectinated rhombs." Foss. 37. f. 1. has two of
these curious markings, which are usually absent in Lower Silurian
genera. The following species have been observed at Dudley, and are
seen in the rich collections of Mr. Gray and Mr. Fletcher of that place
—Pseudocrinites magnificus, fig. 1.; P. quadrifasciatus, f. 2.; P. bifas-

ciatus, and P. oblongus, Forbes *; Apiocystites pentremitoides, f. 4.; Prunocystites Fletcheri, f. 3.; Echinoencrinites baccatus, f. 5., and E. armatus, f. 6.   Mr. Salter has shown, under figure 6., the curious 5-valved ovarian opening, and a pair of pectinated rhombs which always occur opposite each other on neighbouring plates.   Ischadites Königü, Pl. 12.f. 6. may also possibly be an animal of this group.

FOSSILS 37.

CYSTIDEÆ OF THE WENLOCK LIMESTONE.

1. Pseudocrinites magnificus, Forbes. 2. P. quadrifasciatus, Pearce. 3. Prunocystites Fletcheri, Forbes. 4. Apiocystites pentremitoides, Forbes. 5· Echino-encrinites baccatus, Forbes. 6. E. armatus, Forbes.

The last-named genus, Echino-encrinites, but of course with different species, is found in the Lower Silurian rocks of Russia.   In North America, the representative of the Wenlock limestone (the 'Niagara group' of New York) contains some characteristic Cystideæ, Apiocystites, &c.

Passing on to the great group of Crinoid animals, which are numerous but rarely perfect in our lower division, we find the Upper Silurian very rich in forms of this class.   Many of these are yet undescribed, and we can only refer at present to those figured in our Plates 13. to 15., and to a few other species lately published in the works of Austin †, D'Orbigny ‡, &c.

By far the commonest fossil of this class is the Periechocrinus moniliformis (Actinocrinus, Sil. Syst.) pl. 13. f. 1, 2., whose long necklace-like stems cover the slabs of Dudley limestone, and sometimes attain five feet in length.   This and the following species have the arms composed of a double row of plates set side by side:—Dimerocrinites decadactylus, pl. 13. f. 5.; D. icosidactylus, f. 4., a common fossil; three species of the singular genus, Eucalyptocrinites, Gold-fuss (Hypanthocrinus of the Sil. Syst.), which are found in the Dudley limestone; E. decorus, pl. 14. f. 2.; E. polydactylus, M'Coy, a very large species; and the rare E. granulosus, Lewis, as yet de-tected only at Walsall in Staffordshire; also Marsupiocrinites cælatus, pl. 14. f. 1.   Of the last a reduced copy of a perfect specimen is given, Foss. 38. f. 1.; and at f. 3. the same species is drawn without the arms, but showing the long proboscis inserted into the shell of a mollusk — Acroculia haliotis, better figured in pl. 24. f. 9.

From the very frequent occurrence of the same shell, tightly em-braced by the arms of this crinoid, and from the fact, that the mouth of the shell is always turned downwards over the proboscis, it is inferred, and without much doubt, that it was the habitual food of the animal.   This has been long observed, by Mr. John Gray of Dudley, who has dissected many specimens from the stone to illustrate the point.   And it has received further confirmation by the naturalists of America, Mr. Yandell and Dr. Shumard having observed the same feature in several of the beautiful Silurian crinoids of America.

Of those species which have but a single row of joints in each arm, Cyathocrinus tuberculatus is the most common, occurring of all sizes, from that represented, pl. 14. f. 6., to much larger than our f. 5.; C. tesseracontadactylus, (Cyath. simplex, Sil. Syst.) f. 4., and C. Or-bignyi, M'Coy, are rarer forms; the former is found in Gothland. Icthyocrinus? goniodactylus, pl. 14. f. 3.; I. arthriticus, f. 7., and I. capillaris, pl. 15. f. 3., are quite common Dudley fossils.   Icthyo-crinus pyriformis often grows larger than the form represented in pl. 14. f. 8.; and this species extends its range to North America. Actinocrinus (?) retiarius, pl. 14. f. 9., when perfect, has a long stem with many auxiliary arms.

Enallocrinus punctatus, which occurs both in England and Sweden, is frequently found on Wenlock Edge.

Glyptocrinus (?) expansus (Actinocrinus, Sil. Syst.) pl. 15. f. 1., is perhaps one of the most stately species, conspicuous for the size to which it grows and the numerous plates of which its cup is composed.

Lastly may be noted a very common Wenlock fossil, Crotalocrinus rugosus, whose structure is, perhaps, more remarkable than that of any Silurian crinoid.

FOSSILS 38.

CRINOIDEA OF THE DUDLEY LIMESTONE.

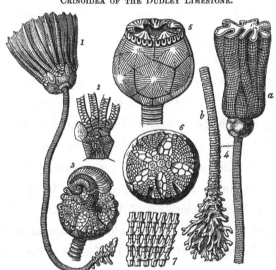

1. Marsupiocrinites cælatus. 2. Magnified base of the arms. 3. Proboscis of do. inserted in the shell of Acroculia Haliotis. 4. Reduced figure of Crotalocrinus rugosus; the bag-like cluster of arms surmounting the small round pelvis. 5. The latter of its natural size, with the stomach plates stripped off, and showing the base of the many-fingered arms. 6. The flat stomachal surface, showing also the branching of the arms from their bases. 7. A part of the reticulate congeries of fingers, each joint being anchylosed to its neighbour on either side. — [J. W. S.]

In most Encrinites the arms start immediately from the edge of the pelvic cup, commencing with a single joint, and soon branching into two, three, or four, the subdivision varying in different species.

But in this wonderful encrinite, the upper edge of the pelvis is seen to be surmounted by at least twenty or twenty-five arm-joints, instead of the usual five; and when the specimens have lost all but the pelvis and these lower joints, the latter are seen (Foss. 38. f. 5.), to have each a perforation in their middle, as in the arm-joints of all other crinoids. These multitudinous arms soon divide, and subdivide again and again, the several branches lying close, side by side, and leaving scarcely any free space between them.

Mr. Salter has pointed out to me that each joint gives off from either side a small lateral process, f. 7., which unites with that of its next neighbour, and this is continued through the whole extent; so that, instead of a star of free and waving arms, there is formed a deep, wide funnel, like a wicker basket; or rather, as the joints are all of equal length, and the lateral processes therefore form continuous transverse lines, like a piece of the coarsest woven serge, or canvas, f. 4. a. This curious funnel of anchylosed arm joints was either flexible, or grew in a lobed and puckered form. But, although the numerous arms seem to start at once from the pelvis, as said before, their real origin is on the ventral surface further inward, see f. 6., where they commence with single joints as in other crinoids, and are clothed with short tentacles to their very base. This surface is but rarely visible, the usual appearance of the cup being that seen in f. 5., and in Pl. 13. f. 3., which are of the natural size. The cup consists of 15 plates. The stem (f. 4. b.) was long ago figured by Parkinson in his Organic Remains.* It is made up of close joints, each with a row of tubercles, which are perforated, the hole communicating with the central canal of the stem. Near the root, these tubercles lengthen out into cylindrical tubular processes, which attach the animal to shells and corals. The cabinets of my friends, Mr. Fletcher and Mr. Gray, furnished the materials to Mr. Salter for the illustration of this remarkable fossil.

Asteriadæ or starfish are not absent: four species, which are here figured, being found in the Ludlow rocks of Kendal. The most common is Uraster primævus, Foss. 39. f. 1. Uraster Ruthveni, f. 3., and U. hirudo, f. 2., are rarer species. Protaster Sedgwickii, f. 4., is believed by Professor Forbes to be closely related to certain species of the group of Euryales inhabiting the northern seas; and

---

* "Turban, or Shropshire Encrinite," vol. ii. Pl. 15. f. 5. p. 193.

not to the Ophiuræ, which it much resembles. And, lastly, there is a starfish in the Dudley limestone, named Lepidaster Grayii * by the same author, more crinoidal in its aspect than any existing species.

FOSSILS 39.

UPPER SILURIAN STARFISH.

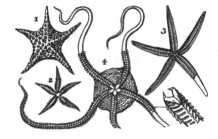

1. Uraster primæ-
vus, Forbes.   2. U.
hirudo, Forbes.   3.
U. Ruthveni, Forbes.
4. Protaster Sedg-
wicki, Forbes.

All from the Upper
Ludlow Rocks in the
neighbourhood   of
Kendal. First found
by Professor Sedg-
wick.

Through the assistance of Mr. Salter, I have dwelt particularly in this chapter on a few of the radiate animals, because they present us with some new or little known characteristics.   The remaining groups, Mollusca, Worms, and Crustacea, will not require so much detail.

Brachiopod shells, though not in such great preponderance as in the lower rocks, nor so numerous in species, are yet very abundant. There are fewer Orthides, and more of the Terebratulæ and Spirifers. The Pentameri are still present, but mostly of distinct species ; and the same may be said of Lingula, of which there are decidedly fewer kinds.

Several characteristic species have been quoted as ranging upwards from the lower rocks, such as Orthis elegantula, O. biforata, and O. biloba: the first of these being equally abundant in both divisions. Strophomena depressa, and S. pecten ; Leptæna transversalis ; Atrypa marginalis and A. reticularis, are far more abundant here than in the lower division ; except in the upper portions of the Caradoc sandstone, where they are very plentiful.   The same may be said of Spirifer plicatellus, (radiatus, Sil. Syst.), pl. 21. f. 2., Rhynconella cuneata, and R. borealis.   These species have a wide geographical range.   In chapters 6 and 7. thirty other species are enumerated from the various

* Mem. Geol. Surv. Decade iii. Pl. 1.

strata of the Upper Silurian rocks, among which Leptæna trans-
versalis, Strophomena euglypha, Spirifer elevatus, S. plicatellus, and
Pentamerus galeatus are most characteristic of the Wenlock strata;
Rhynconella Wilsoni, R. nucula, and R. navicula, Orthis lunata, and
Chonetes lata, chiefly distinguish the Ludlow rocks.

All the species above enumerated will be found in the plates, — 20
to 22.   There are, however, several others which must be noticed.
Among the rarer species may be reckoned, — Discina (or Orbicula)
Morrisi, and D Verneuilii of Davidson, D. Forbesii (Foss. 40. f. 11.),
and D. implicata, pl. 20. f. 4.   These are Wenlock species of a genus existing at the present day, whilst Discina rugata, pl. 20. f. 1, 2.,
and D. striata, f. 3., are common Ludlow forms.

Siphonotreta anglica, Foss. 40. f. 10., is a rare fossil, and of a genus
characteristic of the Lowest Silurian rocks in Russia.*

Foss. 40.

UPPER SILURIAN BRACHIOPODA.

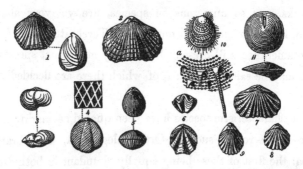

1. Rhynconella nutula. 2. R. Lewisii, Davidson. 3. R. Grayii, Davids. 4. R. Capewelli
Davids. 5. R. Bouchardii, Dav. 6. R. Barrandii, Dav. 7, 8. R. Salterii, Dav. 9. Var. of do.
(R. Baylei, Dav.). 10. Siphonotreta anglica, Morris; a. portion of the surface and spines
magnified. 11. Orbicula Forbesii, Dav.

* In addition to the many proofs given in the preceding Chapters, let me offer
an additional example of the community of certain forms of life throughout the
Silurian æra, which has been obtained even whilst these pages are printing. In
Russia, the inferior portion of Lower Silurian rocks, is strikingly characterized
by the presence of a genus of small horny brachiopods — Obolus; and now, through
the discrimination of Mr. Davidson, two shells, found in the Wenlock shale, have

Spirifer sulcatus, Hisinger, is a small species common on Wenlock slabs; Athyris Circe, Barrande, a Bohemian fossil, occurs, according to that author, with the common A. tumida at Walsall; Rhynconella Lewisii, Foss. 40., f. 2., R. Salterii, f. 7., and its varieties, f. 8, 9., and T. Bouchardii, f. 5., are common Wenlock species; R. Barrandii, f. 6., R. Capewellii, f. 4., and R. Grayii, f. 3. (a strange twisted species), are rare Dudley fossils. R. nucula, f. 1., is the commonest of all the Terebratulæ of the Upper Ludlow rock. Imperfect specimens of it are figured in pl. 22. f. 1, 2.

Three Pentameri only were formerly described from the Upper division, viz. P. linguifer, P. galeatus, and P. Knightii (see pl. 21.). To these may now be added P. liratus, Foss. 8. p. 88., found in the Woolhope limestone; and P. lens, ib. f. 1., which may be considered as belonging to the very lowest beds of this division, since at May Hill, Gloucestershire, the strata containing it alternate with bands of limestone, which contain abundance of Upper Silurian fossils.

Several species of Orthis have been already quoted, Chap. 6., as very common species in the Wenlock and Dudley limestone. O. Bouchardi, Foss. 41. f. 1.; O. Lewisii, Davidson*; O. equivalvis, Davidson, are rare in the same formation; and so is O. calligramma, adverted to at p. 184., if the shell here figured, Foss. 41. f. 2., be not a distinct species (O. Davidsoni), as De Verneuil supposes.

Strophomena antiquata, Foss. 41. f. 8., grows five or six times the size of the figure in my original work (see Pl. 20. f. 18.). S. funiculata, Foss. 41. f. 4., and S. imbrex, f. 6, 7., are common Wenlock

been referred to the same genus, viz. O. transversus and O. Davidsoni, Salter. Mr. Davidson shows the close approach in structure of the Russian and British species, though the latter is much larger; and he remarks, with regard to the Siphonotreta anglica above mentioned, "it has also been found in the same beds (Wenlock shale, Dudley) with the Obolus, and is of a form scarcely to be distinguished from Lower Silurian species." These constantly augmenting data prove that it is quite impossible to divide the "Silurian" into two natural systems.

* Bull. de la Soc. Geol. de France, 2nd series, t. 5. p. 309, 1848. This memoir contains figures of all the British Upper Silurian species then known.

species; S. corrugata, Portlock, is a beautiful species, which has
been already cited in the lower division, where it is plentiful. S. pecten, f. 3., though common both in Upper and Lower Silurian, escaped
notice when the "Silurian System" was published. S. applanata,
Salter, is a small species resembling the last, which occurs sparingly
in both divisions.

FOSSILS 41.

UPPER SILURIAN BRACHIOPODS.

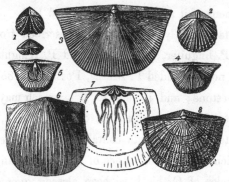

1. Orthis Bouchardi, Davidson. 2. O. calligramma (O. Davidsoni, De Vern.). 3. Stro
phomena pecten, Linn. 4. S. funiculata, M'Coy. 5. Interior of do. 6. S. imbrex (Pander
and Davidson.). 7. Interior of dorsal valve of do. 8. S. antiquata, full grown (half natural size).

Lingula Lewisii, pl. 20. f. 5., is one of the commonest middle Ludlow
or Aymestry species; and L. cornea, pl. 34. f. 2., abounds in the tilestone of the Upper Ludlow. They are both well marked species, and,
with the small but characteristic L. lata of the Lower Ludlow rock,
are the only species of the genus yet described from the Upper
Silurian of Britain.

The Lamellibranchiate bivalves, though numerous in species, have
not yet been fully described. They consist chiefly, as Prof. Phillips
has pointed out*, of one or two families closely related, and which
are represented by the living forms, Mytilus, Arca or Nucula, and
Avicula. The genera Pullastra, Mya, Cypricardia, and Cardium,
to which several of these were formerly referred, do not appear in
these early geological times.

* Mem. Geol. Surv., vol. ii. pt. 1. p. 264.

Pterinea (Avicula) is a most plentiful genus. P. retroflexa, pl. 23. f. 17., is a species subject to great variation, — it has already (p. 191.) been quoted from various strata. P. Sowerbyi, f. 15., (Avic. reticulata, Sil. Syst.) is a characteristic Aymestry species, whilst P. lineatula (A. lineata, Sil. Syst.), f. 16., is very frequent near Ludlow. Besides these common species, Professor M'Coy has enumerated several others, viz., P. Boydi, P. demissa, P. pleuroptera, and P. sub-falcata, of Conrad, and P. tenuistriata, Foss. 42. f. 5. The three last are, it is believed, also found in Lower Silurian rocks in Wales and Westmoreland. Others are Upper Silurian forms only, such as P. hians, M'Coy, an Aymestry rock fossil, and P. asperula, f. 4., a fossil from the Wenlock shale at Builth. A common Wenlock species, doubtfully referable to this genus, is the P. planulata, Conrad, fig. 6.

FOSSILS 42.

UPPER SILURIAN LAMELLIBRANCHIATA.

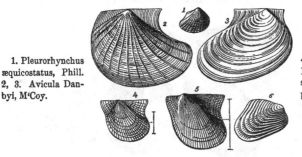

1. Pleurorhynchus æquicostatus, Phill. 2, 3. Avicula Danbyi, M'Coy.

4. Pterinea asperula, M'Coy. 5. P. tenui-striata, M'Coy. 6. P? planulata, Conrad.

There are some fine large species, which seem more like Avicula, or Aviculopecten ; such as A. Danbyi, Foss. 42. fig. 2, 3., and A ampli-ata, Phillips, both occurring in the Upper Ludlow rock; whilst Avicula mira, Barrande, a beautiful, reticulated species, is common at Dudley, and is identical with one from the Upper Silurian of Bohemia.

Next in importance are the mytiloid shells, or Modiolopsis, as the genus is now called *, which contains several species. Thus M. com-

* There does not appear, however, to be any decided character to separate these shells from the Modiolæ of our own day. — J. W. S.

planata, pl. 23. f. 1., is an Upper Ludlow fossil ; M. Nillsoni, Foss. 43.
f. 8., is both a Wenlock and Ludlow species; and M. platyphyllus,
f. 7., is characteristic of the uppermost Ludlow beds, where it
occurs with some other forms, now referred by Professor M'Coy to his
new genus Anodontopsis. M. quadratus and M. ovalis, Salter, M.
angustifrons, M'Coy, M. lævis, pl. 34. f. 7., and M. bulla, Foss. 43.
f. 5., are species of this kind. Mytilus exasperatus, Phillips, and
M. antiquus, pl. 23. f. 14., are Wenlock shale fossils, the latter
being common. Mytilus mytilimeris, Foss. 43. f. 6., is also plentiful
in Wenlock shale. Goniophora cymbæformis, pl. 23. f. 2., seems
nearly allied to Mytilus, and is one of the most abundant Upper
Ludlow shells. Certain forms, less evidently related to this family,

FOSSILS 43.

UPPER SILURIAN LAMELLIBRANCHIATA.

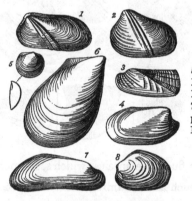

1. Grammysia cingulata, Hising. 2. G. triangulata, Salter. 3. Orthonota angulifera, M'Coy. 4. O.—— sp (O. semisulcata, M'Coy.)

5. Anodontopsis bulla, M'Coy. 6. Mytilus Mytilimeris, Conrad. 7. Modiolopsis platyphyllus, Salter. 8. M. Nillsoni, Hisinger.

have been termed Orthonota by some authors, and Sanguinolites by
others. They are thin shells, without hinge teeth, and outwardly
much resemble Mya, Panopœa, &c., to which they are not at all
allied in reality. Orthonota semisulcata (Modiola, Sil. Syst.) is one
example from the Ludlow rocks, not here figured; and a new species,
(Foss. 43. f. 4.) is one of the most frequent fossils in the tilestones of Westmoreland. Other species referred to this genus in

the publications of the "Geological Survey"* differ still more widely in outward appearance from the Modiola, to which they are nevertheless related. They are abundant in the Ludlow rocks, especially O. amygdalina, pl. 23. f. 6., and its variety, — retusa, f. 7. This species covers the surfaces of the uppermost stratum of the Upper Ludlow rock, wherever it is exposed, and with it, more rarely, occur O. impressa, f. 4., and O. undata, f. 3. Orthonota rotundata (Mya, Sil. Syst.) f. 5., is characteristic of the Aymestry rock; but O. solenoides, f. 9., and O. rigida, f. 8., are more common in the Lower Ludlow. O. angulifera, M'Coy, Foss. 43. f. 3., is an ornamented species, rare in the Ludlow rocks of Westmoreland.

Grammysia, De Verneuil, is a genus very like the shells last mentioned, but easily recognized by its deep furrows on the valves. G. triangulata, Foss. 43. f. 2., is a typical tilestone fossil, and G. extrasulcata, Salter, and G. cingulata, Foss. 43. f. 1., are found with it, both in S. Wales and Westmoreland. The latter fossil is equally common in other Upper Silurian strata (Wenlock), at Dudley, Usk, and near Llandeilo, and is also found in Gothland, Sweden, and Norway. It appears to have flourished best on sandy ground; a condition apparently most favourable to the Lamellibranchiata in general.

Cardiola interrupta, pl. 23. f. 12., is perhaps the most abundant bivalve in the Wenlock and Ludlow shales. On the continent it is also found in Upper Silurian strata, and does not extend its range upward into the Devonian rocks. Cardiola fibrosa, fig. 11., is a Lower Ludlow fossil, and C. striata (Cardium striatum, Sil. Syst.) f. 13., is also to be referred to the same genus (see p. 129.). The shells of the families Nucula and Arca, in which the hinge line is beset with numerous teeth, are very numerous in individuals, though of few species, in the Upper Silurian. Nucula Anglica, pl. 23. f. 10., N. (Arca) primitiva, Phillips, N. Edmondiiformis, M'Coy, and

* Vol. ii. pt. 1. p. 360.

N. (or Ctenodonta*) subæqualis, pl. 9. f. 5., are Upper Ludlow shells; most of these, indeed, had their origin much earlier. Nucula sulcata, Hisinger (Mem. G. Surv. vol. ii.), ranges throughout the Upper Silurian. The genus Cucullella, distinguished by a strong internal ridge, contains several Ludlow species. C. antiqua, pl. 34. f. 16., C. Cawdori, f. 3., and especially C. ovata, f. 17., are common shells in the tilestone. C. coarctata, Phillips, is found in great plenty in the Ludlow rocks of Pembrokeshire, and occurs also in the Wenlock shale.

The genus Cleidophorus, which has no teeth in the hinge, is otherwise much like Cucullella. C. planulatus, Conrad, is found in Wenlock shale. Lyrodesma (Actinodonta) cuneata, Phill., connects these Arca-like shells with the Mytiloid or muscle-shaped forms: it is found in Wenlock shale, and is figured in the "Memoirs of the Geol. Survey," vol. ii.

Lastly, as a representative of the Cardiaceæ, we have one small species of Pleurorhynchus; a genus which, as we have before seen, p. 192, commenced existence in the lower division. P. æquicostatus, Foss. 42. f. 1., is a miniature example of it, found at Wenlock and Woolhope. In America, much larger species of this genus occur in Upper Silurian rocks.

Lamellibranchiate shells are by far more frequently met with in the Upper Silurian strata than in the Lower; a fact indicated long ago in the plates of organic remains given in my former work, and which has been confirmed by subsequent observations. Out of forty-five species quoted in Professor Phillips' memoir †, but five or six are found in Caradoc sandstone or Llandeilo flags. Professor M'Coy, in his catalogue of the Woodwardian Museum, enumerates eleven species from the Lower Silurian, a formation in which that collection is peculiarly rich; and of the fifty Upper Silurian species

---

* Proposed by Mr. Salter, 1851, for the palæozoic Nuculæ, or those of them at least which have the ligament external, and no spoon-shaped internal process to contain it.

† Mem. Geol. Surv. vol. 2. pt. 2. 1848.

described by him, as many as twelve are also found in the lower division. Col. Portlock has also described many Lower Silurian forms of this group in Ireland. But an inspection of the museum of the Geological Survey, and of numerous private collections, will convince any one that the larger proportion in species is still on the side of the Upper Silurian rocks, whilst the individuals in them are unquestionably more numerous.

We ought here to notice Professor Phillips's suggestive remark, that the families to which the Silurian (and generally Palæozoic) Lamellibranchiata belong, are just those which occur at the junction, so to speak, of the Monomyarian and Dimyarian groups.   Mytiloid, Nuculoid, and Aviculoid shells, with few additions, were, therefore, the representatives of this order in the Silurian seas.

Gasteropod mollusca, or univalve shells, are spread thinly throughout the Upper Silurian, seldom forming a conspicuous feature. The Wenlock limestone, however, is neither poor in species nor individuals, though several are yet unpublished. Among them Euomphalus occupies a marked place; three species especially swarming in certain localities, viz., E. discors, pl. 24. f. 12. ; E. rugosus, f. 13. ; and E. funatus, pl. 25. f. 3.   A species also very like E. centrifugus, Wahlenberg, not unfrequently occurs with them.   Euomphalus sculptus, f. 2., which appears to be only a variety of E. funatus, is found as frequently in the Upper Caradoc as in Wenlock rocks.   E. alatus, f. 4., is a Wenlock shale species.   E. carinatus, pl. 24. f. 11., is more characteristic of the Middle Ludlow or Aymestry rock.   Of all these species, E. funatus is by far the most common, and has the greatest vertical range.   Its operculum, a concentric one, is now often met with, and helps to show the near relation of the genus to Delphinula, as above stated (p. 192. note).

Of the genus Turbo, T. cirrhosus, pl. 24. f. 10., often occurs in the Wenlock shale, and the T. Williamsi, pl. 34. f. 14., together with two other species of the genus mentioned in p. 136., are frequent in

the Upper Ludlow, especially in the tilestones. This last rock is also
crowded with the Trochus? helicites, pl. 34. f. 12, 13., a shell which
is referred to that genus with great doubt, and looks more like
a land-snail in outward form. It is probably quite a distinct genus,
in no way allied to the group in which it now stands. T. cælatulus,
M'Coy, is a rare species from the Woolhope limestone; and T.
(Litorina) undifer, M'Coy, is a small Aymestry rock species.

The Turritellæ, Sil. Syst., of the uppermost Ludlow rocks have
been already referred to the genus Holopella, p. 192.; H. obsoleta,
pl. 34. f. 11., H. gregaria, and H. conica, f. 10., being very frequent
fossils in this stratum in Westmoreland, Shropshire, and S. Wales.
Professor M'Coy has distinguished another small species, H. gracilior,
from the Wenlock shale of Llangollen. Loxonema sinuosa, pl. 24.
f. 3., and L. elegans, M'Coy, are Ludlow rock species. The latter
is a fine shell, two inches long, and is frequent both in the Wenlock
and Ludlow shales.

Spiral shells with notched apertures, Pleurotomaria and Murchisonia, are common. Murchisonia corallii, pl. 24. f. 7.; M. articulata, f. 2. and M. torquata, M'Coy, are slender turreted forms in the
Ludlow rocks. M. cingulata, Hisinger, three inches long, is an Aymestry shell found also in Gothland. M. Lloydii, pl. 24. f. 5., abounds
in the middle and lower Ludlow, and is frequent in Wenlock limestone. Pleurotomaria balteata, Phillips (Mem. Geol. Surv. vol. ii. pt. i.
pl. 15.), is another interesting Wenlock form; and Pleur. undata, pl.
24. f. 6., is from the Lower Ludlow, where some other large species,
yet unpublished, are also found. Acroculia (Nerita) Haliotis, pl. 24.
f. 9. and A. prototypa, f. 8., are exceedingly abundant, the first
especially, in Wenlock limestone, and also in the Upper Silurian of
Bohemia. These mollusks seem to have formed the chief diet of
the numerous Encrinites of the period, both in England and America. Natica parva, pl. 25. f. 1.; Cyrtolites lævis, f. 9., and a few
others, complete the list of published species, but many others, I am
assured by Mr. Salter, remain to be described.

One or two Pteropods only have yet been detected, nor do they seem to be so plentiful as in the lower rocks. Theca Forbesii, Foss. 21. in chap. 8., and T. anceps, Salter, are Wenlock shale species. Besides these, Conularia seems to be the only British Upper Silurian pteropod. The beautiful and variable C. Sowerbyi (C. quadrisulcata var. of my former work), pl. 25. f. 10., is often found in Wenlock limestone, and occurs with a rarer species, C. subtilis, Salter, in the Ludlow rocks of Westmoreland.

Bellerophons are frequent. B. dilatatus, pl. 25. f. 6., is one of the largest univalves in the Ludlow and Wenlock rocks. The broad expanded mouth is often three inches wide, and is sometimes furnished with radiating ribs (see the figure),—at other times smooth, Foss. 21. in Chap. 8. B. Wenlockensis, pl. 25. f. 7., is very characteristic of the strata implied in its name. B. expansus, f. 8., is equally so of the Upper Ludlow rock. The latter, and B. Murchisonæ? pl. 34. f. 19. B. carinatus, f. 8., and B. trilobatus, f. 9., generally of small size, are most abundant everywhere in the upper beds of the Ludlow rock.

Of the Cephalopods which chiefly typify the Upper Silurian rocks much might be said; though the species which most abound have already been enumerated (see pp. 112. 121, 122. 129.). It may, however, be noted, that some of these forms are specially characteristic of particular strata. Thus, the thin-shelled species of Orthoceras,—O. sub-

FOSSILS 44.
UPPER SILURIAN CEPHALOPODS.

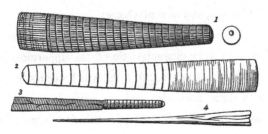

1. Orthoceras filosum.   2. O. Ludense. These two species are figured from specimens two feet in length.   3. O. subundulatum, Portlock (Creseis Sedgwicki, Forbes).   4. O. primævum, Forbes. The two latter are usually 8 or 9 inches long.

undulatum, and O. primævum, here figured, are the most frequent
shells of the Wenlock shale;   O. annulatum, pl. 26. f. 1., and its
variety, fimbriatum, f. 2., is a well-known Wenlock limestone species;
O. filosum, pl. 27. f. 1., here figured (Foss. 44. f. 1.), O. perelegans,
pl. 29. f. 5, 6.; and O. dimidiatum, pl. 28. f. 5., are common Lower Ludlow fossils;   O. Mocktreense, pl. 29. f. 2., is a Middle Ludlow or
Aymestry species; while O. Ludense, Foss. 44. f. 2., and pl. 28.
f. 1.; O. angulatum, f. 4. ; O. Ibex, pl. 29. f. 3., and especially O. bullatum, f. 1., are frequent Upper Ludlow forms.

Orthoceras nummularius, pl. 26. f. 5.; O. excentricum, pl. 27. f. 3.;
O. canaliculatum, pl. 28. f. 3., are rarer Wenlock species; whilst O.
distans, pl. 26. f. 4.; O. subgregarium, pl. 27. f. 2.; and O. imbricatum, pl. 29. f. 7., are not abundant Ludlow fossils.

The same distribution is observable in the other genera; for
Phragmoceras is principally grouped in the Lower Ludlow while Lituites ranges throughout. Phragmoceras ventricosum, pl. 32., which has
been already mentioned, is frequent at Leintwardine, Shropshire, where
also the P. (Orthoceras) pyriforme, pl. 30. abounds. P. arcuatum,
pl. 31. f. 3., and P. intermedium, pl. 30. f. 4., are also from this
stratum. P. nautileum, pl. 31. f. 1, 2., and P. compressum, f. 4., are
found in Wenlock shale, as also are Lituites articulatus, f. 6., and L.
Biddulphii, f. 5.   Lituites giganteus, pl. 33. f. 1, 2, 3., is one of the
finest fossils from Leintwardine, and L. tortuosus, of which a fragment is figured, f. 4., is found in the black nodules of the Wenlock
shale near Welshpool, and also at Dudley.

Orthoceras tracheale, pl. 34. f. 6., and O. (Diploceras?) semipartitum, f. 5., are very characteristic of the uppermost Ludlow or
tilestones, and O. bullatum, pl. 29. f. 1., which occurs in the greatest
abundance with them, must also again be noticed.

To complete the catalogue of the Upper Silurian fauna, a brief sketch
must be given of the annulose animals, such as worms and crustacea,
and of the very few remains of placoid fish which have been discovered.

Cornulites serpularius, see pl. 16. f. 3—9. is still, as in the Lower Silurian, the principal annelid; and though more frequent in the Wenlock limestone, it is not rare in the Ludlow rocks. Tentaculites ornatus, f. 11., abounds in the Dudley limestone; whilst a small species, T. tenuis, pl. 16. f. 12., occurs in the Upper Ludlow rock. The place of the latter is sometimes taken by a form so like T. annulatus of the lower division, that it may be a variety only; in which case we lose another characteristic fossil for distinguishing the two groups. This Tentaculite is plentiful, for example, in Pembrokeshire, in the sandy Ludlow rocks of Marloes Bay and Freshwater. Other annelids (Serpulites, Sil. Syst.) are very common in the Ludlow rock. They are flattened tubes, thick at each projecting edge, sometimes shelly, and at other times corneous, or almost membranous on the sides. Serpulites longissimus, pl. 16. f. 1., of the Upper Ludlow rock, which is characteristic of this formation wherever found, grows to a length of twenty inches, measured round its long spiral curve. S. dispar, Salter, from Kendal, the tube of which has remarkably thin sides, is figured in the late work descriptive of the Woodwardian Museum. Trachyderma coriacea and T. squamosa of Phillips* are wrinkled tubes of these creatures, frequent in the Ludlow rocks of the original Silurian region. Serpulites curtus, Salter, and a few other species, are found in the Wenlock strata.

The crustacea, with rare exceptions, are trilobites. They are exceedingly abundant, forming in some strata, such as the Wenlock limestone, the most conspicuous and characteristic fossils. Many forms have been already quoted in the preceding chapters, and two woodcuts are here given, to represent either the principal forms not illustrated in the original work, or those now more perfectly known. Among the latter we may reckon the elegant and highly ornamented Dudley fossils, Encrinurus punctatus, Foss. 46. f. 5. and E. variolaris, f. 6. These, which are known by the name "Strawberry-headed" by collectors, are found on every slab of Wen-

* Mem. G. Surv. l. c. Pl. 4.

lock limestone, but seldom perfect except at Dudley, where numerous
specimens ornament the cabinets of Mr. Gray and Mr. Fletcher of

FOSSILS 45.

UPPER SILURIAN CRUSTACEA.

1. Lichas Anglicus, Beyrich, and its hypostome or labrum.
2. Calymene tuberculosa, Salter, also with its labrum.
3. Lichas Barrandii, Fletcher.

4. Beyrichia tuberculata, Klöden, natural size and magnified. The other figures are reduced to about half size.

that place. Cheirurus bimucronatus, Foss. 46. f. 4., figured from
fragments only in my old work, is found perfect in the same
collections, and also in the cabinet of the Rev. T. T. Lewis.
Acidaspis Brightii, f. 8., is now better understood than when formerly described, and no less than five other species of this genus
are known in the Upper Silurian rocks, of which we figure the head
of one, f. 9., called A. Barrandii by Mr. Fletcher. The common
Calymene Blumenbachii, Pl. 18. f. 10., is the most prolific of all the
Upper Silurian trilobites, and is, as before stated, known in the lower
division of the system; and C. tuberculosa, which was formerly given
as a variety of it, is a very plentiful species in the Wenlock shale,
Pl. 18. f. 11., and Foss. 45. f. 2.

Some forms are here figured which were not published before.
Sphærexochus mirus, Foss. 46. f. 1., is singularly formed; the head
being so inflated as to resemble a ball. This Wenlock fossil has a very
general range, being found both in America and Bohemia in Upper
Silurian rocks. It appears to have existed earlier in Britain than in
those distant parts, for it is found in the equivalents of the Llandeilo
flags in Ireland. Four or five species of Lichas (a genus already figured
in chap. 8.) are now known in Wenlock limestone. L. Grayii,
L. Salteri, L. Barrandii, L. hirsutus, are all species lately figured
from the rich collection of Mr. Fletcher (Quart. Geol. Journ. vol. vi.

p. 235.)　With them is also represented a small common species, the L. (Arges) Anglicus of Beyrich, Foss. 45. f. 1., of which a very imperfect illustration was formerly given in Brongniart's work. The genus is remarkable for the shape of the head and its lobes. Phacops Downingiæ, Foss. 46. f. 3. and pl. 18. f. 2—5., P. caudatus, pl. 18. f. 1., with its variety, P. longicaudatus, pl. 17. f. 3—6., have been already quoted (p. 122.) as very common forms.

FOSSILS 46.
WENLOCK LIMESTONE TRILOBITES.

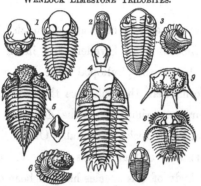

1. Sphærexochus mirus, Beyrich, with a coiled-up specimen. 2. Cyphaspis megalops, M'Coy. 3. Phacops Downingiæ, both extended and coiled-up. 4. Cheirurus bimucronatus, and its hypostome.

5. Encrinurus punctatus, and its hypostome, from a Dudley specimen. 6. E. variolaris. 7. Proetus latifrons, M'Coy. 8. Acidaspis Brightii, in part restored. 9. A. Barrandii, Fletcher and Salter, MSS.

Proetus Stokesii, pl. 17. f. 7., and P. latifrons, Foss. 46. f. 7., are also far from rare in Wenlock strata.　The Bumastus Barriensis (Illænus auct.), Foss. 9. and pl. 17. f. 9—11., ranges, as before said, from the lowest Wenlock beds to near the top of the Upper Silurian.　Two species of Homalonotus also have been already quoted, pp. 108, 136.

Cyphaspis megalops, Foss. 46. f. 2., and a small species, C. pygmæus, Salter, are the only species of that genus known in England.　The former began life in the Lower Silurian epoch.　Deiphon Forbesii, Barrande, is a rare, globular-headed trilobite, common to the Dudley limestone, and its equivalent rock in Bohemia.　And lastly, there is a species of Bronteus, B. laticauda, Wahlenberg, found by my deceased friend, Dr. Lloyd, in the Wenlock limestone.　It is figured by Professor Phillips under the name of B. signatus (Pal. Foss. f. 254.).

A goodly list of British trilobites of the Upper Silurian rocks

has thus been given, though many other interesting forms will probably reward research. The small bivalve crustaceans Cytheropsis
and Beyrichia remain to be noticed. Of the latter genus, the
Upper Silurian species is B. tuberculata, Foss. 45. f. 4. It is very
abundant from the base of the Wenlock shale to the highest
Ludlow stratum, and is a good index of Upper Silurian rocks,
though found sometimes in the upper division of the Caradoc; (the
Lower Silurian form, it will be recollected, is B. complicata, Ch. viii.
p. 201.).

These bivalve crustacea, which are of much larger size in the
Upper Silurian limestones of Sweden, are, in the opinion of good
naturalists, probably of the phyllopod tribe, and much more nearly
allied to living forms than the trilobites that accompany them. So
also are some larger, shrimp-like forms, described by Prof. M'Coy
from the Upper Ludlow rock of Kendal. One of these, the Ceratiocaris inornatus, is three or four inches long. The carapace, or front
part only of the body of this species has yet been described; but in
all probability, says Mr. Salter, the entire creature was very like the
fossil figured in Ch. ii. Foss. 3. The same Ceratiocaris, or a closely
allied species, occurs at Dudley, and portions of the jointed body have
been found there also.

Such were also, according to M. de Barrande's discoveries, those
curious striated spines, which are represented in pl. 19. f. 1, 2., and
which have been of late distinguished, by Prof. M'Coy, from fish
defences, with which they were formerly confounded. (See Chap. vi.
p. 137.) That naturalist, indeed, supposed them to be the slender
pincers of some large crustacean, and has called them Leptocheles
from this circumstance: — L. leptodactylus and L. Murchisoni, pl.
19. f. 1., from the Lower Ludlow rock. But M. Barrande having
found perfect specimens in Bohemia, has indicated, that these long
pointed spines are the trifid tail of a crustacean, something like that of
the genus Dithyrocaris of the carboniferous rocks.

The largest, if not the most highly organised crustacean in the Silurian list, is one allied to the living king-crab or Limulus of the Indian seas.   It is Pterygotus problematicus, before mentioned. The fragments of the carapace, which alone are represented, Plate 19. f. 4, 5., were formerly considered by Agassiz to be fish scales *; but that author soon afterwards † corrected this statement, and assigned the animal to its true place among the Crustacea.   A gigantic species of the genus will hereafter be spoken of as occurring in the Old Red sandstone of Scotland.   The dimensions which P. problematicus attained are unknown; but from fragments of the pincers lately described by Strickland and Salter ‡, it was probably less than that of the great fossil, the " Seraphim " of the Scottish quarry-men, which must have been four feet in length. §   Prof. M'Coy has very satisfactorily referred these monsters of their order to the group above mentioned, the Xiphosura, of which the large king-crab is the modern type.   There are two other Ludlow rock species of smaller dimensions probably to be referred to the same group, which, as we know, existed, and apparently under forms very like the living Limuli ‖, in the carboniferous epoch (see Ch. xii.).

Lastly, in the ascending order of animal life, and in speaking of fishes, we find that even in the " bone-bed" of the upper Ludlow rock, their remains are very scanty, some of those few which were formerly referred to the class, having been recently removed, as stated above, p. 137., to crustaceans.   Still there are, in this stratum ¶,

* Sil. Syst. p. 606.   † Poissons du Vieux Gres Rouge, pl. 1.
‡ Quart. Geol. Journ. v. viii. p. 386. pl. 21.
§ Mr. Strickland recently found this fossil in the Upper Caradoc of Malvern.
‖ Prestwich, Geol. Trans., 2d series, vol. v. pl. 41.
¶ In a recent survey of the May Hill district, in Gloucestershire, made whilst these pages were passing through the press, Mr. H. E. Strickland and myself, re-examining the cutting of the Gloucester and Ross railroad, near Flaxley which my friend had already described (Quart. Jour. Geol. Soc. vol. ix. p. 8.), discovered two thin bone-beds, each little more than an inch thick, and separated by about fifteen feet of fossiliferous Upper Ludlow rock.   On the same occasion we

true remains of fish, such as defences of the fins of sharklike species, with shagreen or skin, jaws and teeth, and other minute fragments less easily determinable.

Onchus tenuistriatus, Pl. 35., f. 15—17., and O. Murchisoni, f. 13. 14., are true bony fin defences, such as were possessed by many placoid fish of the old rocks.   Sphagodus, f. 2., is the prickly skin of some such animal, and may have belonged (as suggested by M'Coy) to one of these species.   The small cushion-like bodies, f. 18., called formerly Thelodus parvidens by Agassiz, and which occur by myriads in the stratum, often forming large portions of its thin layers, are certainly the granules of the skin, or shagreen, of one or other of these two common species.   The remarkable jaws and teeth, f. 3. to 8., first figured in my former work, and named by Agassiz, Plectrodus mirabilis and P. (Sclerodus) pustuliferus (f. 9.) must be regarded still as the jaws and anchylosed teeth of some of these small but predaceous fish.   Prof. M'Coy, however, is inclined to refer these bodies to the order of crustaceans, an opinion in which Messrs. Sowerby and Salter do not concur. Their texture, say they, is solid, bony, and retains in the fossils the jet-like lustre which the other bony fragments exhibit.   Then, as additional evidence of the predaceous habits of these fish, there are the coprolitic bodies (f. 21.—28.), which, by Dr. Prout's analysis, contain the due admixture * of phosphate and carbonate of lime, with other matters, and retain imbedded in them the various small mollusca and crinoids which inhabited the sea bottom in company with the fish.   It thus appears, that Bellerophon, Holopella, Lingula, Orbicula, and Orthis, were all preyed upon by these minute but dominant creatures;

observed, that the only remains of small land plants, one of which seemed to be a stem, as well as the numerous seeds ("spore-cases of Lycopodiaceæ," Dr. Hooker), occurred in beds above the uppermost fish-layer, *and therefore at the very top of the Ludlow formation, just beneath the lowest beds of the Old Red*, in which we found a Cephalaspis; the two deposits being there conformably juxta-posed.   See Ch. X.

* Sil. Syst. pp. 199. 607.

and the half-digested shells remain here, as in the coprolites of fish of many later formations, to attest the character of their food, and the extent of their depredations.

Fourteen years have now elapsed since I proclaimed that these fishes of the Upper Ludlow rock appeared before geologists for the first time as *the most ancient beings of their class* * ; and all the subsequent researches in the various parts of the world over which Silurian rocks have been found to extend, have failed to add to or modify this generalization.   In other countries, indeed, besides our own, as in America and Bohemia, one or two ichthyolites have been discovered within the pale of the Silurian rocks; but there, as with us, they are merely found on the outer threshold of the system, and very sparingly.   We may, therefore, fairly regard the Silurian system, on the whole, as representing a long, early period, in which no vertebrated animals had been called into existence.

But here we must recollect that, when first created, the Onchus of the uppermost Silurian rock was a fish of the highest and most composite order; and that it exhibits no symptom whatever of transition from a lower to a higher grade of the family, any more than the crustaceans, cephalopods, and other shells of the lowest fossiliferous rocks; all of which offer the same proofs of elaborate organization. In short, the first created fish, like the first forms of those other orders, was just as marvellously constructed as the last which made its appearance, or is now living in our seas.

In closing, then, this sketch of the first great or Lower Palæozoic system of life, as exhibited in the most typical region of Britain, it may truly be said, that the geologist who stands on the summit of the Silurian rocks of Shropshire and Herefordshire, and casts his eye westward over the mountains of Wales, sees nothing but ancient masses in which, though replete with copious animal remains of pre-existing shores and deep sea bottoms, no traces of

* Sil. Syst. c. xlv. p. 606.

vertebrata have ever been detected. In looking eastward, on the
contrary, he has before him hills of red sandstone, the lower strata
of which immediately overlap the fish beds of the Ludlow rocks, and
in those he finds that fishes are, on the contrary, the characteristic
fossils.

Examining upwards from that first great *piscina* of antiquity, the
Old Red Sandstone or Devonian, which will be considered in the next
chapter, the geologist has ascertained, that all the other superjacent
and younger strata, whether of primary, secondary, or tertiary age,
are more or less characterized by containing ichthyolites. In other
words, the one or two small fishes which we have just been considering, may be viewed as the heralds which announced the close of the
Silurian æra, and the introduction of the numerous other families
of this class, which thenceforward are found in sediments of every
succeeding age. The name Silurian marks, therefore, the first series
of fossiliferous deposits, throughout the great mass in which no vertebrated animals have anywhere been discovered.

CHAPTER X.

THE DEVONIAN ROCKS, OR OLD RED SANDSTONE, AS EXHIBITED IN THE
BRITISH ISLES.

HAVING explored at some length the lowest known burial-places of former beings, a shorter consideration only can be devoted to the younger races which successively occupied the higher tiers in the vast Necropolis of primeval life.

During the accumulation of nearly all the Silurian deposits, as characterized by a certain fauna, the bottom of the sea was, to a very wide extent, successively occupied by deposits of dark grey-coloured mud. At the close of that period a great change occurred in the nature and colour of the sediment, over large areas of those ancient seas. In and around the Silurian region of Britain, for example, the grey mud was succeeded by red deposits, for the most part sandy; their colour being caused by the diffusion of iron oxides in their waters.* These physical changes were accompanied by the disappearance of those tenants of the deep, whose records we have been tracing, and by the creation of other animals suited to the altered conditions.

The passage upwards from the highest strata of the Silurian rocks into such red deposits of England and South Wales, has already been alluded to †, and illustrated by several diagrams.

In the lower junction beds, as seen within the Silurian region,

* See some excellent observations on the influence of iron oxides on marine life, by Sir H. De la Beche, Mem. Geol. Surv. of Great Britain and Ireland, vol. i. p. 51.

† Pp. 49. 73. 75. 126. 140. &c.

it is only by the detection of Upper Ludlow fossils in thin beds or
tilestones, which are often of reddish colours, that the limit can be
defined, so gradual is the mineral transition from the strata of
the one era into those of the other.   Still, the change is on
the whole well marked; and nothing can be more in contrast, in
that region, than the overlying yellow and red masses, and the
underlying grey Silurian rocks.

Good evidences of this succession are to be seen near Ludlow
and between that town and the Clee Hills, and thence all along the
eastern edge of the Upper Silurian rocks in Radnor and Brecon, as
well as on the west flank of the Malvern and May Hills, and
around the valley of Woolhope.   But the grandest exhibitions of
the Old Red Sandstone in England and Wales appear in the escarpments of the Black Mountain of Herefordshire, and in those of the
loftiest mountains of South Wales, the Fans of Brecon and Caermarthen; the one 2860, the other 2590, feet above the sea. (See Map.)

In no other tract of the world which I have visited, is there seen
such a mass of red rocks (estimated at a thickness of not less than
8000 to 10,000 feet) so clearly intercalated between the Silurian
and the carboniferous strata.   For, whilst in the fore and middle
grounds all the rocks are Lower and Upper Silurian, as before
described (pp. 75, 76.), the distant mountains are entirely composed
of Old Red Sandstone (as represented in the highest of the opposite
diagrams), the observer has only to repair to their southern slopes,
beyond the line of vision in the sketch, and he will there see the
uppermost beds of the red rocks conformably overlaid by the carboniferous limestone of the Great South Welsh coal basin.   (See
lower diagram.)

Whilst then, the opposite pictorial sketch represents the whole region from the Lower Silurian slates to the summit of the Old Red, the
other diagram, taken from one of the coloured sections of the Silurian
System, indicates the copious succession of red strata, which are

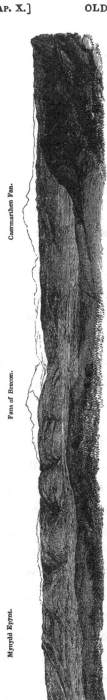

Mynydd Eprnt.   Fans of Brecon.   Caermarthen Fan.

VIEW FROM THE SLATY LOWER SILURIAN ROCKS (*a*) NEAR LLANWRTYD, BRECON, OVERLOOKING ROUNDED HILLS (*b*) OF UPPER SILURIAN, WITH MOUNTAINS OF OLD RED SANDSTONE IN THE DISTANCE, THE LATTER BEING DEFINED BY THE UNTINTED OUTLINE.

(From a coloured drawing, Sil. Syst. p. 346.)

N.W.   S. E.

Lip of the S. Welsh Coal Field.

Black Mountain.

The Hay.

Trewerne Hills.
R. Wye.

SECTION FROM THE TOP OF THE SILURIAN ROCKS, ON THE N.W., ACROSS THE WHOLE AREA OF THE OLD RED SANDSTONE, TO THE BOTTOM BEDS OF THE S. WELSH COAL FIELD, ON THE S.E. (*Horizontal distance, 25 miles.*)

a. Upper Ludlow Rock and Tilestones.   *b.* Red Marls and Flagstones.   *c.* Whitish Sandstones.   *d.* Red and green Marls, Sandstones, and Cornstones. *e.* Chocolate coloured Sandstones.   *f.* Conglomerate and Sandstone forming the summit of the Old Red.   *g.* Lower Limestone Shale.   *h.* Carboniferous Limestone. *i.* Millstone grit.   [To save space, the wide and deep valley of the Usk, between the Black Mountain and the lip of the S. Welsh coal-field, is omitted.]

exposed between the Upper Silurian, *a*, of the Trewerne Hills, and the carboniferous rocks, *g*, *h*, *i*, of the Great South Welsh coal field near Abergavenny.

Consisting of red and green shale and flagstone, *b*, with some whitish sandstone (*c*) in its lower parts, the central and largest portion of the deposit is composed of spotted green and red clays and marls, *d*, which afford on decomposition the soil of the richest tracts of Brecknock, Monmouth, Hereford, a large portion of Salop, and small parts of Gloucester and Worcester. These argillaceous beds alternate, indeed, with sandstones, occasionally of great thickness. They also contain many irregular courses of mottled red and green, earthy limestones, termed " Cornstone," and which, usually consisting only of small concretionary lumps, expand here and there into large subcrystalline masses, as on the western face of the Brown Clee Hills, Salop. The highest member of the series is composed of red and chocolate-coloured, fine-grained, micaceous sandstones and flagstones, which, after alternations with other mottled marls and cornstones, pass upwards into pebbly beds and conglomerates, and in that state form the encircling underlying edge (*g*) of the Great South Welsh coal basin. These last-mentioned rocks are as well exposed in the lofty escarpments of the Fans of Caermarthen and Brecon above alluded to, as in the range of hills near Abergavenny on the south bank of the river Usk. Similar hard sandstones and conglomerates form in like manner a symmetrical girdle to the coal basin of the Forest of Dean, Gloucestershire (see Map), where, and in the gorges of the Wye, as well as in the adjacent and much larger basin of South Wales, they are everywhere conformably overlaid by the shale and limestone which form the base of the carboniferous rocks. In short, the above generalized section gives an idea of the prevalent distribution of the strata in and around the original Siluria.

Now, whilst in all that region, the only organic remains which have been found in this great red and green series are fragments of fossil fishes, we shall presently see, that under mineral conditions to a great extent similar, the widely spread Old Red Sandstone of Scotland, which is of like age, has also yielded scarcely any other class of animals. This fact is in striking contrast to all that has been said of the contiguous and inferior Silurian rocks; in which no trace of a vertebrated creature has been found, except in those highest beds which usher into the natural group under consideration, — or the earliest known period of vertebrate life.

In Herefordshire and Brecknockshire, as well as in Shropshire, remains of ichthyolites were formerly described as occurring chiefly in those strata where calcareous matter is most diffused; fragments and scales being often found in the cornstones, though the best specimens have been procured in the finely laminated flagstones and marls which are in the vicinity of such limestone. These fishes, none of which were known to geologists, in this region, before the publication of my larger work, consist of the genera named by Agassiz, Cephalaspis, Onchus, Ptychacanthus, and Holoptychius; figures of which, as taken from the "Silurian System," are given in plates 36, 37. Others will presently be noticed in treating of the Scottish rocks of similar age.

The lowest strata of the Old Red, in which I have personally observed the remains of the peculiar fish Cephalaspis, is at the western slope of the Silurian ridge of May Hill.* In recently examining the cuttings of the Ross railroad, near Flaxley, Mr. H. E. Strickland† and myself detected a fragment of that ichthyolite in the very lowest beds of the Old Red, which, containing also certain land plants, there lie

---

* See Sil. Syst. Pl. 36. f. 13.

† The passage of this sheet through the press having been delayed until I returned from a tour in Germany, I learn with profound grief, upon my arrival, of the lamentable catastrophe by which, in examining another railroad cutting, my distinguished friend, Mr. H. E. Strickland, has been taken away from those sciences of which he was so ardent a promoter and so great an ornament. The correction of these pages by his hand, during my absence, was one of the last memorials of a friendship which I truly cherished. Sept. 22, 1853.—R. I. M.

in apparent conformity upon the uppermost Ludlow rocks. This close superposition of the lowest member of the Old Red Sandstone to the fish beds of the Upper Ludlow, each stratum containing distinct ichthyolites, is a fact confirmatory of the view of Agassiz, that those animals are very exact indicators of the age of rocks.

When we follow this series of red beds from Caermarthen into Pembrokeshire, considerable lithological changes are seen to occur. Though the prevailing colour is still red (the "red rab" of this county), some portions of the rocks consist of dull green and yellow flags or sandstones, and others of hard grey grits, which even resemble certain inferior Silurian deposits; whilst no true corn-stone or limestone occurs, nor have traces of fossil fishes been there detected. Yet, the strata in question occupy precisely the same geological horizon as in the tracts of England and Wales previously mentioned, and lie just as clearly between the Silurian and carboni-ferous rocks, as in the section before given. In fact, no better sections of the Old Red Sandstone of Britain, in relation to the rocks beneath and above it, can be offered than are exposed around the magnificent land-locked bay of Milford Haven. (See Map.)

We shall presently show the order of succession in vast masses of strata of the same age under a different and more slaty aspect on the south side of the Bristol Channel in Devonshire, and indicate that, with a great increase in calcareous matter, they contain a much larger number of imbedded animal remains.

Let us previously glance at that country, where, preserving much of the same mineral character as in the parts of England just mentioned, the group is more grandly displayed as a red sand-stone than in any other part of northern Europe.

Old Red Sandstone of Scotland.—Dwindling away from its copious development in Brecon, Hereford, and Salop, the Old Red Sandstone becomes in the north of England, as before stated, little more than a single band of coarse conglomerate, which lies unconformably, as a small skirting mass between the Silurian and carboniferous deposits of the lake region of Cumberland. This mere fragment of a series so copious elsewhere, there reposes on the truncated edges of many of the older strata.

In the South of Scotland where the lower Silurian strata are so fully developed and the uppermost Silurian (Ludlow rocks) have not been recognized, the great hiatus between the lower rocks and the coal fields is also occasionally occupied by old red conglomerates only, which range transgressively across the tops of the inclined layers of the older Silurian rocks. Now, although in that southernmost portion of the Scottish region, it may ever be impossible systematically to connect the Silurian series with the next superior formation, on account of the absence of a large portion of its upper members, more close and accurate researches have already been made, which, by marking the range of fossils, still more clearly define the upper limits of the Old Red Sandstone, and thus exhibit an ascending order into strata of sandstone, limestone, and coal, charged with true carboniferous fossils. This has been shown in Fifeshire, by the Rev. Dr. Fleming *, and partially, indeed, on the east coast of the Lothians. Hereafter the same upward succession will doubtless be also ascertained on the opposite or Ayrshire coast, whence specimens of fishes of the species of Holoptychius, which occur in the Old Red, have been sent to me.† It is, however, in the North of Scotland that the Old Red Sandstone attains its full development. Yet even there it wants the conformable fossiliferous Upper Silurian base which has been described in Wales and England.

I have, however, alluded to indications of the upper member of the Silurian rocks that seem to exist in Forfarshire, and particularly on the flanks of the Sidlaw Hills, where the grey, Arbroath paving-stones repose upon clay slate, and contain forms of the lobster-like crustacean Pterygotus, which approach very nearly to, if they

* See Edin. Journ. Nat. Science, vol. 3.
† By Mr. Alexander McCallum, of Girvan, who found them near the Point of Ayr.

are not identical with a fossil of the uppermost Ludlow rock.* The general section over a part of that region, given by Sir C. Lyell in the last edition of his Manual of Geology, seems to me so indicative of the clay slate which underlies the Old Red being also an equivalent of some Silurian rock, that it is here repeated.

S.W.　　Whiteness.　　　　　　　Sidlaw Hills.　　　　　　Strathmore.　　　　　N.E.

g　　c　　b　　a　　　　　a　　　　a　　b　　c　　d　　　c　　b　　a

SECTION ACROSS THE OLD RED OF FORFARSHIRE.

a. Clay slate. b. Grey paving stone and tile stone, with green and reddish shale, containing Pterygotus and fruit-like bodies. c. Red conglomerate. d. Red sandstone. g. Red shale or marl, unconformable to the lower rocks.

I have further suggested theoretically, that Lower Silurian rocks, like those of the South of Scotland, may, in large portions of the Highlands, have been converted into the chloritic and micaceous schists.† (See p. 163. antè.)

Thus, we know that the granitic nuclei of the Grampians and their western prolongations, as well as those of the North Highlands, are often enveloped by such apparently metamorphosed, crystalline schists, and that around their flanks, flagstones and coarse conglomerates are arranged. In general, the latter form the base of the great Scottish system of Old Red Sandstone, which in some parts is divisible into separate, thick bands, and is seen to graduate upwards, particularly in Fifeshire, as above stated, into the lowermost beds of the carboniferous system.

Professor Sedgwick and myself first pointed out the true ‡ general features of this vast deposit in the Highlands of Scotland, and showed that the flagstones of Caithness, with their numerous ichthyolites (Dipteri and Diplopteri), formed a part of it. To Mr. Hugh

* Quart. Journ. Geol. Soc. Lond. vol. vii. p. 169.

† Ibid. vol. vii. p. 168.; vol. viii. p. 386.

‡ Trans. Geol. Soc. Lond. N. S., vol. 3. p. 125. with map and sections. Dr. M‘Culloch had described these rocks as primary sandstones.

Miller, however, science is indebted, for expounding the wonderful organization of many of the fossil fishes of these deposits, and for throwing the clear lights of zoology upon what was formerly a very obscure page in geological history.*

The Old Red of the north of Scotland is justly termed by Miller "the frame" in which the crystalline rocks are set. In other words, it is the rough mantle which has been thrown over their shoulders and sides. That this deposit was of enormous thickness and occupied a very long period in its formation, is manifest to every one who surveys either the east or west coasts of the Highlands. In the latter the deposit is chiefly known as a coarse conglomerate or a hard red sandstone, which rests in layers, more or less horizontal, on low and gnarled bosses of crystalline gneiss; out of which and other ancient rocks the conglomerate has been formed.

This view, sketched by the graphic pencil of the late Duchess Countess of Sutherland, exhibits the three insulated mountains of Suil-vein, Coul-beg, and Coul-more.

MOUNTAINS OF THE WEST COAST OF THE HIGHLANDS
(as sketched from the sea by the Duchess Countess of Sutherland).

* The British reader is, or ought to be, familiar with the "Old Red Sandstone" of Miller; but as my volume may fall into the hands of foreigners, perchance unacquainted with that work, let me urge them to refer to it, not only as an

I have gazed in wonder at these mountains from a boat off the western shore, not only as proofs of the long period during which the boulders, pebbles, and sand were gathered together in former seas, but still more as evidences of the enormous subsequent furrowing out or excavation of the strata which at one period filled up such deep intervening depressions to the depth of upwards of two thousand feet, and united the more widely separated masses.* These mountains, which extend their rugged range from Ross into Sutherland, may be taken as samples of many similar rocks that occur in other parts, of the Highlands; as the Ord of Caithness, parts of Easter Ross flanking Ben Wyvis, the slopes of Mealfourveny, Invernesshire, and the sides of the great valley extending from S. W. to N. E., which is occupied by the Caledonian Canal.

Such conglomerates and sandstones, having often derived their red colour from pre-existing granitic rocks in which hematitic iron is much diffused, are overlaid (and particularly in Caithness) by dark grey micaceous flagstones and schists, in which fossil fishes abound; and these again by other sandstones, usually of a lighter red colour than those beneath, in which ichthyolites also occur. This succession of flagstone and red sandstone is copiously developed in the Orkney

eloquent and original treatise, but also as singularly instructive and well calculated to incite the general reader to the study of geological science. (See address of Lord John Russell to the Mechanics of Leeds. Public Journals of 1853.)

* In an able memoir on the former glacial condition of the surface of the Highlands, by Mr. Robert Chambers, (Edin. Phil. Journ. April, 1853) this enormous denudation, first described by Dr. Macculloch in his Western Islands, vol. ii. p. 93., is hypothetically referred to the furrowing action of ice and glaciers. In this volume it is impossible to go into the discussion of that very recent geological cause—ice-action, of the effects of which there are no signs whatever in the primary or secondary or even in the older tertiary rocks. Admiring the ability and research displayed in the memoir of Mr. R. Chambers, I cannot, however, admit that this vast denudation of the Western Highlands can have been due to any operation of glaciers or icebergs; though very powerful debacles, due in great measure to the melting of ice masses, may have somewhat contributed to the final result.

Islands, and is partially seen in Rosshire* and Invernesshire, counties in which the lower conglomerates are most abundant. It was in one of the most dislocated and therefore most obscure of those tracts, the coast of Cromarty, that Mr. Hugh Miller detected his remarkable nondescript fishes, and, from a few fragments, so admirably assigned to them their peculiar attributes and their proper place in the animal kingdom.

On the south side of the Murray Frith, the mineral succession is somewhat varied by the interpolation of large masses of concretionary, greenish and reddish limestone or cornstone, in which, as well as in the conglomerates below,—and in the shale, flagstones and sandstones above them, — ichthyolites have been found, and in some places very largely.†

Now, among the fishes which most abound in the lower group, whether composed of conglomerate, sandstone, flagstone, schist, or clay, with calcareous nodules, or cornstone, one of the most remarkable is the Pterichthys, that singular winged fish ‡, of which a woodcut is here given.

FOSSILS, 47.

OLD RED SANDSTONE FISH.

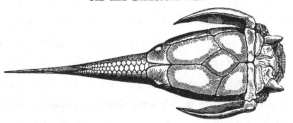

Underside of Pterichthys cornutus, Agassiz, from Morayshire.

* See Trans. Geol. Soc. 2d series, vol. iii. p. 125. with map of the Highlands.

† The late Lady Gordon Cumming, of Altyre, was the discoverer of many of these ichthyolites, and the exquisite manner in which that lady and her eldest daughter sketched and coloured them, is duly recorded in the pages of Agassiz's classic work. (See *Poissons du Vieux Grès Rouge, passim.*)

‡ See the striking description of this ichthyolite by its discoverer, Hugh Miller, " Old Red Sandstone," p. 46.

Another is the Cephalaspis, Ag., which, occurring, as before stated, in the English Old Red, abounds also in the Scottish rocks of the same age (see Pl. 37.). This fish, with its large buckler-shaped head and its thin body, jointed somewhat like a lobster, is perhaps the most remarkable example of a fish of apparently so intermediate a character, that the detached portions of its head when first found were supposed to belong to crustaceans.*

Then, there is the very singularly shaped creature the Coccosteus, Ag. One sgeciesis represented in the present woodcut,

FOSSILS, 48.

OLD RED FISH.

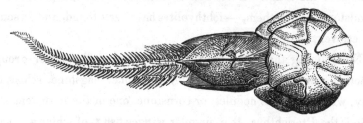

Coccosteus decipiens, Agassiz, somewhat restored.†

which also through the ingenuity of Hugh Miller was first put together from small detached pieces. That author has well said it required all the skill of Agassiz to determine that the uncouth Coccosteus and the equally uncouth Pterichthys, with their long articulated tails and tortoise-like plates, were bonâ fide fishes.

At the same time, no one could observe even fragments of many other fossils, such as those which were first published by myself from the flagstones of Caithness, and not recognize that they are parts of fishes. Such, for instance, are the Dipterus and Diplopterus

* Mr. Miller has requested his readers to compare the head of the Asaphus (now Phacops) caudatus, a well-known Silurian trilobite, with that of Cephalaspis Lyellii, to illustrate how the two orders of crustaceans and fishes *seem* here to meet,—in the view of persons who have not mastered the subject.— "Old Red Sandstone," p. 54.

† A perfectly restored head of this fish is given in Mr. Miller's eloquent work, the "Footprints of the Creator," p. 50. Edinburgh, 1850.

described by Cuvier, — such are the subsequently discovered genera, Osteolepis, Glyptolepis, Cheirolepis, Diplacanthus *, and Cheiracanthus, Agassiz, most of which have several species.† A drawing of one of the Caithness fishes is here given.

FOSSILS, 49.

OLD RED GANOID FISH.

Dipterus macrolepidotus, Ag. — from the black schists of Caithness.

The upper division of the Scottish Old Red, whether consisting of red and mottled marls, and sandstone beneath, or of overlying yellow and whitish sandstone, which, as before said, immediately supports the coal fields of Fifeshire, is chiefly characterized by fishes of the genus Holoptychius — the finest specimen of which ever seen is the H. Nobilissimus, first represented in Pl. 2. bis. of the ' Silurian System,' now in the British Museum, and named by Agassiz at my request after the discoverer, Mr. Noble.

In short, the Scottish Old Red Sandstone, as a whole, has afforded not less than sixty-five genera and species of fishes. Yet with this abundance of one class, we have in Scotland no traces of those marine mollusca and zoophytes which occur elsewhere, and which, in

* The vigilant eye of Hugh Miller detected a land plant, probably the oldest Conifer ever yet seen, in the lower Old Red at Cromarty, associated with scales of the Diplacanthus. "Footprints of the Creator," pp. 197. 199. The supposed fern (*ib.*, p. 194.) is from Orkney (see *ib.*, p. 254.).

† These Caithness fishes were first noticed in my memoir on the Brora coal, Trans. Geol. Soc. Lond. N. S. vol. ii. p. 314. They were afterwards partially described by Cuvier, and placed in their proper geological position by Professor Sedgwick and myself, Trans. Geol. Soc. Lond., vol. iii. 2d Ser. p. 142, Plates 15, 16, 17.

some countries, as will hereafter be shown, are commingled with fishes like those of Scotland, in deposits of the same age.

Shells, though very scarce, have, however, been found, and were indeed alluded to many years ago by Hugh Miller, as certain undescribed mollusks. The occurrence of better specimens and a minuter attention to them has led to the opinion that they had a freshwater origin, and may possibly be considered species of Cyclas. Again, the researches of the last two years have brought to light a still more remarkable and truer denizen of former lands in the small air-breathing reptile, Telerpeton Elginense, Mantell, of which a figure is here given.

FOSSILS, 50.
REPTILE OF THE UPPER OLD RED.

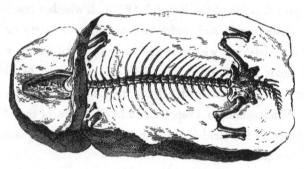

Telerpeton Elginense, Mant. (*Somewhat less than natural size.*)

Now, this the oldest known reptile, was found* in the light-coloured sandstones of Elgin, on the south side of the Murray Firth — strata which were classed by Professor Sedgwick and myself as the upper division of the Old Red Sandstone, and as being superior to the central flags with many ichthyolites. The same rock occupies a similar position in Ross, Caithness, and the Orkney Islands†, and the

---

* See Mantell's description of this animal, as found by Mr. P. Duff, of Elgin, and brought into notice in reference to the strata and to certain tracks of an animal in this sandstone, by Capt. Brickenden, Quart. Journ. Geol. Soc. vol. viii. p. 97.

† Trans. Geol. Soc. Lond. vol. iii. 2d Ser. p. 130. 1828.

same order of sequence, as in the mainland, has even been observed in
the Shetland Isles.   Hence it is inferred, that certain land plants *
recently brought from the sandstone of Lerwick in Shetland, and
which Dr. Hooker has referred to Calamites of species differing from
those of the carboniferous rocks, belong to the same formation as
the small reptile of Elgin and the freshwater shells alluded to.
Again, in the uppermost old red or Devonian of Ireland, the government surveyors have recently found at Knocktopher, in the county
of Kilkenny, large freshwater bivalves, and certain species of tree
ferns differing from any known to botanists in the overlying coal
formation.   In this Irish case there is no more ambiguity than in
certain Devonian plant-bearing beds, which will be described, in
Devonshire and other countries.   In fact, the Irish strata in question
rise out from beneath the very lowest strata referred to the carboniferous system, and besides the plants contain fossil fishes (Coccosteus), which unquestionably pertain to the Old Red in Scotland.
One of these plants, the Cyclopteris Hibernicus, E. Forbes, is
here figured, as belonging to one of the oldest of the clearly defined
tree ferns which have been discovered in the crust of the globe.

FOSSILS, 51.

OLD RED SANDSTONE PLANT.

Cyclopteris Hibernicus,   From specimens in the
Forbes. Proc. Brit. Assoc.   collection of the "Geol.
1852.   Survey."

* The plants in question were brought from Lerwick by the Right Hon. H
Tuffnell, and when presented by him to the Geological Society, I confirmed my
own views of their position, not only by reference to the work of Dr. Hibbert, but
by consulting Dr. Traill and the Rev. Dr. A. Fleming, both of whom have visited
Shetland, and are well acquainted with its rocks.   The latter had observed plants
many years ago in the sandstones of the Shetland Isles.   (See Quart. Journ. Geol.
Soc. Lond. vol. ix. p. 49., and H. Miller, on Plants, *antè*, p. 253.)

SECTION ACROSS NORTH DEVON.

a. Lowest beds of schist and red micaceous sandstone. *b.* Red sandstone and conglomerate. *c.* Grey schists and *Stringocephalus* limestone. *d.* Quartzose schists. *e.* Arenaceous rocks, red and brown. *f.* Upper Devonian bands, with *Clymenia* limestone. *g.* Trough of culm or coal, with *Posidonomya* limestone near the base. This limestone is overlaid by grits, schists, and culm; the whole representing the carboniferous rocks.*

The lowest of these Devonian rocks (*a*) are hard, close grained, greenish and reddish, silicious and slightly micaceous sandstones, by no means unlike parts of the old red of Pembrokeshire, on the opposite side of the Bristol Channel (see p. 246.). Among these are chloritic schists, occasionally calcareous, and enclosing corals and some shells, such as *Spirifer ostiolatus*, *S. apertwratus*, and *Orthida*; they are usually much distorted by slaty cleavage.* Then follow thick coarser beds, *b*, red and purple, which here and there pass into red conglomerates; whilst some beds are flag-like and spotted, and others are white sandstones. Some of them contain fragments of the older schists, and many of them are highly impregnated with red oxide of iron and hæmatite. "Considered as a whole, and from mineral characters," said Sedgwick and myself, "we might compare some parts of this group with the most characteristic portions of the Old Red Sandstone." (See Trans. Geol. Soc. Lond. N. S. vol. v. p. 145.)

These red rocks, *a*, *b*, which may be compared with the lower shelly greywacke of the Rhine (Coblentz, &c.), are of considerable thickness, and are distinctly overlaid on the south by grey coloured slaty schists, *c*, in which calcareous geodes are frequent, occasionally ranging through some thickness. One band of limestone at Combe Martin has the dimensions of sixty feet, and is overlaid by eight or nine other calcareous courses of smaller dimensions. They contain remains of fossils, including *Stringocephalus* and many corals, some species of which, such as the *Cyathophyllum cæspitosum*, and *Favosites polymorpha*, are well known elsewhere. Most of the fossils, however, are so disturbed by cleavage and pressure, that we can only speak confidently of the species when these remains are found in rocks of the same age in South Devon, where the limestones are purer, and the schists in a less crystalline state.

The calcareous group of Ilfracombe is succeeded on the south by hard, slaty grey, and chloritic schists, without limestone, *d*, and with many veinstones of quartz; these again are surmounted by softer schists, with reddish sandstone, and greenish grey and purplish flagstone. Then follow the arenaceous flagstones of Baggy Point and Marwood, *e*, which are much more earthy than the underlying rocks, and contain casts of fossils and remains of land plants. Here again we have a *partial return to the Red Sandstone character*; for the brown and grey flagstones really often pass into a red sandstone, hardly to be distinguished from the inferior red rock, *a*, *b*, of Linton; whilst others are of greenish grey and brownish sandstone, resembling varieties of the Upper Old Red Sandstone of Scotland. The calcareous group (*f*) of Pilton and Petherwin completes the Devonian series, and is surmounted by the carboniferous schists (*g*) with the *Posidonomya* limestone, the representative of the carboniferous or mountain limestone.

* See the able memoir of Mr. D. Sharpe, on the effects of slaty cleavage on the form and outline of several of the Devonian shells, Quart. Journ. Geol. Soc. vol. ii p. 74.

Devonian Rocks (the equivalents of the Old Red) in Devon and Cornwall. — The crystalline and slaty condition of most of the stratified deposits in Devon and Cornwall, and their association with granitic and eruptive rocks and much metalliferous matter, might well induce the earlier geologists to class them as among the very oldest deposits of the British Isles. In truth, the south-western extremity of England presented apparently no regular sedimentary succession, by which its grey, slaty schists, marble limestones, and silicious sandstones could be connected with any one of the British deposits the age of which was well ascertained. The establishment of the Silurian System, and the proofs it afforded of the entire separation of its fossils from those of the Carboniferous era, was the first step which led to a right understanding of the age of these deposits. The next was the proof obtained by Professor Sedgwick and myself, that the " culm measures" of Devon were truly of the age of the carboniferous limestone, and that they graduated downwards into some of the slaty rocks of this region. Hence, in the sequel it became manifest, that the rocks now under consideration, were the immediate and natural precursors of the coal era, and stood therefore in the place of the Old Red Sandstone of other regions. The highly important deduction, however, of Mr. Lonsdale, that the fossils of the South Devon limestones, as collected by Mr. Austen and others, really exhibited a character intermediate between those of the Silurian system and of the carboniferous limestone, was the most cogent reason which induced Professor Sedgwick and myself (after identifying North and South Devon) to propose the term Devonian.* The inference that the stratified rocks of Devonshire and Cornwall, though of such varied composition, are really the equivalents of the Old Red

---

* See Reports of Brit. Assoc. for the Advancement of Science, 1836. Bristol Meeting. Sedgwick and Murchison, Trans. Geol. Soc. Lond. vol. v. p. 633., and Phil. Mag. vol. xi. p. 311. Lonsdale, *supra*.

Sandstone in the regions alluded to, has since, indeed, been amply
supported and extended by the researches of Sir Henry De la
Beche, Professor Phillips, and many other good geologists.*

The most instructive of the sections published by my colleague
and myself to illustrate the general structure of Devonshire, is that
of which the diagram in p. 256. is a compiled reduction.† It is a
section across North Devon from the Foreland on the Bristol Channel, to the granitic ridge of Dartmoor on the south, and exhibits a
copious succession of the Devonian rocks between Linton and Ilfracombe on the north, and Barnstaple on the south; the whole
dipping under strata of the carboniferous age‡, on the opposite side
of a wide trough of which, or on the north flank of Dartmoor, the
Upper Devonian strata again rise to the surface.

North Devon has thus been selected as affording, on the whole, the
best type of succession of the rocks to which the name Devonian was
applied; because it offers a clear ascending section through several
thousand feet of varied strata, until we reach other overlying rocks,
which are undeniably the bottom beds of the true carboniferous
group.

For, whether we advance to the south of Barnstaple from the
north, or to the north of Petherwin from the edge of Dartmoor on the
south (section, p. 256.), we find ourselves in a wide §, overlying trough
of much softer nature, and in which the slaty character does not

* See Report on the Geology of Cornwall, Devon, and W. Somerset, by De la
Beche, 1839, and the Palæozoic fossils of the same region,byProfessor Phillips, 1841.

† See Trans. Geol. Soc. Lond. 2nd Ser. vol. v. pl. L. figs. 1. and 2.

‡ The coal-field which is bituminous in Monmouth, Glamorgan, and Carmarthen,
becomes anthracitic in Pembroke. There the stone coal, much fractured by dislocations, differs from that of Devon only in being more productive in broken
culm or stone-coal.

§ The thickness of these lower carboniferous strata must not be estimated by the
breadth which they occupy on a geological map: for, owing to countless convolutions, the very same beds are repeated over and over. The bottom beds only of
these undulations or small portions of each side of the culm trough (g) are represented in the section.

exist. And, although this overlying series is in mineral aspect as much unlike the carboniferous series of most other parts of Britain, as the rocks of North Devon are unlike the ordinary Old Red Sandstone of Central and North Britain, we have clear proofs, besides the analogy with Pembrokeshire before spoken of, that the Posidonomya black limestone, or the calcareous band (in *g* of the previous section), does represent, on a miniature scale, the mountain or carboniferous limestone; that the next series of white grit and sandstone stands in the place of the millstone grit; and that the overlying courses of stone coal or culm, with many remains of plants, are consequently the equivalents of some of the coal-bearing strata of other tracts described in the next chapter.

The objections, therefore, (which have, however, been only very partially made) to the view taken by Professor Sedgwick and myself, that the Devonian rocks of the above section are the true representatives *in time* of the vast deposits of Old Red Sandstone of Scotland and England, seem to me to be quite untenable. In truth, the long period which was occupied in developing those enormously thick deposits must have produced elsewhere equivalent accumulations; and the vast series of North Devon thus occupies precisely the position of a succession of strata synchronous to that of the Old Red Sandstone. And, as every geologist knows, that the crystalline feature of slaty cleavage has been impressed upon these Devonian rocks long after their formation, so must he also admit, that the change from the red sand and mud of Hereford, Brecknock, and Caermarthen to the red and grey mud of Devon, is by no means abrupt, but resembles the gradual change shown in Pembroke. Nor, is there any difficulty in supposing, how by a less diffusion of iron, and under dissimilar submarine conditions, the southern portion of the area of the very same sea should have less of the red and sandy character than the northern.

The great eruption of the granite of Dartmoor, which affected both the Devonian and carboniferous strata in contact with it, has so

usurped the place of the regular deposits in South Devon, that in
vain do we look, either in that district or in Cornwall, for the same
clear order as in North Devon. In fact, the derangement in the
western part of South Devon and the adjacent parts of Cornwall is
so great, that the Lower Silurian rocks are seen, as already stated,
to overlie true Devonian rocks!* The metamorphism and mineralization of some of the schists has indeed been carried to such an
extent, that they often resemble the oldest primary rocks.

Still, there are adequate means of bringing the disjointed and
occasionally inverted masses of South Devon into comparison with
the clear order of North Devon.

This is in great measure accomplished through the fossils† of its
extensive lower limestones, being much more numerous and better
preserved than those of like age in North Devon. The lowest limestones, for example, which are on the parallel of those of Combe
Martin and Ilfracombe (*c* of the long section), and rise in great
masses near Plymouth and Ogwell, ranging at intervals to Newton
Bushell and Torquay ‡, are laden with corals and shells, many species

---

* See p. 145. and Quart. Journ. Geol. Soc. vol. viii. p.13.

† The differences between the *mineral* succession in North and South Devon,
on the opposite sides of the great granitic axis of Dartmoor, are explained by
Professor Sedgwick and myself, Geol. Trans. vol. v. p. 3. p. 635.

‡ One of the finest collections of the S. Devon fossils was made by Mr. R. A. C.
Austen, whose researches in the field, and whose study of the organic remains,
have materially contributed to a correct acquaintance with the stratified rocks
of Devonshire. See Lonsdale, Trans. Geol. Soc. Lond. vol. v. p. 721.

In a recent Memoir on the Palæozoic rocks of the Boulonnais (Quart. Journ.
Geol. Soc. vol. ix. p. 244) ; Mr. Austen classes the Petherwin and Pilton beds,
like myself, with the Devonian rocks, whilst he separates the S. Devon limestones
of Newton and Ogwell from those of Ashburton, Bickerton, and Chudleigh. If
in this popular work I retain the older view, and group together the S. Devon
limestones, I would in no respect detract from the value of such a subdivision.
My present belief is, indeed, that the lower sandstones, conglomerates, and
slates (or the *a b* of the previous section), are truly the equivalents of the
Lower Rhenish or Coblentzian, shelly, greywacke sandstone ; thus completing
the parallel between the British and Rhenish Devonian rocks, and giving to each
a similar base. In respect to the Upper Devonian division, all foreign geologists

of which occur in rocks of the same age in various parts of the continent of Europe, and notably in the limestones of the Eifel, the Rhenish provinces, and Belgium.   Now, many of these fossils are quite peculiar: for whilst they exhibit an intermediary character, approaching in the lower beds of this series to those of the Silurian System, and in the upper strata to those of the Carboniferous era, there can be no doubt that on the whole they constitute an independent group.   Among the most typical of these peculiar fossils, and which, wherever they may be found, mark the strata in which they occur as being of Devonian age, are the large mollusks Stringocephalus giganteus, Foss. 52. f. 4., the Megalodon cucullatus, f. 2., and Calceola sandalina, f. 1., represented in this woodcut, together with Pleurodictyum problematicum, Cyathophyllum cæspitosum, and Murchisonia bilineata, f. 3.

FOSSILS, 52.

SHELLS OF THE LOWER DEVONIAN LIMESTONES.

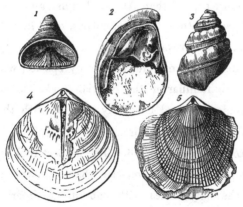

1. Calceola sandalina, Linn.   2. Megalodon cucullatus, Sow.   3. Murchisonia bilineata, Goldf.
4 Stringocephalus giganteus, Sow.   5. Atrypa desquamata, Sow.

who classify by organic remains, have placed the Clymenia and Cypridina limestone in the Devonian system, and I must therefore entirely dissent from a recent proposal of Mr. D. Sharpe, to remove it to the base of the carboniferous rocks.   But, if an *intermediate* band like this may in one region present more of a carboniferous type than in another, and therefore admit of much liberty in drawing the line

All the species of Trilobites, known in the Silurian system, have disappeared, and their places are taken by others, of which Brontes flabellifer, Goldfuss, and Phacops latifrons, Bronn, are striking types. In short, Trilobites, which swarmed in the Silurian era, are comparatively very scarce in the Devonian; two or three only of the very numerous genera of the former era being known in it.

Among the mollusca, nearly all the species of Atrypa, Orthis, and Spirifer, differ from those of the Silurian era. One or two exceptions only occur. Great changes in the proportional number of species in such genera also take place, and certain genera which were common in the older period are no longer traceable. Thus, the genus Orthis becomes infinitely less abundant, whilst the Spirifer, comparatively rare in the older rocks, augments much in number of species, and especially in the size of the shell—the large broad winged species being singularly characteristic of rocks of this age, and particularly in the Lower Devonian of several foreign countries. But no form in Britain, is more typical than Atrypa desquamata, Foss. 52. f. 5.

One shell, however, the Atrypa reticularis, must here be mentioned as an exception to the prevalent rule of each great group being characterized by distinct forms; for this hardy species, with which the reader became so familiar in the Silurian rocks (see p. 188), lived on to the Devonian era, and is equally common with the last in the limestones and shale of Devonshire.

To return to the section, p. 256. The highest rock which is there classed as Devonian, and is, on the whole, a mere upward continuation of the slaty series with subordinate inferior sandstone,

of separation, I can still less admit the comparison sought to be established by the same author, between a few unfossiliferous beds of red sandstone, shale, and conglomerate, which at Pepinster, in Belgium, form a *very small* portion indeed of the series of Devonian rocks, and the enormous deposit of the Old Red Sandstone of the British Isles, which, without any break, occupies the vast interval between the Silurian and Carboniferous Rocks. This point will be further explained in describing the Rhenish Devonian Rocks. See Sharpe, Quart. Journ. Geol. Soc. vol. ix. p. 23.

is occasionally also calcareous, presenting thin courses of limestone, in which many fossils occur.   This is the band of Pilton and Barnstaple.   Now, whilst the lower sandy strata of this division, as at Marwood and Baggy Point, contain Dolabra (Cucullæa) Hardingii, Foss. 53. f. 2., and D. trapezium, Sow.; Bellerophon subglobatus, M'Coy, with casts of Stigmaria and other land plants, — the upper or calcareous part is charged with small, true, Devonian Trilobites, such as Phacops latifrons, Bronn; together with Orthis, Pleurotomaria, Spirifer, Terebratula, and Productus, few of which occur in the Carboniferous Rocks.   Here again, though with considerable mineral variations, we see the same upward succession as in Scotland and Ireland, and in approaching the summit of what has been classed as Devonian or Old Red, we are gradually introduced to the land vegetation of the Carboniferous era.

This uppermost course of impure Devonian limestone is represented on the south side of the great trough of culm or coal measures, *g*, as expressed in the previous section, by the band of Petherwin, *f*, in which a great number of fossils have been found, and these clearly identify it with the Upper Devonian of many parts of the continent of Europe.   Such are the Phacops granulatus, Foss. 53. f. 5.; Clymenia linearis, f. 1.; C. lævigata, Münster; Spirifer Verneuili (disjuncta, Sow.), f. 4. &c.

Fossils, 53.

Shells of the Upper Devonian Limestone.

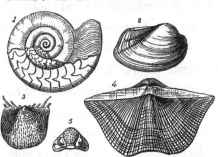

1. Clymenia linearis, Münst.   2. Dolabra Hardingii. Sow.   3. Strophalosia caperata, Sow.
4. Spirifer Verneuili, Murch.   5. Phacops granulatus, Münster.

This Upper Devonian of South Britain, which occupies, in my opinion, the same place in the geological series as the uppermost red sandstones flanking the Silurian region, — the yellow sandstone of Fifeshire with shells and plants,— and the sandstones with the Telerpeton in the N. of Scotland — is, I repeat, well exposed in the beds of Petherwin, Pilton, and Marwood, N. Devon (f of the section). In addition to Phacops granulatus, Clymeniæ, and other fossils, above mentioned, this band is also marked by the presence of a minute crustacean, Cypridina serratostriata, Sandb., of which I shall have to speak in treating of the Devonian rocks of Germany. The discovery in Britain of this small crustacean was not made until after the distinguished palæontologists, the brothers Sandberger, had found it in myriads occupying the upper schistose and calcareous rocks of the Rhine, which Professor Sedgwick and myself had long ago shown to pass immediately under the lowest carboniferous deposits, and to be the true equivalents of our rocks of Devonshire. The observation of M. F. Sandberger (for it was he who first detected this Cypridina in British rock specimens sent to him) has therefore been peculiarly valuable, as, through the presence of this minute but very characteristic crustacean, we now learn more conclusively, that the limits of the Devonian rocks in S. Britain have been correctly defined; their equivalents in Germany being similarly so distinguished.

In subsequent chapters we shall see that analogous divisions of the Devonian Rocks are also developed in different parts of Europe, and that in the Rhenish provinces especially, a triple division may even be preferred. It will further appear, how, in other tracts, but more particularly in Russia, the *ichthyolites of the Old Red Sandstone of Scotland and the marine mollusca of Devonshire are found united in the same strata:* thus demonstrating the simultaneous accumulation of deposits, which, although they differ considerably in mineral aspect, occupy precisely the same place in the series of deposits.

In having previously alluded to a shell found in a given tract of the Old Red of Scotland, and to others in the S. of Ireland, as being of fresh-water origin, we must, however, only consider them as exceptions to the general character of deposits which are essentially marine in other parts of the world. It is, in fact, chiefly the upper member of the series, which affords evidence of the spoils of the adjacent land having been swept into estuaries and seas.

In quitting this brief consideration of the highly diversified and important group of the Devonian or Old Red rocks of the British Isles, it must not be forgotten, that whilst some of its lowest members have rarely afforded traces of land plants, its uppermost members in Ireland, as in Devonshire and Shetland, contain a greater number, including tree ferns and calamites. We must equally remember, that the latter plants are associated, in Scotland and Ireland, with ichthyolites of the Devonian epoch, and occur in strata which rise out from beneath the lowest beds of the carboniferous rocks. Nay, more, it is in strata of this age that we also first meet with that assemblage of natural products which might be expected; for the most ancient fresh-water shells and the oldest air-breathing reptile which have been discovered, are the legitimate associates of this early land vegetation.

Certain terrestrial plants of this period will be afterwards adverted to as occurring in Germany. In fact, the uppermost member of the Devonian and the lowest division of the Carboniferous are sometimes so linked together, that where the carboniferous limestone is wanting, it is no easy matter to draw any positive line of separation between these groups. Such results must, indeed, have invariably followed from the ordinary operation of natural causes, wherever great subaqueous deposits were quietly and successively accumulated.

In terminating this chapter the reader must be reminded, that although there is no Upper Silurian in the south-western extremity

of England, Lower Silurian strata have been adverted to, as ranging
along a few of the southern headlands of Cornwall, and as being at
once flanked by Devonian strata (p. 144.). In subsequent chapters it
will be shown, how Lower Silurian rocks are similarly followed by
Devonian formations over very large portions of the continent of
Europe, to the almost entire exclusion of the Upper Silurian.

My friend, the Rev. John Fleming, D.D., to one of whose original papers
on the organic remains of the Old Red Sandstone I have already referred, has
recently assured me, that the fruit-like body found in Fifeshire and in the Arbroath
paving-stones of Forfarshire, (see p. 247.) is unquestionably a vegetable, and
cannot be classed as the egg of a mollusk, as suggested by Lyell, or as the egg of a
batrachian, as more recently proposed by Mantell.   The figure which Dr. Fleming
gave in the year 1831, Edin. Journal of Nat. Science, vol. 3, pl. 2. fig. 5., is in
itself pretty good evidence in favour of his opinion, that it is an aggregate fruit.
Since then he has examined both the upper and under surfaces, and has quite
satisfied himself that it is a receptacle, and that the round bodies covering it are
carpels or fruits.   In the above-mentioned Memoir, Dr. Fleming first pointed out
certain terrestrial and freshwater conditions of the Upper Old Red Sandstone of
Scotland.

CHAPTER XI.

CARBONIFEROUS ROCKS.

GREAT PRIMEVAL FLORA THE SOURCE OF THE OLD COAL DEPOSITS. — GENERAL
VIEW OF THESE DEPOSITS AND THEIR ORGANIC REMAINS IN THE BRITISH ISLES.

ASCENDING in the scale of deposits, we have now reached another
grand accumulation of strata, which is not only replete with many
types of animal life peculiar to it and unknown in antecedent periods, but specially characterized by the earliest abundant remains
of a terrestrial vegetation. For, the reader will remember, that
there are no traces of land plants in the great mass of the Silurian
rocks; and that it is only where their uppermost strata unite with
the Devonian, that certain traces of land plants have been sparingly
detected. Even in the Devonian rocks such plants, as before stated,
are rare, and only begin to prevail as we pass upwards and are surrounded by the spoils of this first great woody era.

Now, as these primeval plants were the substances out of which
the great mass of the mineral termed coal has been formed (all naturalists and chemists admitting the fact); so, for the first time in mounting up from the basement rocks, do we meet with those copious accumulations of that mineral in conjunction with the impressions and
casts in stone of the numerous plants out of which it has been formed.
Some idea of the characters of the rank, luxuriant vegetation which
must in this age, have overspread very wide areas of land, from
polar to nearly equatorial latitudes, may be formed by inspecting the
annexed woodcut, in which an ideal representation is given of a portion of the earth's surface, as clothed with plants, the fragments of

which bespeak the presence of rich, vascular cryptogamia, whose
fossilized stems and leaves are frequently to be detected even in the

IDEAL VIEW OF THE VEGETATION OF THE CARBONIFEROUS ERA.

coal itself. In the standard work of Bronn, von Meyer, and Göppert, which gives the most complete, general tabular view of ancient
nature hitherto published*, Professor Göppert estimates the total
number of known species of fossil plants of what is here considered
the carboniferous æra as 934, which are thus distributed: —

PLANTÆ 934.
{
Cellulares, including the Fungi, Algæ, &c., 19.

Vasculares, 915; of which 772 are cryptogamous
plants, or Ferns, Calamites, Asterophyllites, &c.;
and 94 are dicotyledonous plants, such as Cycads,
Conifers, and Club-mosses.
}

The result, arrived at by this fossil botanist, agrees generally with
that of his precursor in this line of inquiry, M. Adolphe Brongniart,
who first gave to the world a general and philosophical view of the
distribution of former vegetation. On his part M. Göppert has not

---

* Bronn's Geschichte der Natur, vol. iii. part 2. The so called "Transition"
plants of Göppert are included in this list; because it has been ascertained, that
nearly all of them occur in strata which, formerly viewed as ancient "*grauwacke*,"
are now known to be of no higher antiquity than the lower division of the carboniferous rocks.

only added considerably to the number of species, but also to the number of dicotyledonous or forest trees.

Both, however, of these eminent men, as well as many other botanists are agreed in the opinion, that as the great mass of the plants belong to the vascular, cryptogamic class, the conditions of climate under which such vegetables grew in very various latitudes, is a phenomenon *per se*, and especially characteristic of that period which geologists call "Carboniferous." These plants, marking, as above stated, the earliest great vegetable period, are, in fact, as decisive of the Carboniferous as the great abundance of Trilobites and Graptolites is of the Silurian era. But, whilst the earlier system is marked only by its submarine contents, the carboniferous deposits owe their chief features to the actual presence or contiguity of lands covered with a peculiar vegetation, which, disappearing with the younger primeval strata, was never afterwards reproduced upon the earth. For, no one of the floras of subsequent geological periods possessed those characteristic features of rank, gigantic cryptogamia, indicative of an inter-tropical climate, which so prominently marked the vegetation of the epoch which geologists term *the* Carboniferous.

In most of their lithological characters the successive strata of the Carboniferous rocks do not differ essentially from many which have preceded them. Like the Silurian and Devonian, they contain beds of shale, sandstone, pebbles or conglomerate, and limestone, though they seldom exhibit a true slaty cleavage.* But, when examined more closely, they vary considerably in their nature and contents in different parts of the world. In subsequent chapters a few allusions will be made to these variations in other countries; but for the present we will only take a cursory view of some of their features in Britain.

In the region represented in the map annexed to this work the Car-

---

* In parts of France the carboniferous rocks are very crystalline and slaty, as I have shown in a memoir on the environs of Vichy. Quart. Jour. Geol. Soc. Lond., vol. vii. p. 14.

boniferous rocks are most fully developed in the great South Welsh basin of Glamorgan, Caermarthen and Monmouth, where an ascending order from the summit of the Old Red Sandstone, through limestone, shale, sandstone and grits, upwards into an enormously thick coal field, is clearly exhibited in lofty escarpments, particularly around the northern, eastern, and western edges of that grand basin. (See lower section, p. 243.) The same succession, though on a smaller scale, is seen around the smaller basin of the Forest of Dean, and again, with certain mineral variations, in the county of Pembroke.

Towards the north, however, the calcareous and lower members of the series are more developed, than in South Wales, as observed in the Oswestry coal field, and in Shropshire and Flintshire. This expansion of the inferior strata of the system towards the north, becomes still more striking in Derbyshire, Yorkshire, and Northumberland.

In some of the coal tracts within or adjacent to the Silurian region, as at Dudley and Wolverhampton, the true base of the group, or the carboniferous limestone, is wanting; the productive fields being there seen to repose at once on Silurian rocks (see woodcut, p. 465.). In those districts, however, and along the banks of the Severn, good evidences are obtained of the relations of the coal strata to the overlying red deposits now termed Permian, and of which hereafter.

When viewed, therefore, as a whole, and *in the region of the coloured map*, the Carboniferous group (*b* to *g*) may be stated to lie between the Old Red *a*, and the Permian rocks *h*, thus:—

GENERAL RELATIONS OF THE CARBONIFEROUS ROCKS IN THE CENTRAL AND SOUTHERN PARTS OF ENGLAND.

a. Upper beds of the Old Red Sandstone (Devonian).   *b*. Limestone shale.   *c*. Carboniferous limestone.   *d*. Millstone grit.   *e*. Coal and ironstone.   *f*. Main coal fields.   *g*. Upper coal with a peculiar limestone.   *h*. Red sandstone (base of Permian rocks).

As on this occasion it is impracticable to attempt to treat in detail of the various aspects of these carboniferous deposits, strictly so called, in different British districts, a very slight sketch only of some of their striking peculiarities can be attempted.

Separated from the upper band of the Old Red Sandstone by beds of dark and party-coloured shale (*b* of the section), the limestone (known as the mountain limestone of geologists *c*,) is the dominant rock of the lower division.   The massive nature of these calcareous rocks and their vast development in Derbyshire, Yorkshire, and Northumberland, are well known to geologists.   Even within the region of my original Silurian map, this limestone when traced along the rim of the great South Wales coal field, or from Caermarthen into Pembrokeshire, is seen to be there exhibited continuously, in bold coast cliffs, particularly in the promontory of Stackpole, where it is much contorted, as in these two sketches.   The first of them represents the contortions of the rocks, with their numerous clefts

STACKPOLE CLIFFS OF CARBONIFEROUS LIMESTONE.
Sketched by Lady Murchison.   From Sil. Syst. p. 382.

and caverns, as seen from Bull Slaughter Bay.   The next is a view of one of the detached folds of the same limestone, known as Stackpole Rock.   Though these cliffs seldom exceed 150 feet in height,

yet, being precipitous and abrupt, they present a very rugged, wild,
and picturesque barrier to the sea.

STACKPOLE ROCK.*

Sketched by Lady Murchison.   From Sil. Syst. p. 370.

Whilst the carboniferous limestone is separated, as before stated,
from the Devonian, or Old Red, by shales (which in Pembrokeshire
are in parts both sandy and calcareous), it is there, as in most other
districts of England, surmounted by light-coloured sandstones of considerable thickness, which are known under the name of millstone
grit.   The following section of the general succession of the strata in
Pembrokeshire exhibits, at one view, the whole order from the Silurian rocks up to the productive coal above the millstone grit.

GENERAL ORDER IN PEMBROKESHIRE.

Coal or Millstone Carb.   Old Red Sandstone.   Silurian.
culm.   grit.   limest.

* These striking cliffs are the property of my esteemed friend the Earl of
Cawdor, whose chief residence, Stackpole Court, is situated on the carboniferous
limestone, and whose other residence, in Wales, Golden Grove, is on the Llandeilo
formation of the Lower Silurian rocks, (see distant edifice, sketch, p. 69.).

To pass, however, from a diagram of the general order to a natural section, the English student may examine with ease and in a small compass the Carboniferous series from its base upwards to the productive coal inclusive, on the south-east slope of the Titterstone Clee Hill in Shropshire.* Ascending from a depression in the Old Red Sandstone near Cleobury to the eastern summit of that hill, he will pass successively over the three divisions of limestone and shale, millstone grit, and productive coal; the whole being capped by basalt, which is seen to have been erupted through the entire series, and to have overflowed on the top of the hill.

SECTION ACROSS THE CORNBROOK COAL BASIN OF THE CLEE HILLS.

(From Sil. Syst., p. 113. pl. 30. f. 6.)

a. Upper beds of Old Red or Devonian. *b.* Carboniferous limestone and shale. *c.* Millstone grit. *d.* Coal measures. * Eruptive basalt, which has risen through and overflowed the coal.

The lower Carboniferous members, or the shale, limestone, and grit (*b, c*), are, however, of comparatively small dimensions in and around the Silurian region, when compared with their representatives in Derbyshire and the north of England. In that territory, particularly in the west of Yorkshire and Lancashire, they swell out into a vastly thicker series. In ascending order those strata consist of the great scar limestone, estimated to vary in thickness from 500 to 1,000 feet; and that mass is followed, in the west of Yorkshire, by considerable alternations of shale and limestone, considered by Pro-

* Any one who is desirous of understanding the peculiarities of this coal-field may refer to the "Silurian System," p. 113., where the proofs of the existence of a vertical mass of basalt (*) rising up through the strata are explained, as well as the extraction of coal by shafts from beneath the overlying table of basalt.

fessor Phillips to be about 1,000 feet thick; while the whole is surmounted by 800 feet of millstone grit.*

Again, besides its grand protrusions of basalt and greenstone, the
bold coast of Northumberland and South Berwickshire exhibits these
lower carboniferous limestones opening out into different courses and
interlaced by copious masses of schist and several coal seams. One of
the calcareous bands, near to the centre of the group, is especially
characterized by Posidonomya Becheri (Bronn), a shell which
marks the thin courses of black or culm limestoné in Devonshire.
This fossil, which is also of frequent occurrence in the schists and
'kiesel-schiefer' of the Rhenish provinces of Prussia, both with and
without limestone, is therefore a good type of the lower carboniferous
age; though in tracts where the limestones expand, it occurs in
one of their upper bands. It was by such proofs, and by the relations
of the strata, that Professor Sedgwick and myself showed, that the
Rhenish schists with Posidonomyæ were the exact equivalents of
the British carboniferous or culm limestone of Devon, and that
the sandstones which overlie them (the 'Jungere Grauwacke,' or
'Flötz-lehrer Sandstein' of the Germans), are the representatives
of the British millstone grit.† It is in this lower division of the
Carboniferous rocks, as composed of schists, and some limestones with
Producti and other fossils, and a great underlying sandy series, that
the chief Scottish coal-fields are developed. They are truly a continuation of those beds of North Northumberland, which extending
into Berwickshire, seem there to graduate downwards into the Old
Red Sandstone.

Having alluded to the convolutions of the Pembrokeshire limestone, it is right to explain, that the violence of the movements which

---

* See the Memoirs of Professor Sedgwick, "On the General Structure of the
Cumbrian Mountains," and "On the Carboniferous Chain from Penyghent to
Kirkby Stephen." (Trans. Geol. Soc. Lond. N. S., vol. iv. pp. 47. *et seq.*; also
Phillips, Geology of Yorkshire, vol. ii.)

† See Trans. Geol. Soc. London, N. S., vol. vi. p. 228. *et seq.*

produced them, affected likewise most remarkably all the Carboniferous rocks of that region, and particularly the coal.   Thus the coal
which is there nearly all in the state of stone-coal, culm * or
anthracite, has been for the most part shivered into small fragments,
and is frequently accumulated in small troughs or hollows, the
' slashes ' of the miners.   Of the great lateral pressure and violent
fractures to which the strata have been subjected, the following
woodcut may convey some idea.

Slash of Culm.

(From Sil. Syst., p. 377.)

a. Contorted culm strata with stone coal. *b.* Fault. *c.* Slash of finely triturated culm between
violently contorted strata, and probably upon a great line of fracture.

These great disturbances were produced after the accumulation
and solidification of the deposits under consideration.

Lower Carboniferous Rocks in Ireland.—There is, perhaps, no
country of the same size in which the carboniferous rocks are so
widely extended and at the same time so little productive of coal as
Ireland.   This is doubtless in some measure due to the fact, that by
far the greater portion of the Irish Carboniferous series of strata
pertain to those lower divisions, which, though very rich in coal
in Scotland, exhibit little or none of it in those parts of England
which lie in the same latitudes as the mass of the Irish rocks of

* The variations in mineral character of the anthracitic or culm trough
of North Devon, which was doubtless at one time a mere extension of the
Pembrokeshire strata of the same age, have been partially adverted to in the
previous chapter. (See p. 260, and section, p. 256.)   One thin course of black
limestone with Posidonomyæ is there the true representative of the great massive calcareous cliffs of the opposite coast of Pembroke, and of the diversified
North British Series; so much for great mineral changes of the same rocks in
our own isles! (See pp. 256. *et seq.*, and 271. *et seq.*)

this age. The great overlying and productive coal-fields of England, therefore, either never existed in Ireland, or have been removed by denudation. Exclusive of the strata which are to be separated as being of Devonian age (see p. 255.), the 'yellow sandstone' of Griffith, so defined where it is inter-stratified with courses of limestone and shale with Carboniferous fossils, is the oldest member of the system. This is followed by shale or schist, which assumes a slaty aspect, particularly in the Cork district, and alternates with other courses of limestone. The middle portion of the system is composed of the great lower limestone, followed by the dark grey, earthy limestone, known under the name of 'Calp,' whilst the 'Upper Limestone' of the midland and southern districts of Ireland is covered by the 'Millstone Grit.'

This great succession of limestones and shales, the latter occasionally affected by a slaty cleavage, is seldom, indeed, seen in one clear and consecutive section, but undulating over very wide areas, is usually much obscured by gravel, shingle, and clay. When, however, the limestones form natural escarpments, as, for example, on the south bank of Lough Erne, and in the hills near Florence Court to the south of Enniskillen, where they exhibit a splendid succession and numerous fossils, including large Producti in the lower and Posidonomyæ in the higher strata, they contain no seams of coal worthy of notice. In the Cork district, indeed, (where the lower schists are so slaty and crystalline, that they might be mistaken for much older rocks,) some coal has been found; but the quantity and value of it are little known.

The limestone series of Ireland is proved to be of exclusively marine origin, from the multitude of well-preserved fossils it contains; and of these Mr. Griffith (with the assistance of Professor M'Coy) has already prepared a list of about 500 species *, a consider-

---

* A Synopsis of the Characters of the Carboniferous Limestone Fossils of Ireland, 1844.

able number of which are stated to be identical with those of the
underlying and Devonian rocks, but altogether different from those
of the Silurian system.   These lower members of the system, which,
unluckily for Ireland, constitute, as before stated, by far the larger
portion of all that is called carboniferous, are surmounted (but very
partially) by grits and sandstones, in which, at three or four localities
(Kilkenny in the south, and Ballycastle, Dungannon, and Coal Island
in the north), a few thin beds of coal are situated; but as they have
as yet proved to be of comparatively slight value, and have no special
bearing on the object of this work, I forbear to say more respecting
them.*

Overlying productive Coal-fields. — The lower Carboniferous series
hitherto spoken of, or the shale, limestone and millstone grit, is all
over Europe and North America essentially of marine origin.   It has,
indeed, affinities to the upper portion of the underlying Devonian
rocks, into which it graduates.   It is, however, distinguished from
them by its chief animal remains, including many remarkable fishes;
and, above all, by the intermixture of terrestrial plants in an infinitely
greater quantity than are known in the preceding epoch.   With
the repeated evidences of thin seams of coal being intercalated with
bands of limestone of exclusively marine or estuary character, it is
indeed a fair inference, that the vegetables (often in great matted
bodies) out of which such lower coal was formed, were transported
from the mouths of great rivers, or broke away from the shores of
broad jungles, into the adjacent waters, and so together were associated with marine remains.

---

* I must not, however, omit to do justice to a recent spirited endeavour of the
Marquis of Downshire, who, discovering a fine mass of rock-salt and gypsum on
his estates near Carrickfergus, sunk a shaft through it, which, when I visited the
spot in 1852, had traversed a considerable thickness of the New Red formation; it being expected that coal in a good and unbroken state might be found
underlying this, the only extensive mass of New Red Sandstone which had then
been found in Ireland.

When, however, we examine the nature of the great overlying coal strata, particularly those of England, another order of things is opened out to us.

In most of these, whether in the South Welsh basin, or the northern and Durham fields, we lose all traces of marine life, and can recognize only huge accumulations of terrestrial, lacustrine or fluviatile origin. Thus in South Wales, where the coal measures are estimated to attain the great thickness of 12,000 feet, and one hundred coal beds are intercalated at various levels, we have undeniable evidence of successive terrestrial conditions—each of these coal-seams having immediately beneath it a band of sandy shale, called under-clay, and abounding in Stigmaria, or the *roots* of Sigillaria, one of the plants out of which coal has been generated. For this important fact, science is chiefly indebted to Mr. Logan, who demonstrated that the under-clay of the miner was the real soil of a primeval marsh or jungle. (See also Mammatt's Geological Facts, &c. 1834, p. 73.)

The diagram in the beginning of this Chapter is intended to convey a general idea of the nature of the wet and swampy tracts in which the vegetation of this period flourished.

The comparative rarity of true dicotyledons or forest trees in this flora, is as remarkable as the extraordinary uniformity in the families of plants of which it was composed. These consisted chiefly, as before said, of Lepidodendra, Equisetaceæ, Asterophyllites, and Sigillariæ, with some curious extinct pine-trees, a few of which resembled the living Araucaria of Norfolk Island. As many of these are found in the roof of the coal, and sometimes even in the coal itself*, there can remain but little doubt, that, in most of these cases,

---

* Few persons, who are attentive observers of the contents of their cellars, will have failed to detect the forms of plants in the coal itself. Göppert has, indeed, recently demonstrated, by microscopic examination (and Witham in part anticipated him, "Observations on fossil Vegetables," 1833), that the vegetable fibres and tissues of all the families of the plants of this era are to be detected in the coal itself. In some layers all the plants are calamites, in others, ferns. (Quart. Journ. Geol. Soc. Lond., vol. v.; Mem. p. 17.)

the mineral resulted from the fossilization of extensive ancient jungles or marshes. It may fairly be inferred, that this conversion of vegetables into coal, took place at a period in the formation of the crust of the earth, when very different physical conditions prevailed, and when a warmer and more equable (though probably not a hot) climate pervaded our islands, as well as latitudes far to the north and south of them. The supposition of many and successive subsidences of vast swampy jungles beneath the level of the waters, best explains how the different vegetable masses became covered by beds of sand and mud, so as to form the sandstone and shale of such coal-fields. But this theory of oscillation, or of the subsidence *en masse* of ancient marshes and their re-elevation (with occasional sand-drifts), though good in such examples as those of the South Welsh and Newcastle coal-fields in England, as also of the large coal-fields in British North America, to which Sir C. Lyell has recently called attention *, can have little application to those other seams of coal which, as before mentioned, are interstratified with beds containing marine shells, the animals of which, such as Producti and Spirifers, must have lived in comparatively deep sea-water.

In such examples (and nearly all the older coal beds come into the category) we may, on the contrary, endeavour to explain the facts by the supposition, that ancient streams like the present Mississippi and other large rivers, which flowed through groves on low lands and mud banks, transported great quantities of trees, leaves, and roots entangled in earth, and deposited them at the bottom of adjacent estuaries, or that they were carried *en masse* into the broad, open sea. (See Postscript, p. 288.)

The coasts of Northumberland and Berwickshire exhibit fine proofs of such conditions; but the most remarkable confirmation

* See an excellent general sketch of the chief carboniferous deposits of Europe and America, with illustrations of the prevailing plants, in the 24th and 25th chapters of Lyell's Manual of Geology, 1851. See also Ansted's " Ancient World."

of this view of the method by which some of the older coa has
been formed, which ever fell under my own observation, occurs in
the Russian field of the Donetz, in the Southern steppes. There,
besides numerous beds of shale and sand, with remains of plants,
bands of limestone, charged with species of Productus, Spirifer,
Bellerophon, Nautilus, and other marine shells and many corals,
alternate several times with grits, sand, and shale charged with
coal *, as well as numerous terrestrial plants, including tree ferns,
Sigillariæ, Equiseti, Calamites†, &c. &c.

In geology, less indeed, than in other sciences, can we ever
hope to account for certain results by one *modus operandi* only.
For, nature appears to have worked out phenomena apparently similar, through distinct paths and by devious processes; and of this no
clearer proof can be given, than that coal was formed out of vegetables which were accumulated by at least the two methods above
described.

General Observations on the Organic Remains. —In treating of
the general physical relations of the Carboniferous rocks of the British
Isles, and of some changes which they have undergone beyond the
region of the annexed map, it has been stated, that in North
Northumberland, as in Berwickshire and other parts of the South of

---

* See Russia in Europe and the Ural Mountains, vol. i. p. 89. *et seq.*, Pl. i. and
large woodcut showing the vertical succession, p. 111.

† The old or primeval coal is here spoken of in contradistinction to the
secondary and tertiary coals, which were formed at periods long afterwards, and
do not exhibit those uniform geological characters or that wide and equable
spread which characterize the old coal. The enquirer who wishes to study
the British fossil plants of this period must consult the Fossil Flora of Lindley
and Hutton. The same subject is developed by M. Adolphe Brongniart, who
has given his views in an admirable sketch of the successive floras imbedded
in the crust of the earth. The more recent publications on the fossil plants of
Germany, by Göppert and Unger, are productions of high merit. In comparing
these works with those on the coal plants of America, we find that the same
species inhabited very distant regions, or had a wide range throughout many
latitudes.

Scotland, the greater part of the coal is either inferior to lime-stones charged with the remains of marine animals, or interstratified with them. Most of the fossils, indeed, of those tracts, such as Nautilus, Productus and Spirifer, with Crinoidea and Crustacea, were unquestionably inhabitants of the sea.

Such was also the case as respects the large sauroid fish, Me-galicthys Hibberti, Agass., whose remains occur both in the lower and middle coal measures, and the shark-like fishes, Gyracanthus formosus, and G. tuberculatus; the Ctenacanthus; Hybodus, &c.

It is in the lower or purely marine limestones of the Carboni-ferous epoch, that the geologist takes his final leave of Trilobites. Abounding in the Silurian era, these crustaceans had, we see, dwindled to a comparatively small number in the Devonian, whilst, in the lower Carboniferous, the three genera, Phillipsia, Griffith-ides and Brachymetopus, mark the last appearance of any indi-viduals of this family in the ascending scale of natural deposits. On the other hand, it is among the superior coal strata where Trilobites became extinct, that the Limulus appears for the first time, a genus of crustaceans which has lived on to the present day —the great King Crab of the Indian seas being the existing type.

Fossils, 54.

Limulus rotundatus,
Prestwich.

From the Coalbrook Dale
Coal Field.

One example of this primeval Limulus is here given. It is a rarer species than the L. trilobitoides of Buckland.* Limulus anthrax,

* See the Bridgwater Treatise, t. 46″, fig. 3.; — that very remarkable work of my eminent friend the Dean of Westminster.

Prestwich, is another rare form from Coalbrook Dale.* We may, however, remark, that the Pterygotus, found in the Silurian Rocks and the bottom beds of the Old Red Sandstone, pertains to the same group of crustaceans. (See p. 237.)

Here, also, we last meet with any large Orthoceratites. Those scavengers of the ancient seas had now begun to diminish rapidly, and with them most of the genera of Cephalopods, which have a simple form of air-chamber, their office being taken, in the Triassic and later secondary strata, by other groups of Cephalopoda, such as Ceratites, Ammonites, &c., in which the air-chambers are minutely foliated at their edges.

The true characters of the very numerous genera and species of fossils which occur in this group, must, indeed, be studied in other works; but in support of the opinion, that the corals of the lower division often lived where they are found, a woodcut is here given of a gigantic specimen of Lithostrotion floriforme †, (Sil. Syst.) the

FOSSILS, 55.
CORAL OF THE MOUNTAIN LIMESTONE.

Lithostrotion floriforme, Fleming, in its natural position in the rock.

lower parts of which are rooted in the shale *f*, whilst the superior portion is imbedded in a limestone, *e*, with a mass of red con-

* Trans. Geol. Soc. Lond. vol. v. pl. 41. f. 1—4.

† My friend, Mr. Lonsdale, who, as before stated, described all the corals in the Silurian System, does not admit the generic word *Lonsdalia*, as applied to this form by Mr. Milne Edwards; he, Mr. Lonsdale, having first defined and limited the genus Lithostrotion. See Russia and the Ural Mountains, vol. i. p. 602., and Annals of Nat. Hist., November, 1851.

cretions, *d d.* This coral, when in its native bed, appeared, therefore, to be precisely in its original position, and conveyed to me
the impression that it had remained undisturbed beneath the sea,
whilst fine red sand at one time, and mud with calcareous matter
at another, were deposited around it. The small figure, in the
woodcut, represents the large shell Productus hemisphericus, which
in nature is four inches broad, and shows by comparison the very
great size of the coral, which has a width of two feet five inches.*

The large Producti are, however, of all its fossils, the most
characteristic of the lower carboniferous group, the same species
being found in this rock through many degrees of latitude, in
Europe, Asia, and America, and even in India and Australia. A
few of the characteristic fossils of these formations are exhibited in
the next woodcut.

FOSSILS, 56.

SOME FOSSILS OF THE CARBONIFEROUS LIMESTONE.

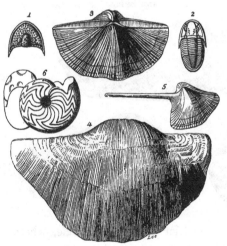

1. Brachymetopus Ouralicus, De Vern. 2. Phillipsia pustulata, Schloth. (P. gemmulifera, Phill.)
3. Spirifer striatus? Mart, sp. 4. Productus giganteus, Sow. 5. Pleurorhynchus aliformis,
Sow. 6. Goniatites crenistria, Phillips.

* See Silurian System, p. 107. The coral is from Lilleshall, Shropshire.

Associated with these species, which are found wherever the limestone exists, are fishes of the placoid order—chiefly Cestracionts,
their hard bony palates and fin defences being the only portions
preserved to us; but these are sufficient to show, that they were
both very numerous, and of many genera. Psammodus, Cochliodus,
Ctenacanthus, as well as Megalicthys, Gyracanthus, Hybodus, &c.,
are familiar types, which, with many others, must be studied in the
great work of Agassiz, or in the subsequent publications of Egerton
and other authors.*

But besides these numerous evidences of the presence of the sea,
fresh water shells (Unionidæ) every now and then occur in thick
layers, as we ascend in the series, and are among the most characteristic of the upper portions of the formation.

An estuarian intermixture is also observable in coal-bearing
strata of parts of Shropshire, Staffordshire, and Lancashire.   It is

FOSSILS, 57.

INSECT AND SHELLS OF THE COAL.

1. Wing of (Cory-
dalis ?)

2. Productus scab-
riculus, Sowerby.

3. Unio acutus,
Sow.

(From. Sil. Syst. p.
105.)

well seen in the district of Coalbrook Dale, where remains even of
insects allied to the Corydalis of America, are found associated

* See Palæichthyologic Notes, by Sir Ph. de M. Grey Egerton, Bt., M.P.,
Quart. Jour. Geol. Soc. vol. iv. p. 302 ; vol. v. p. 329 ; vol. vi. p. 1. and Nos. 4, 5.
in vol. ix. The collections of ichthyolites made by Sir P. Egerton and his associate
the Earl of Enniskillen, are, it is believed, unrivalled.

with marine and fluviatile shells and land plants. The figures in
cut 57. represent this case of intermixture as formerly pointed out
by myself. A fuller account, however, of it is given by Mr. Prestwich, in his elaborate memoir on the strata and fossils of that
district.*

FOSSILS, 58.

FERN FROM COAL OF COALBROOK DALE.

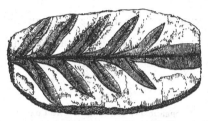

Pecopteris lonchitica, Brongn. (From Sil. Syst. p. 105.)

Other probable examples of such associations, indicating the contiguity of land on which plants grew, and from which they were
transported to marine or brackish water estuaries, are seen near

FOSSILS, 59.

UPPERMOST LIMESTONE OF THE COAL.

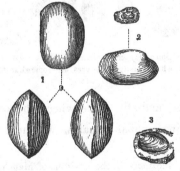

1. Cypris inflata, natural
size, and magnified 12
times. 2. Cyclas (or
Edmondia?) 3. Mo-
diola, from Ardwick,

Manchester, in a some-
what similar band of
limestone. (From Sil.
Syst. p. 84.)

* Trans. Geol. Soc. Lond., vol. v. p. 413. One of the plants, *Pecopteris lonchitica* (Brongn.), which is given in the upper woodcut, is associated in these
tracts with forty or fifty species of terrestrial plants belonging to Calamitaceæ,
Polypodiaceæ, and Lycopodiaceæ. Of these, the Stigmaria ficoides (considered
to be the root of Sigillaria), Neuropteris cordata, Odontopteris obtusa, and the
Pecopteris here figured, are perhaps the most common.

Shrewsbury, Manchester, and other localities, among some of the younger coal strata. In the environs of Shrewsbury, small shells of the genus Cyclas (?) with the minute crustacean Cypris (or possibly the marine genus Cythere) are commingled in the same limestone with a minute shell, originally termed Microconchus carbonarius, but now referred by naturalists to the marine or estuary genus Spirorbis.*

The preceding and following figures (cuts 59, 60.), formerly published in the Silurian System, explain more particularly the contents of this peculiar limestone of the coal strata.

FOSSILS, 60.

UPPERMOST LIMESTONE OF THE COAL.

(From Sil. Syst., p. 84.)

Microconchus (now called Spirorbis) carbonarius, Sil. Syst. The real size is given in the minutest of these figures, whilst the upper figures are somewhat magnified, and the lower very greatly so.

In all such cases, and still more where the coal is intercalated among purely marine animal remains, we must believe, that the

* Some of the coal sandstones in the environs of Manchester exhibit on their surfaces the clearest indications of having been shore deposits; certain tracts being marked on them, of animals which must have crawled at ebb tides. Mr. Binney states also, that the shells of the above Spirorbis occur throughout the whole of the thick series of the Lancashire coal-field, and thus indicate long-continued marine conditions. See "On some Trails and Holes in the Carboniferous Strata, with Remarks on Microconchus Carbonarius," by Mr. E. W. Binney, Trans. Lit. and Phil. Soc. Manchester, vol. x., 1851-2.

vegetables from which it was formed were carried down into seas or estuaries fed by freshwater affluents.

Whilst researches in all parts of the world have only produced one small reptile from the Upper Devonian rocks, it was stated that the same period afforded comparatively few land plants. The appearance, therefore, of land plants and air-breathing reptiles was, as far as our evidences go, nearly simultaneous.

With the proofs in our possession of the large quantities of terrestrial vegetables which occur .in the Carboniferous era now under consideration, accompanied as they are in numerous cases by river and lake shells, and, in other instances, by insects, and land crabs, it might be expected that reptiles would also be found in them. And such has proved to be the case. The first discovery of this nature was made in the coal-field of Saarbrück in Rhenish Bavaria, wherein two species of Archegosaurus, Goldfuss, have been found, — a reptile which Hermann Von Meyer supposes to be a connecting link between Batrachians and Lizards.

Since then, the footsteps of a large reptile allied to the Cheirotherium, have been discovered in the Carboniferous strata of Pennsylvania. Recently, Professor M'Coy detected in the Museum of Lord Enniskillen, the remains of an allied reptile from British coal-fields, which Owen has described under the name of Parabatrachus Colei. Sir C. Lyell has also communicated the still more interesting account of a discovery made by Mr. Dawson and himself, in the coal-field of South Joggins, in Nova Scotia *, of a reptile called Dendrerpeton Acadianum, which Owen and Wyman consider to belong to the perennibranchiate Batrachians. With this reptile were associated many land plants, and the shell of a mollusk, which is supposed to have been air-breathing: if this be so, it is the first true land-shell which has been found in strata of such high antiquity.†

* Quart. Journ. Geol. Soc., vol. ix. p. 55.   † Ibid vol. ix. p. 58.

These discoveries afford us proofs of associations of organic remains which might, indeed, have been anticipated from what we know respecting other palæozoic deposits.

In the earliest wide and general diffusion of a copious and peculiar vegetation, we further recognize the prevalence of that equable temperature and of similar conditions over various latitudes, which must, in my opinion, have also existed in the preceding periods. In the Carboniferous deposits, however, we are for the first time surrounded by proofs of the existence of vast plant-bearing lands. Of these ancient plants, the great mass of the Silurian sediments afford, I repeat, no traces in any part of the world, though they occur in just the same latitudes as the Carboniferous strata, and are very generally in juxta-position, often with, but sometimes without the interposition of the Devonian rocks. The first feeble traces of such vegetation are observable at the close of the Silurian era; and if the Devonian rocks exhibit here and there proofs of an increase of land plants, it is only in the Carboniferous period that we are presented with the complete materials for the elaboration of vast coal-fields.

P. S. — Referring to the preceding pages, let me state, that the reader will find an admirable and succinct sketch of the methods by which coal was accumulated and formed, in a Report by Professor H. D. Rogers of the United States, as published in the Transactions of the American Association for the Advancement of Science, 1842, p. 433. Having explained the views of various European geologists, among whom he justly assigns praise to Mr. Mammatt, for having been one of the first persons who sustained by physical evidences the previous theory of De Luc, M'Culloch, and others, of the formation of coal out of terrestrial vegetation *in situ*, Professor Rogers, having under consideration the enormous area occupied by the carboniferous strata of North America, endeavours to reconcile conflicting hypotheses. He has, indeed, so ingeniously applied the opinions of his brother, W. Rogers, and himself, respecting the influence of great proxysmal earthquakes which affected the earth's crust in ancient periods, as to suggest how the rapid alternations of terrestrial and oceanic remains, to which attention has been called in this work (p. 277, *et seq.*), can be best accounted for. This subject will be in part reconsidered in subsequent chapters.

CHAPTER XII.

PERMIAN ROCKS.

CHANGES OF THE SURFACE BEFORE THE PERMIAN DEPOSITS WERE ACCUMULATED.—
ORIGIN OF THE TERM PERMIAN AS APPLIED TO THE HIGHEST GROUP OF PRIMEVAL
DEPOSITS. — THE PERMIAN ROCKS OF RUSSIA, GERMANY, AND ENGLAND. — THE
ORGANIC REMAINS OF THE GROUP.

IN the two previous chapters, slight allusions only have been made
to eruptions of igneous or volcanic materials, which were ejected and
spread out on the sea bottom, together with the ordinary sedimentary
deposits; for such volcanic ejections, during the Devonian and carboniferous eras, were not by any means so abundant in the British
Isles as when the Silurian rocks, and especially their lower divisions, were accumulated. (See *antè*, pp. 54. 57. &c.)

The great upper coal-fields which we have been considering, were
probably formed under conditions which geologists would consider
comparatively quiescent. This is supposed to have been particularly
the case with all that portion of them, in which, as we have seen,
coal beds must, in many cases, have been formed by repeated
downward movements of low lands beneath the waters, followed
by many elevations of the same into the atmosphere. The close,
however, of that period was specially marked by convulsions and
ruptures of the crust of the earth, which, from the physical evidences placed before us, must have extended over distant regions of
Europe, as well as of America. Whatever may have been the

previous changes, it was then that the coal strata and their antecedent formations were very generally broken up, and thrown, by grand
upheavals, into separate basins, which were fractured by numberless
powerful dislocations.   The order of ancient sedimentary succession
was thus very widely, though not universally, interrupted; and all
the strata previously spoken of were so disturbed, as to produce a
prevalent break or discordance between the carboniferous and inferior
formations and those which succeeded to them.

That this disturbance was not, however, general, is demonstrated
by the fact, that both in Russia on the east, and England on the
west, there are exceptional tracts in which the carboniferous and
overlying Permian deposits are apparently in conformable relations
to each other.

Still, the general effect of the great disseverment alluded to, in de-
termining the outline of the earth, is obvious ; for whilst the Silurian,
Devonian, and Carboniferous rocks were at that period so heaved up
as to constitute portions of mountain chains, the strata which we are
now considering, though allied to the carboniferous by their imbedded plants and animal remains, have been usually spread out
in western Europe in less highly inclined positions; constituting,
therefore, for the most part, countries with a lower level.   Whilst
their fossils also are, on the whole, different from those of all pre-
existing palæozoic rocks, they are yet far more connected with
the fauna and flora of the carboniferous deposits, than with the
organic remains of any secondary or mezozoic rock subsequently
formed.

The strata, which were accumulated after the great deposits of coal,
are those which in England have been severally termed Lower New
Red Sandstone, Magnesian Limestone, and marl slate.   Similar
rocks have been long known in Germany under the names of Rothe-
todte-liegende, Kupfer-Schiefer, Zechstein; and with these I have

united an inferior portion of what had been commonly classed with the Bunter Sandstein. Having become satisfied that these strata, so different in mineral character, constituted one natural group only, which must be distinguished from, yet connected with, the carboniferous series beneath, and which, from its organic contents, must be entirely separated from all formations above, I proposed in 1841*, to my associates, de Verneuil and von Keyserling, that the group should receive the name of Permian, as taken from an enormous region which composed the ancient kingdom of Permia. To that vast country, my illustrious friend the Baron von Humboldt particularly called my attention when about to revisit Russia with my friend de Verneuil; and there we found, that all the characteristic features of an independent assemblage were elaborated on a very grand scale.

After two explorations of Russia with my companions, I re-examined (1844) the portions of Germany, where the Zechstein and its associated strata, underlying as well as overlying, are best displayed; and, by placing them in parallel with the Russian and English rocks, my views were confirmed, and thus all the deposits above mentioned were included under the term " Permian."

The value of this euphonous name, for a series of strata which constitute one natural mass, soon became so obvious, that, although proposed only in 1841, it has already been adopted by geologists of different nations, including my own.

In Germany, where the Rothe-todte-liegende and the Zechstein had long held their head quarters, and where their characters had been described by numerous geologists, from the days of Werner and

* The term was first proposed in a letter addressed by myself at Moscow to the venerable and accomplished Russian palæontologist Dr. Fischer, October, 1841. See Bronn and Leonhardt, Journ. an. 1841, and Phil. Mag., vol. xix. p. 417.

Schlotheim, to those of the great geologist Leopold von Buch, whose recent loss we deplore, these rocks have been recognized in the works of Naumann of Leipsic, and of Geinitz, and Gutbier of Dresden, as the "Permische System." *

In France, M. Alcide D'Orbigny, in his systematic work on Palæontology, and other authors, have given currency to a name which would indeed have had little or no value without the close palæontological † comparisons of my colleagues, de Verneuil and von Keyserling, as worked out with me in the distant eastern regions of Russia in Europe.

In the north of England, where its most varied and richest fossiliferous members were first described by Sedgwick ‡, its organic remains have recently been illustrated by King, under the term Permian; and the word is now used by Sir Henry de la Beche and the Government surveyors in the construction of the geological map of England.

As many persons may not have access to my large work on the geology of the Russian empire, in which a full description is given of the leading characters of the group, and, as I wish to impress the reader with the reasons for the adoption of the name Permian, a few passages from the writings of my colleagues and myself are here reproduced. By the first of these extracts it will be perceived that, in describing the whole structure of Russia, we began, as on the present

---

* See Geinitz and Gutbier, "Permische System in Sachsen," 1848. Dresden and Leipsic.

† See Cours Elémentaire de Pal. et Geol. Strat., par Alcide D'Orbigny, vol. ii. p. 370. I am happy to learn that the French Government have rewarded the labours of M D'Orbigny by naming him Professor of Palæontology in the Jardin des Plantes. My friend Professor Philips had the merit of having first suggested that, on account of its fossils, the "magnesian limestone" of England should be classed with the palæozoic rocks. Treatise on Geology, p. 189.

‡ Trans. Geol. Soc. Lond., vol. ii. new series, p. 37. et seq.

occasion, with the lowest rocks in which any traces of life could be detected. We then proceeded :

"Having worked our way upwards through Silurian, Devonian, and Carboniferous rocks, we have now to describe the next succeeding natural group. Spread out over a larger surface than any other in Russia, the rocks in question, with certain overlying red deposits, which we cannot separate from them, occupy the greater part of the Governments of Perm, Orenburg, Kazan, Nijni Novogorod, Yaroslavl, Kostroma, Viatka, and Vologda, a region more than twice the size of the whole kingdom of France !"*

After showing to what an extent opinions varied respecting the age of these Russian deposits, most authors referring them to the New Red Sandstone or Trias, others to the carboniferous era, we next observed † :

"Such was the state of the question when we entered upon the survey of Russia. To arrive, therefore, at a sound conclusion respecting the age of these rocks, it became essential to traverse, as far as possible, the countries over which they extended, and compare the phænomena which had led to such contradictory opinions. The result has been, that though these deposits are of very varied mineral aspect, and consist of grits, sandstones, marls, conglomerates, and limestone, sometimes inclosing great masses of gypsum and rock salt, and are also much impregnated with copper, and occasionally with sulphur, *yet the whole group is characterized by one type only of animal and vegetable life.*"

"Convincing ourselves, in the field, that these strata were connected by their organic remains with the carboniferous rocks on the one hand, and were independent of the Trias on the other, we ventured to designate them by a geographical term, derived from the ancient kingdom of Perm, within and around whose precincts the necessary evidences had been obtained."

"With the highest respect for the labours of German geologists upon the Zechstein, and for the researches of those authors who have placed the Magnesian Limestone of England on the same parallel, we are con-

---

* Russia in Europe, vol. i. p. 137.   † Russia in Europe, p. 138.

vinced, that neither in Germany nor in Great Britain do the same accumulative proofs exist, to establish the independence of a geological system. If mineral characters be appealed to, no German writer will contend, that the thin course of 'Kupfer schiefer' is of like importance with the numerous strata which in Russia constitute many bands of various structure; rendering, in fact, the Zechstein itself a mere subordinate member of a vast cupriferous series. Subordinate, however, as it is in some tracts of Russia, the Zechstein is so magnificently displayed, in others, in masses of both limestone and gypsum, that it more than rivals the finest sections of that deposit, whether to the south of the Hartz, or in Thuringia. We object, however, to a lithological name, hitherto reserved for one portion only of a complicated series; and as the Germans have never proposed a single term for the whole group, which is based upon the Rothe-todte-liegende, and surmounted by the Trias, we have done so, simply because we first found in Russia the requisite union of proofs."

Occupying the enormous area before mentioned, the Permian deposits of Russia are flanked and underlaid on the west, east, and north, by upper members of the carboniferous rocks, but with little or no coal. These Permian strata of Russia seldom exhibit a mineral succession similar to that of rocks of the same age in Western Europe; and in different tracts of the vast region explored, they exhibit, as explained in the preceding quotation, many variations in their contents and relations. In some places, as on the river Kama, near the Volga, cupriferous, red grits with plants underlie the chief limestone, to which succeed marls; but along the eastern limits of the system, as flanked by the Ural Mountains, gypseous limestones form the base, followed by the red copper grits, sands, marls, and pebble beds, which extend on all sides around the city of Perm. On the whole, indeed, whether we appealed to the sections on the banks of the great Dwina, above Archangel, or to the western flank of the Ural Mountains, or to the banks of the Lower Volga, near Kazan, localities removed from each other by vast distances, we found that limestones, often interstratified with much gypsum, prevailed towards the base of the Russian deposits.

In some parts of the region salt springs occur. These may, doubtless, rise from bodies of rock salt in older palæozoic rocks, since the mineral is known to occur in the Devonian Old Red Sandstone or Devonian of Western Russia; but in the steppes south of Orenburg, it is certainly subordinate to true, red Permian deposits.* These salt beds range up to the foot of the older palæozoic and crystalline rocks of the South Ural Mountains to the east of Orenburg, the strata of Permian age being alone visible in the low wooded slopes at the foot of the rocky chain of older rocks represented in this vignette.

THE GURMAYA HILLS OF THE SOUTH URAL MOUNTAINS.
As seen from the Steppes of Orenburg. (From "Russia in Europe," vol. i. p. 450.)

---

* Russia in Europe, &c., vol. i. p. 184.

SECTION ON THE WEST FLANK OF THE SOUTH URAL. (From "Russia in Europe," p. 146.)

Gurmaya Hills.

Giriakaya.

a. Sandstones, limestones, gypsum and grits. (Permian Rocks part of.)
d. Red sands and copper ores.
c. Conglomerate and sandstone.

a. Carboniferous limestone.
b. Goniatite flags.

Along certain portions of the west flank of the same chain the Permian strata occur in almost apparent conformity to the carboniferous rocks. There they have manifestly undergone a movement impressed on them by great forces, directed from north to south, or parallel to the Ural mountains; all the strata, whether carboniferous or Permian, having been raised up and thrown off sharply to the west, as represented in the woodcut on the side of the page.

At the imperial baths of Sergiefsk, and on the banks of the river Sok, a tributary of the Lower Volga, magnesian limestone and marl are surmounted by gypsum, copper ore, native sulphur, with sulphureous and asphaltic springs in the middle masses, whilst other marlstones and white limestones form the summit. A considerable volume of gaseous, sulphureous water, which forms a large pool, issues from these rocks. (See diagram, Russia in Europe, &c., p. 158.)

Near Kazan, about 150 miles to the north of Sergiefsk, huge masses of gypsum (*a*), rising high above the level of the river Volga, are surmounted by limestone cliffs with Zechstein fossils (*b*), and the latter by red, green, and white marls (*c*), as here represented.

PERMIAN DEPOSITS NEAR KAZAN.

(From "Russia in Europe," p. 162.)

W   E.

Volga R.   Kazan.

a. Thick deposits of gypsum.   *b.* Limestone with fossils.   *c.* Red and green shelly marls.

On the other hand, in the central tracts between the Ural
Mountains and the Volga, as on the Dioma, and Kidash, tributaries
of the great river Kama, the limestone, which in some tracts assumes
a definite horizon, and is underlaid by coarse grits, is repeated at
various levels in a succession of beds, interlaminated with sandstones,
and yellow, white, and greenish marls, occasionally containing plants,
and small seams of impure coal, — the whole being surmounted by
red grits and conglomerates, with copper ore.

Now, the calcareous and gypseous deposits which interlace this
series are throughout characterized by an analogous group of fossils,
and even by some of the same species, as the Zechstein of Germany
and the magnesian limestone of England; whilst the beds of copper
grit, with most of the red conglomerates, which in Russia are *uppermost*, contain bones of reptiles (Thecodont-saurians), belonging to a
genus which in Germany occurs in the copper slate *beneath* the
limestone.

In the exploration of Russia, therefore, geologists were taught, by
this diversified Permian group, not to dwell on the local mineral distinctions of Central or Western Europe, but to look to the wide
spread of certain fossil remains, which, in vastly distant countries,
occupy the same general horizon, between the carboniferous rocks beneath, and all those overlying strata which can be called secondary.

The survey further proved, by the extension of some of these
fossils upwards into red and green marls and sands far above the chief
bands of limestone, that the Zechstein of Germany cannot be properly considered the summit of this natural group.

Revisiting the districts of central Germany *, which I had often
traversed, and which had been rendered classical by the writings of
many native geologists, I ascertained that the Flora of the Rothe-
todte-liegende, as collected by Colonel Gutbier and Professor Geinitz,
has essentially the same supra-carboniferous type as the Permian
Flora of Russia, and that certain fishes of the todte-liegende of

* In the year 1844, and after my journeys in Russia.

Silesia are closely analogous to, if not identical with, Russian forms.
Further, seeing that the whole of the vast red deposit (which according to M. Credner, of Gotha, swells out in Thuringia to a maximum thickness of about 2500 feet) was conformably succeeded by
the Zechstein, and was transgressive to the coal beneath it, I naturally placed it (this lower red sandstone and conglomerate) as the
base of the Permian group in Germany; the more so as by this
means the successive strata of that country and England were
brought into parallel positions. Next, it was evident, that all the
copper slates, marls, and limestones both plain and dolomitic, whether
in Russia, Germany, or England, were occupied by a similar fauna.
Lastly, I became convinced that the analogy, with Russia on the one
hand, and with England on the other, might be completed by
grouping with the Zechstein a certain lower portion only of the vast
series of red sandstones which overlies that formation in Germany.

Any one who looks at the natural succession of those overlying
rocks in Hessia, Thuringia, and Saxony will see, that the Bunter
Sandstein is divided into two parts, as faithfully, indeed, laid down
on maps by several good native geologists. It is the lowest of these
only that I have grouped with the Zechstein, of which it frequently
forms the natural and conformable cover. As yet scarcely any organic
remains have, I admit, been detected in this Lower Bunter Sandstein
of the Germans. Still, in several places, the deposit is characterized
by the presence of the plant published by M. Adolphe Brongniart as
Calamites arenarius, which has quite a palæozoic aspect, and differs
from all known plants of the mezozoic or secondary series.*

---

* In a recent excursion to central Germany, when accompanied by Professor J. Morris, I had
an opportunity of confirming the Permian classification, by appeals to nature and to the collections of different individuals, as well as by reference to maps. In Saxony we examined the grand
sections of the Rothe-todte-liegende under the guidance of Professor Geinitz, and ascertained
from him how symmetrically that vast, old, coarse formation of conglomerate, sandstone and
interpolated porphyries, is there succeeded by the Zechstein, the fossils of which that author
has so ably described. (See Permische System in Sachsen.) He further pointed out to us a
thin-bedded limestone in the Plauen-Grund W. of Dresden, and considered by him to be subor-

If therefore this view be sustained, it may oe said that the Permian of Germany is a lower or palæozoic *Trias*, the central mass of which is the Zechstein limestone, and base the Rothe-todte-liegende; whilst the Upper or Secondary Trias is marked by its middle limestone, the Muschelkalk, with its underlying red Bunter Sandstein, and its overlying red Keuper marls.

But whether this lower " Bunter " be abstracted from the Trias or not, let me again call attention to the fact, that whilst the Permian and Trias are conformable to each other, and exhibit nowhere any of those signs of disseverment which so often mark the close of the coal deposits, their respective fauna and flora are entirely dissimilar; the one exhibiting the last traces of primeval life, the other being charged with the exuviæ of plants and animals entirely distinct from those which preceded them.

dinate to the copious lower red sandstones of that tract, which are penetrated to such great depths by shafts that reach productive coal seams.

In the south-east parts of Thuringia (Saalfeld), where the higher members of these strata appear' we saw how the Weiss-liegende or upper member of the Lower Red Sandstone, is succeeded by the Kupfer Schiefer and the Zechstein with its subordinate schists and overlying dolomites and gypsum, as described by M. Richter (Einladungs-Programm der Realschule Saalfeld, 1853); and again how all these dip conformably under the noble escarpments of Bunter Sandstein which range from Rudolstadt eastwards.

At Berlin, through the kindness of M. Beyrich, we inspected many fishes and plants from the Rothe-todte-liegende of Silesia (Friedland, Ruppersdorf, &c.), rocks, which after a former examination, I classed as Permian; and I was more than ever impressed with the belief, that the ichthyolites which there occur in bituminous schists or micaceous red flagstones (Palæoniscus, Holacanthodes, and Xenacanthus), pertain to the same system of life as the ichthyolites of the Kupfer Schiefer. Again, in the numerous fossil plants of this red sandstone which were laid before us, Professor Morris, who assisted me materially in describing the Permian plants of Russia, was of opinion, that like them, this Silesian flora was to be distinguished from the ordinary carboniferous flora; for among these plants we could perceive no Stigmaria, Lepidodendron, nor any form abundant in the old coal; but on the contrary, species of Neuropteris and Odontopteris, and also those coniferæ (Araucarites or Voltzia) which first begin to appear in the Permian era. (See Russia and Ural Mountains, vol. i. p. 218., and Plates, vol. ii.)

At Gotha we had the advantage of examining the collections of that good geologist and miner, M. Credner, who communicated to me the new part of his geological map of Thuringia, " Geogn. Karte des Thüringer-Waldes." The double division of the Bunter Sandstein to which I formerly adverted, as made by M. Althaus, of Rothenburg, and others, has also been clearly laid down in an useful and instructive map by M. Ludwig, ' Petrog. Karte, der gegend zwischen Frankfurt, Giessen, Fulda, und Hammelburg; Darmstadt, 1853.'

The general succession, which for the most part applies to Western Europe, is expressed in this diagram, though, as before said, there are tracts both in Russia on the east, and in England on the west, where these Permian rocks repose with apparent conformity on the subjacent coal strata.

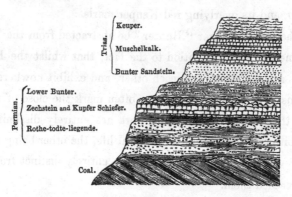

Permian Rocks of Britain.—Let us now see of what the British Permian rocks (No. 9. of map) consist.

In one district of the original Silurian region, or that which lies to the south-west of the Staffordshire coal-field (near the Severn), this group seemed to me occasionally to exhibit, in its lowest part, a sort of transition upwards from the coal measures. Calcareous shale with thin courses of coal is there followed by sandstone with trappean tuff, and quartz pebbles, with fragments of plants, and concretions of iron stone, all of which ought probably to be grouped with the carboniferous rocks. But the overlying ·members, — red shale, argillaceous marl, and incoherent yellowish sandstone with calcareous laminæ, some very thin layers of poor coal, may be viewed as the base of the Permian. These strata are succeeded, in apparent conformity, by dark red sandstone and shale, here and there charged with concretions of dark mottled limestone. These sandstones are surmounted by conglomerates, which in parts are so calcareous as to be burnt for lime, and in others, as at Alberbury near Shrewsbury, become almost compact, yellow, magnesian limestones.

In an adjacent part of Staffordshire and Worcestershire, on the contrary, the older rocks of the coal-field have been heaved up to the surface from beneath a former cover of the red ground or Permian rocks; and hence the formations, as in many places around the Dudley coal-field, are separated by a powerful fault, thus:

RELATIONS OF PERMIAN ROCKS TO THE COAL NEAR HALES OWEN.

c. Trappean conglomerate.   b. Red sandstone with calcareous concretions.   a. Coal.

In the tract near Hales Owen calcareous matter so abounds in the red Permian rocks, that it constitutes zones of earthy, concretionary limestone (*b* of this diagram), which are perfectly undistinguishable, as formerly pointed out, from some of the cornstones of the Old Red Sandstone. Thus, the coal of this part of England was shown to lie between two similar red deposits, Sil. Syst., p. 55.

In following this succession upwards, the trappean breccia and conglomerate of the great Lickey Hills above Hagley Park is met with, but the order is there no further very distinctly traceable. Fragments of a few other rocks, besides the prevalent trap which I described, having been found in these breccias since my examination, the government geologists, Ramsay and Jukes, view them as stratified aqueous deposits, forming an integral part of the Permian, and as the equivalents of the ordinary conglomerates of the group.*

In one portion only of the annexed map do the Permian conglomerates and sandstones contain any traces of fossils except plants; and this is to the north of Bristol, where, in the form of a magnesian or dolomitic conglomerate, they have afforded remains of sauroid reptiles, which were first described by Messrs. Riley and Stuchbury

* Professor Ramsay and the Government Surveyors are now employed in deciphering the real limits and contents of the Permian rocks of central England.

under the name Thecodonto-saurus, and to which the Rhopalodon, Fischer, is considered to be nearly allied by Professor Owen.*

The same conglomerate is seen in patches, accumulated on the edges of the carboniferous limestone and old red sandstone, around parts of the southern rim of the South Welsh coal basin.†

This rock has usually been placed on the same parallel as the magnesian limestone of the north of England; and the analogy of the succession in Shropshire, where the lower red sandstone is inter-polated between the coal-fields beneath and the magnesian conglo-merate above, favours the view. There, the upper coal is surmounted by a considerable thickness of red sandstone and shale (the Lower New Red), which, in their turn, are covered by a yellow magnesian limestone — in some parts highly brecciated, in others a conglomerate, and this rock is again overlaid by red sandstone, marl, &c. The general order in the portions of Shropshire, near Cardeston and Alber-bury is as here given.

POSITION OF THE PERMIAN ROCKS IN SHROPSHIRE.

d. Red sandstone and marl.
c. Magnesian conglomerate and yellow limestone. ⎫
b. Dark red sandstone (Lower Red).   ⎬ Permian Rocks.
a. Coal measures — upper division with thin courses of limestone.

In their range to the north of England, these rocks become more diversified in mineral character and much richer in fossil contents, and therefore more completely represent the Permian of the continent. Thus, at Manchester, the red shale, subordinate to sandstones of this age, is already charged with casts of fossil shells of the genera Schizodus, Avicula, Turbo, Rissoa, &c.‡

* Russia in Europe, &c., vol. i. p. 637.

† This dolomitic conglomerate is also finely exhibited near Thornbury, on the east, and Bridgend on the west, of the Severn. (See map.)

‡ See the Fossils of these Manchester rocks, as published by Captain Brown and Mr. Binney, Manchester Geol. Trans., vol. i. p. 63.

Extending northwards, or from Nottinghamshire into Yorkshire, red marls alternate with sandstones, and, as shown by Sedgwick, overlie the magnesian limestone; thus offering an ascending series similar to that of the Russian and German examples before cited.*

The inferior member of the group, or the "Rothe-todte-liegende" of Germany, where it occurs in Yorkshire, is known as the " Pontefract rock" of Smith.† It consists of yellowish sandstones and conglomerates, containing various plants. Reddish sandstones, re-occur, however, near Ripon, and are not unlike some of the same age in Shropshire and Staffordshire: they there occupy the same position, being covered by the magnesian limestone or Zechstein of the Bedale district, whence they expand over a large tract of central Yorkshire.

The succession of strata in the north of England, as originally described by Professor Sedgwick, in his excellent " Memoir on the Magnesian Limestone," and there paralleled by him with the Zechstein of Germany and its associated beds, consists, in the lower part, of flaglike marls, the equivalents of the " Kupfer Schiefer," and in the upper of the yellow magnesian limestone (" Zechstein") of Yorkshire, Nottingham, and Durham.   The whole of the series, together with the above-mentioned overlying red beds, are included in the term " Permian."

Recently Professor King, who had long studied the fossils of the calcareous members of the group, has placed the detailed component parts still more closely in comparison with the corresponding beds of Germany.   Thus, whilst, with Sedgwick, he considers that the lower sandstones of Yorkshire and Durham, whether red, white, or yellow, which lie between the coal and the magnesian limestone, are the true equivalents of the German Rothe-liegende, and that the marl slate, with its fishes, stands in the place of the copper slate

* See Sedgwick on the Magnesian Limestone, Geol. Trans. 2nd ser., vol. iii.p. 37.

† So named by William Smith, the 'father' of English geology, whose lessons in the field along the Yorkshire coast (1826) were of great service to me.

of Germany — he also shows, that the fossiliferous beds of compact limestone represent the lower, — and the brecciated, concretionary limestones of Durham the upper Zechstein of the Germans, with its overlying beds of dolomite, rauchwacke and " stinkstein."

In England, there is, perhaps, no other yellow limestone so charged with magnesia as to form a true dolomite; and hence in the early days of geology, it was natural to define this rock as *the* Magnesian Lime-stone*, and to associate with it certain subordinate strata. But now that yellow magnesian limestones are known to occur, on a stupen-dous scale (in the Lower Silurian rocks of North America†, and are also of some magnitude in the Devonian and Carboniferous series of Russia) whilst the Jurassic masses of the Alps are to a great extent crystalline dolomites, it becomes necessary to abandon the term " magnesian," and to place the English formation under a name which establishes its synonymy, through fossil contents and position, with rocks of other countries.‡

The Permian deposits, as developed in Russia, and as applied to Germany and England, do not sto, as before said, in the as-cending order, with the Zechstein. They also include another over-lying red sandstone, which, in many parts of Germany constitutes the conformable roof of the Zechstein, and contains the plant Calamites arenarius, Brongn., which has a *quasi* carboniferous aspect. In general language, therefore, the Zechstein of Germany, or mag-nesian limestone of England, may be viewed as the calcareous

---

* I am, however, disposed to think that some of the yellow beds of the carboni-ferous limestone of the Clee Hills are exceptions.   (See Sil. Syst. p. 119.)

† See Dale Owen's Geology of Wisconsin, Iowa, and Minnesota, with Map, on which vast magnesian limestones are laid down as the Lower Silurian of that author. 1853.

‡ True dolomite, whether crystalline or earthy, is known by its containing 45 per cent. of magnesia.   Another proof of the inapplicability of mineral terms to designate the age of strata, is the use of the term "oolitic " for rocks, which in England have a structure that is scarcely ever found in their continental equi-valents.   It is therefore merely an insular lithological name which misleads.

centre of an arenaceous group, or, as before said, a lower " Trias ; "
the upper red marl and sand of Yorkshire, described by Sedgwick,
being as much a parcel of it as those red sandstones and conglomerates,
or "rothe-todte-liegende," which in Western Europe lie beneath it.

There are, indeed, parts of Europe, as in the Ardennes of Belgium,
where these rocks, so diversified elsewhere, are represented by a mere
band of pebbly, silicious conglomerate, which being sterile of fossils
was named " *Penéen* " by M. D'Omalius D'Halloy.[*]   In France, *i. e.*
in the Vosges and at Lodève in Dauphiné, this group has not the
distinctive characters, lithologically or zoologically, which it exhibits
in Russia, Germany and Britain, being simply a red sandstone.   But
even there it is distinguished, as in other regions, by a peculiar flora,
containing certain ferns and conifers which M. Adolphe Brongniart
classes with Permian plants.

Permian Fossil Remains.—The fauna and flora of the Permian
rocks are, as before stated, essentially palæozoic; for whilst in great
measure specifically distinct from that of the carboniferous system, the
amount of their agreement is surprising when we reflect upon the
phenomena, adverted to in opening this chapter, of a great physical
revolution which pretty generally affected the known surface of the
earth at the close of the preceding or carboniferous era.   That disruption, therefore, however violent and extensive, was not universal, but
was, we may suppose, so productive of new conditions as to occasion
the destruction of many genera and species of plants and animals.

The Permian flora has not been yet so developed in the British
Isles as to show in what degree it differs from that of the carboniferous group, to which, in fact, it every where bears a resemblance.   We know, indeed, that some of the species found in the
yellowish sandstones of Pontefract, and in the lower red sandstones
which may be classed as Permian, are identical with plants of the

---

[*] Elémens de Geologie, p. 276. "Penéen," means sterile.

adjacent coal-fields; and at Ashby de la Zouch, *Sternbergia* has lately been detected, together with silicified wood, by the Rev. W. H. Coleman. Prof. Sedgwick* long ago pointed out the traces of Calamites in the lower New Red Sandstone; and we here reproduce figures of two fossil plants, found in the marl slate of Durham, from Professor King's elaborate work.†

FOSSILS, 55.

PERMIAN PLANTS, &c.

1. Caulerpa ? selaginoides, Sternberg.   2. Neuropteris Huttoniana, King. Both are from the Marl Slate of Durham.

3. Fenestella retiformis, Schlotheim; "Magnesian limestone" of Humbleton Hill. The figures, except f. 2, are much reduced.

In certain foreign tracts, however, which have been closely examined, the lower Permian strata contain many plants and some thin beds of coal. The sandstone of Lodève, before adverted to, affords, according to M. Ad. Brongniart, ferns of the genera Sphenopteris, Pecopteris, Neuropteris, Alethopteris, Callipteris, with Annularia floribunda (Sternberg), and several conifers of the genus Walchia.‡ The sandstone and conglomerates also beneath the magnesian limestone near Zwickau in Saxony, have rewarded the researches of Colonel Gutbier § with about sixty species of plants. One-third only of these are known in the carboniferous epoch; the other two-thirds are forms which specifically are unknown in any

* Proc. Geol. Soc., vol. i. p. 344.

† Monogr. Perm. Foss. 1848, pl. 1., &c.

‡ See Russia in Europe, &c., vol. i. p. 219.

§ Geinitz und Gutbier, Gæa von Sachsen, 1843. "Permische System in Sachsen," 1850.

other deposit. Among the plants most characteristic of the group, may be mentioned the Sphenopteris erosa, S. lobata, and the great, reed-like Calamites gigas. The large silicified stems of tree ferns called Psaronius, so much admired by collectors for the exquisite conservation of the fibre of the plant, and for the beautiful polish they take, occur abundantly in the lower red (Permian) beds of Saxony.* Whilst in Russia these species of plants are found in beds above all the limestone courses, they are confined to the lower strata of Saxony, and are there commingled with some species known in the upper coal deposits.

On the whole, it was the opinion of M. Adolphe Brongniart, to whom the plants brought from Russia were referred†, that they exhibit a continuation of vegetable life of the same general characters as those which prevailed in the carboniferous era. But whilst the genera are the same as those of the coal period, the species, with two or three exceptions, are different. On the other hand, no one of the Russian or Permian genera, Neuropteris, Odontopteris, Lepido-dendron, Sphenopteris, Noeggerathia, &c., occur in the overlying Trias, in ascending into which the geologist has before him an en-tirely different flora.

The chief Permian animal remains bear also a strong resemblance to their congeners of the carboniferous era. When the work on Russia was written, my companions and myself were not aware of more than 166 fossil Permian species, but, by the publication of the works of Geinitz in Saxony‡ and King in England, this number has been increased to 230; a small proportion, therefore, when compared with that of the preceding or carboniferous epoch, in which about 1050 species of animals are already described.

---

* The finest collection of these "Psaronites" extant, with which I am acquainted, is that of my valued friend, Mr. Robert Brown, the eminent botanist.

† Russia in Europe, &c., vol. i. p. 218. et seq.

‡ Geinitz, Die Versteinerungen des Deutschen Zechstein gebirges, 1848.

The corals and Bryozoa, twenty-six in number, have the palæozoic type, the small cup-coral Polycœlia; apparently showing the
quadripartite arrangement of the lamellæ characteristic of the cup-
corals of the old rocks.   The abundant Fenestella retiformis, and
Synocladia virgulacea, Foss. 61. f. 3., are also of a type peculiar
to the primeval fauna.   Crinoids are rare; but there is in England a
species of the genus Cidaris, the earliest known representative of the
modern sea-urchins.*   The Brachiopods, of which about thirty-
eight species are known, offer many Producti, analogous to those
of the coal period;   and about five species, including P. horridus, the most common Permian form in Britain and Germany
(Foss. 56. f. 2.), and P. Cancrini, are identical with carboniferous
forms.   Eleven species of Spirifer also occur, one of which, S. undulatus (f. 1.), is the most characteristic, and S. cristatus is supposed
to be common to the carboniferous and Permian rocks.   Of the
genus Orthis, which prevails so greatly from the Lower Silurian to
the carboniferous inclusive, we here find no trace; nor does it
occur in any superior formation.   Its place is possibly taken, in
the Permian rocks, by the new genus Strophalosia, King (Orthothrix
of Geinitz)†, or Aulosteges of Helmersen, which is but rarely found
in older strata.   Of the few species of Terebratula which are known,
two or three are exceedingly abundant, and one, T. elongata of
Schlotheim, can scarcely be distinguished from a common carboniferous species.   The Athyris Roissyi, too, is apparently identical with
the Devonian and carboniferous shell.

The genus Pentamerus, so characteristic of the Silurian epoch,
and so rare in the Devonian strata, but which has also been found

* The species of Cidaris often quoted from carboniferous rocks, belong, as has
been well shown by M'Coy, to a different family.   They are now termed *Archæo-
cidaris.*   See Annals Nat. Hist. 2 ser. vol. iii. p. 352.

† Composed of those species of Producti which have a distinct hinge area and
hinge teeth.

in the carboniferous rocks of England, has not yet been detected in the Permian deposits. In unison, however, with the prevailing evidences of the development of nature, which, in modifying beings at successive periods, often retained some features of the preceding types, the Pentamerus was represented in the Permian division, by the curious terebratuloid shell, named Camarophoria by Prof. King, which offers in its internal arrangement an approach to the structure of the Pentamerus. Two of the most frequent Permian shells, indeed, are C. (Terebratula) Schlotheimi, von Buch, and C. superstes, de Verneuil, both figured in the work on Russia, vol. ii. pl. 8. figs. 4, 5. These singular brachiopods also disappeared at the close of the Permian era, and their places were occupied by genera (Terebratula, Rhynchonella, &c.) which exist even in the present day.

The annexed woodcut represents a few of the typical Permian fossils; the figures being about half the natural width.

FOSSILS, 56.

PERMIAN SHELLS.

1. Spirifer undulatus, Sow.
2. Productus horridus, id.
3. Schizodus (Axinus) obscurus, id. All from the county of Durham.

4. Strophalosia Morrisiana, King. These species, except, perhaps, the Schizodus, are equally common in Germany and Britain.

In the lamellibranchiate shells (the Dimyaria and Monomyaria of most geological works), we note the same diminution of genera and species in reference to deposits of previous eras; and, whilst not less than 288 species are known in the carboniferous, about fifty-three only have been obtained in the Permian deposits. Modiola is one of the most characteristic genera; M. Pallasi, and M. septifer, being common forms. Schizodus, King (*olim* Axinus), is also a characteristic genus: S. obscurus and S. truncatus are from western Europe; both of them are however common in England and Russia. S. Rossicus, de Verneuil, is a Russian species only. The genus Avicula is also abundant; the small spinose A. Kazanensis being common in Russia. The Avicula speluncaria of Schlotheim is one of the most characteristic species of western Europe, whilst A. keratophoga and A. antiqua, Münst., are frequent. Arca tumida, Sowerby, and species of Edmondia, Myacites, &c. are British shells of this age.

" The Gasteropods appear to have undergone great diminution during the formation of the Permian strata, and to have had great difficulty in accommodating themselves to new conditions. For, if we pass over the seven minute species of Turbo and Rissoa, occurring in one locality near Manchester, the number of Gasteropods known throughout England, Germany, and Russia, in rocks of this age, amounts to but fifteen (*now twenty-one*) species, a number which must appear still more insignificant, when we reflect that as many as 225 species of this class are known in the carboniferous.* These fifteen Permian species are almost all new, three only having been able to live on from the carboniferous to the Permian epoch. The rarity of individual Gasteropods, which are met with in the strata, seems to combine with the paucity of species to make us presume that the causes which were opposed to their free development, produced very extensive effects."

" The Cephalopods, which under the forms of Goniatites, Nautilus, and Orthoceras, were so numerous during the carboniferous period, that 160 species (now many more) have been already described from its strata, were almost entirely annihilated previous to the commencement of the Permian era."

At least it is remarkable, that, with the exception of a Nautilus, N.

---

* In the work on Russia, the Devonian, Carboniferous, and Permian rocks were each designated as " Systems; " but, as explained in this work, they are now viewed as groups, that constitute the " Upper Palæozoic System; " the Silurian being the " Lower Palæozoic." (See also the last chapter.)

Frieslebeni, Geinitz, no well characterized cephalopod has been detected
in the whole extent of the formation.

"Trilobites are entirely wanting. In the study of the palæozoic succession we see that the disappearance of this race is regularly announced by a
gradual diminution of its numbers during the preceding epochs. Appearing among the earliest forms of life, and having their maximum of development in the Silurian period, trilobites decrease very sensibly in the
Devonian strata, and in the carboniferous deposits are reduced to some
few small species, of which Phillipsia and Griffithides are the last expiring forms. And here we are presented with one of those beautiful
links in natural history, of which the strata, forming the earth's crust,
have afforded so many proofs. For, with the final extinction of a family
destined never more to reappear, its place is taken by an allied crustacean,
the Limulus, the earliest form of which was created during the formation
of the great coal-fields, and was followed, in our Permian system, by the
large and remarkable species as yet peculiar to Russia, the Limulus
oculatus (Kutorga). Unlike the trilobite, the Limulus has survived all
the numerous revolutions which have followed its creation, and some of its
species are coexistent with our own race." *

On the whole, therefore, if there are only a few species of
mollusks which are common to the older rocks and the Permian,
the latter still retains its connection with the former through the
genera Productus, Spirifer, Strophalosia, its peculiar group of corals,
and of Bryozoa (Fenestellæ, &c.). It is the constant diffusion of
many individuals of such forms, which induced my associates and
myself to hold firm in retaining the term Permian, as marking
the close of primeval life, and in separating it from the Trias and
secondary deposits, with which it has no form in common.

The fossil fishes of this era, of which fifty-three species are
known, all belong, like their precursors, to the division with
heterocercal tails; a distinction that becomes more and more
evanescent in all the succeeding secondary and tertiary formations,
and in our era is confined to the shark and sturgeon families;

* Russia in Europe, vol. i. p. 209, &c.

nearly all other species now living, about 8000 in number, having
homocercal tails.

Of true heterocercal fishes of the Permian group, two figures
only are here given, by which the reader will see that the backbone

FOSSILS, 57.

PERMIAN FISH.

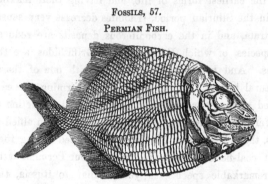

Platysomus striatus, Agassiz ; from the " Marl Slate," Durham.

is prolonged with a bend into the end of the upper lobe of the
tail; whilst in the homocercal fishes, or those common in the younger
rocks and in the present time, it terminates in the middle, between
the equally balanced fins of the tail.

The Palæoniscus, Pygopterus, Cœlacanthus, Acrolepis, and Platysomus, which are the prevailing genera of the Permian era (the
same generic forms being found in England, Germany, and Russia),
are all known in the Carboniferous epoch, but the species of the
younger deposit are wholly distinct.

FOSSILS, 58.

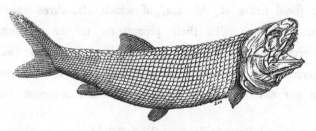

Palæoniscus Frieslebeni, Agassiz ; from the " Kupfer Schiefer " of Mansfeld, in Thuringia.

Lastly, in enumerating the chief features of the Permian fauna, we have to recollect, that whilst a very few reptiles have been found in the next subjacent formations (Carboniferous and Devonian), the strata of this newer era contain the relics of Thecodont Saurians, which Owen refers to a higher order of reptiles than any of the older fossils of this family, by showing that they are even allied to the living monitor.* The Palæosaurus and Protosaurus were, according to him, Sauri which occasionally walked on *dry land*.

Struck with this fact of the Permian deposits containing Saurians, Agassiz thought that the group ought to be placed at the base of the secondary rocks, which mark the great era of reptiles. But the recent discovery of Saurians and Batrachians in the Carboniferous rocks of Europe and America, and of one reptile in the Upper Old Red, have satisfied my colleagues and myself, that we were right in being guided by the general facies of the mollusks and plants, and in grouping the Permian with antecedent deposits, with which they are naturally united by many links.

In reviewing the few reptiles of the Devonian, Carboniferous, and Permian rocks, which have been discovered, we have yet scarcely sufficient materials to enable us to decide on the relative value of such animals in the scale of creation. As far, however, as the above mentioned evidences go, Professor Owen is of opinion, that the solitary, small reptile of the Upper Old Red Sandstone, found in the vicinity of Elgin (see p. 254.), belongs to the Lacertilia, a higher order than the Batrachia, to which the reptiles of the coal are referred. At the same time the latter possess characters in their skull and teeth which bring them nearer to the Saurian order than any modern Batrachian.

* See Russia in Europe, vol. i. p. 213, and Professor Owen's description, p. 637, App.

The same eminent comparative anatomist affirms, that the The-
codonts of the Permian are more nearly allied to the Crocodilians of
the Lias, and so to the higher order of reptiles, than any living
monitor; and " just," says he, " as the coal Batrachia are more
sauroid, so the lacertilian sauroids of the Trias (New Red) are more
Crocodilian than existing frogs and lizards are." In a letter to
myself, from which the above view is derived, he adds, " So far as
Old Red (Upper), Coal, and Permian reptiles are known, they support
the general fact, that the more ancient forms of vertebrata adhered
closer to the general archetype of that sub-kingdom than the more
modern forms."

Thus we learn, that in the oldest rocks (Upper Palæozoic), in
which they first appear, the reptiles are as wonderful and elaborate
in structure as the primeval fishes which accompanied them, or as the
corals, crinoids, shells, and crustacea which preceded them during
the Silurian or Lower Palæozoic epoch.

In concluding this chapter, it may indeed be re-asserted*, that the
mass of the organic remains of the Permian system constitute a mere
remnant of the earlier creations of animals, the various developments of which we have followed in the preceding pages. They
exhibit the last of the partial and successive changes which these
creatures underwent before their final disappearance. The dwindling away and extinction of many of the types, which were produced
and multiplied during the anterior epochs, already announce the end
of the long palæozoic period.

In ascending above the highest of the Permian deposits, the
geologist takes, indeed, a sudden and final leave of nearly everything in nature to which the words primary, primeval, or palæozoic
have been or can be applied.

In short, the two greatest revolutions in the extinct organic world

* See Russia in Europe, vol. i. p. 205.

are those which separated the primeval or palæozoic rocks from the mediæval or secondary strata, and the latter from the tertiary and modern deposits.

To the consideration of these two remarkable and salient features in the history of former races, we shall briefly advert in the concluding chapter.

P. S.—A good additional palæozoic form has been discovered in the Zechstein of Ilmenau, and is described by Professor Geinitz as *Conularia Hollebeni*. Zeitschrift der Deutschen-Geologischen Gesellschaft, vol. ii. p. 465.

I may also mention here a fine specimen of the *Xenacanthus Dechenii*, Beyrich, which, with other Permian fishes (Holacanthodes, &c.), I saw lately in the Museum at Bonn. The fish in question has the fins and tail expanded, and measures 15 inches in length. It is from Ruppendorf. (See p. 339. See also Leonh. and Bronn, Neues Jahrb. 1849, 118.)

CHAPTER XIII.

GENERAL VIEW OF THE SILURIAN, DEVONIAN, AND CARBONIFEROUS ROCKS
OF SCANDINAVIA AND RUSSIA.

IN the last chapter only, which explained the nature of the Permian Rocks, has a sketch been given of the development of any palæozoic deposits upon the Continent. An effort must now therefore be made, to delineate, if only in the most general manner, the equivalents in other countries of the Silurian, Devonian, and Carboniferous groups of Britain.

Throughout large portions of Western Europe, *i.e.* Germany, France, Spain and Portugal, the deposits under consideration have been in some tracts so much metamorphosed and crystallized, in others so penetrated by igneous rocks, and even so dislocated, that notwithstanding the researches of numerous good geologists and mineralogists, the task of reducing them to a normal order of succession is far from being completed. Deferring, therefore, to a subsequent chapter, such explanation of these complicated regions as may be practicable, let us first take a broad view of the succession of primeval life in Scandinavia and Russia in Europe, where, on the contrary, the series of the older sediments is exhibited, over very wide areas, in the clearest and most symmetrical order, and usually uninterrupted by the intrusion of igneous or volcanic rocks.*

---

* The limits of this work do not permit any attempt to delineate the mineral structure of the Ural Mountains, except to indicate by the way how clearly they exhibit the metamorphism of the palæozoic deposits of Scandinavia and Russia in Europe. An acquaintance with their structure must be sought in the works of many authors, from the time of Pallas to the days of the most illustrious of all travellers Humboldt, who explored these regions accompanied by his eminent friends Gustaf Rose and Ehrenberg, and extended the lights of physical science to

In Scandinavia and Lapland, ancient crystalline rocks occupying the chief mass of that territory, and consisting to a great extent of granite and gneiss, with many varieties of schist and quartzose strata in which nothing organic has been discovered, are covered at intervals by true representatives of the Silurian rocks.

The lowest beds which are charged with organic remains, are clearly the equivalents of the lowest Silurian strata of the British Isles, Bohemia, and other countries. In some places these strata repose upon the above-mentioned crystalline schists, which are probably altered or metamorphosed primary sediments. The first of the following diagrams represents the succession of Lower Silurian deposits from a base of crystalline gneiss and associated strata near Kinnekulle in Sweden.*

the frontiers of China. Besides the "Reise nach dem Ural, dem Altai, &c.," of those authors, the reader will find a great body of information in the Archiv für Wissenschaftliche Kunde Russland, conducted by M. Adolf Erman, the explorer of North-eastern Siberia and Kamschatka. Among the authors who have written on the mineral structure of Siberia, Helmersen and Hoffman also stand out conspicuously, as will be seen by those who consult their publications in the " Annuaire des Mines de Russie." The splendid work also of M. Pierre de Tchihatcheff, on the Altai Mountains, and many others, would have to be noticed; but as this volume is chiefly an outline of the nature and succession of the older sediments, I cannot here expatiate upon the labours of many of my contemporaries.

In the work "Russia and the Ural Mountains," references are given to the authors, both anterior and contemporary, who have illustrated the older sedimentary deposits and their fossils in the Russian Empire. Special allusion is there made to the original sketch, by the Hon. W. Fox Strangways, of the environs of that metropolis, in Trans. Geol. Soc. Lond. 1 ser. vol. v. p. 392., to the work of Pander on the fossils of the same district, and to the volume of Eichwald, ' Système Silurien de l'Esthonie.' Since the publication of our work, my friends and self have been gratified by seeing it translated (1849) into the Russian language by Colonel Osersky, who has added some important data from his own observation and other sources, including corrections of our general geological map.

* A detailed geological map of all the crystalline and metalliferous rocks of Sweden and Norway has for some time been in preparation, but is not yet published. This task was undertaken by MM. Forsells, Erdman, Franzen, and Troilius.

d. Black graptolite schists. *c.* Orthoceratite limestone. *b.* Alum slates. *a.* Lowest or fucoid sandstone.

gn. Gneiss.   *t.* Trap or eruptive greenstone.   *bl.* Erratic blocks.

In other parts of Sweden, as near the Billingen Hills, the two lower strata *a, b,* of the above section are seen to rest upon granitic rocks without any intervening crystalline schists. In such cases the lowest sandstone, which is used as a millstone, is simply a re-aggregated granite, the *arkose* of Brongniart; the materials being derived from the subjacent rock.

b. Alum schist.   *a.* Fucoid sandstone.   *a'.* Millstone.   *gr.* Granite.

Now, whether the subjacent rocks be composed of granite, granitic gneiss or flinty slates, quartzose or other crystalline masses, the fundamental strata in which any traces of former life can be detected, are sandstones and schists, which stand in the place of the lowest fossiliferous beds of Britain, and of that primordial zone in Bohemia which will presently be noticed. The bottom beds contain fucoids or casts of sea-weeds only; but the lowest band of schist is well characterized by its fossils in several parts of Sweden, and particularly at Andrarum, where it contains trilobites of the genera Paradoxides, Olenus, Conocephalus, Agnostus, &c.

The limestones which next overlie (*c*, of the opposite page), and which abound in Orthidæ, large Orthoceratites, Trilobites, and Cystideæ, are distinctly the equivalent of the British Llandeilo formation of schists, flags, and limestone, &c. Above these, but connected with them, is a considerable mass of shale or schist, chiefly characterized by graptolites, and this is covered by other limestones, which in Gothland are profusely charged with many species of shells and corals that are found in the limestones of Wenlock and Dudley. The whole of these are capped, in the south of Gothland, by certain sandy strata, which are meagre equivalents of the Ludlow Rocks.*

In Norway, all these Silurian rocks, very clearly divisible into ' Lower' and ' Upper' through their characteristic fossils, are regularly exhibited in a very small compass in the Steensfiord, between the river Drammen and Krokleven, to the west of Christiania, where their uppermost member is conformably overlaid by a great accumulation of Old Red Sandstone with conglomerate, as expressed in this diagram.†

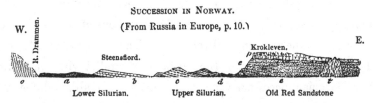

SUCCESSION IN NORWAY.

(From Russia in Europe, p. 10.)

Lower Silurian.   Upper Silurian.   Old Red Sandstone

o. Azoic crystalline rocks.   *a.* Lower Sandstones, Schists, and Flags.   *b.* Limestone, with Pentameri.   *c.* Coralline limestone (Wenlock).   *d.* Calcareous flagstones (Ludlow?).   *e.* Old Red Sandstone.   (*p.* Rhombic porphyry.   *t.* Other eruptive rocks.)

The same strata of Lower and Upper Silurian prevail in the Bay of Christiania; but there they are at many points contorted and penetrated by syenites, greenstones, and hypersthenic rocks, whereby

* See details of this succession, by Murchison and De Verneuil, Quart. Journ. Geol. Soc. Lond. vol. iii. p. 1.

† For the particulars of these phenomena, consult "Russia in Europe," vol. i. p. 10. *et seq.*, and Quart. Journ. Geol. Soc. Lond. vol. i. p. 467. (Murchison "On the Palæozoic Deposits of Scandinavia, &c., and their Relations to Azoic or more Recent Crystalline Rocks "), 1845.

the lower alum schists are crystallized, and the limestone, *b*, converted into marble. The next section (a continuation to the east of
the preceding diagram) explains their relations, and shows how the
undulating and broken masses, regaining their order, fold over and
dip under a great mass of Old Red Sandstone, the eastern extremity
only of which is represented in the preceding diagram.

W.   Christiania. Egeberg.   E.
      Ringerigge.

e. Old Red Sandstone. *d*. Calcareous flagstones. *c*. Coralline limestone and shale (Wenlock), beautifully exposed in many islands of the bay. *b*. Limestone, in parts a marble (Paradis
backen). *a*. Lower Silurian Schists, &c., in parts altered. *o*. Azoic Gneiss. *t*. Various
eruptive rocks. (The *t* of this section represents most of the intrusive rocks, whether syenites,
greenstones, porphyries, or younger granites, which have been forced through the strata.
p refers to the large rhombic porphyry on the summit of the plateau.)

Referring the reader for the general description of Scandinavia
and for the detailed description of its fossils to the works of native
authors*, let me here only call attention to the general comparison
of the lower palæozoic strata of this region with those of Britain. In
doing so, I specially advert to the above sections of a tract which I
examined in the company of Professor Keilhau, in the year 1844, when
I placed the older strata of Norway in exact parallel with the Silurian
and Devonian succession of my own country. For, although the many
thousands of feet of the Lower Silurians of Britain are there represented by a few hundred feet only of thickness, these black Norwe-
gian schists, with their overlying limestones, are as copiously charged

---

* In this brief sketch, no justice can be done to the numerous works on the
rocks and fossils of Sweden and Norway, countries which, from the time of Linnæus to the present day, have contributed so much to the progress of Natural
History. In the memoirs of Nilsson of our time, in those of Wahlenberg and
Gyllenberg of past years, and in the works of Löven and Angelin, which have
recently appeared, the fossil organic remains have received much illustration, in
addition to the knowledge formerly communicated in the "Lethæa Suecica"
(1837) of Hisinger. The "Gæa Norvegica" of Keilhau has great merit for its
description and delineation of the various rocks of Norway.

with characteristic trilobites and other fossils, as the enormously expanded British rocks of the same age.   The latter, as seen in North
Wales, have, in fact, been swollen out by a much more profuse and
active deposit of materials; the bottom of the sea, in which they
were accumulated, having been subjected to a turbulence produced
by contemporaneous submarine volcanos, from which this northern
region was exempt.

The primordial fauna of Scandinavia, which I long ago characterized as "the earliest zone of recognizable life," consists, in its lowest
stage, of fucoids only.   The next stage contains a number of trilobites, which have been since elaborately described by M. Angelin*,
who has added much to our acquaintance with these fossils.   Among
the crustacea of these lowest shales (*Regiones* A, B, of Angelin), we
recognize several of the genera which characterize the lowest fossil
beds of the Silurian region of Britain.   The most plentiful fossils are
the Agnostus pisiformis, and many species of Olenus, one of which
probably also occurs in the lowest schists of our Malvern Hills.

Then, the Swedish *Regiones* C and D of Angelin are nothing
more, as respects their fossils, than masses of Lower Silurian Rocks,
as originally described in England and Wales, and which reappear in
great force in the Bay of Christiania, Norway.   They are specially
identified by the abundance of trilobites, of the same genera; —
Asaphus, Ogygia, Remopleurides, Trinucleus, Illænus, Ampyx, &c.,
as our British strata of this age.   Several species are identical.

Having examined these Lower Silurian Rocks over various tracts
of Sweden and Norway, I can assure geologists that, rich as they
are in fossils — so rich that they have enabled M. Barrande, as will
soon be seen, to parallel every one of their stages, from the lowest
upwards, with the primordial and succeeding zones of Bohemia—the
whole of this Lower Silurian of Scandinavia never exceeds in vertical
thickness 1000 feet!   And yet this mass is nearly as complete in the
development of life, as the 30,000 feet of strata of the same age in

* Palæontologia Suecica, pt. 1., 1852.

Britain ! Nay, the general succession is essentially the same as in our islands. For, the Lower Silurian of Sweden and Norway is overlaid by shale and limestone, which completely represents the Wenlock of England; whilst M. de Verneuil and myself have endeavoured to show, that in the south of Gothland still higher rocks are exhibited, which, though of feeble dimensions, are on the parallel of the Ludlow formation of Britain.

And here, in respect to nomenclature, I may observe, that the geologist who should seek to separate the Lower Silurian of Norway from the Upper, and call the higher part only Silurian, would make an effort which could not be applied to any map. For these two groups are there often exhibited in the space of an English mile, and roll over in such frequent and parallel undulations, as to *constitute one united mass of very small dimensions.*

The remarkable resemblance of the fossils of the chief limestones of Gothland, and of certain isles in the bay of Christiania, to those of Wenlock and Dudley, a very great number being common to the two localities, is well known to geologists and collectors. I have already spoken of some of these—the corals—on page 215. Those persons who may wish to understand how the calcareous masses of Gothland repose on graptolite schists, and these again upon the lower Silurian rocks of Öland, or again pass upwards into schists and sandstones (in the South of Gothland) which represent the Ludlow rocks of England, must refer to the published descriptions of M. de Verneuil and myself.*

Primeval Rocks of Russia. — Following the older rocks from Scandinavia into the Russian Baltic provinces, the hard crystalline nucleus subsides, and the lowest Silurian rocks being depressed beneath the sea-level, the absolute bottom of the fossiliferous strata is no longer visible, as in Scandinavia.

Extending from the bed of the Neva at St. Petersburg to the cliffs west of Narva, the lowest strata are shales, often unconsolidated,

* Quart. Journ. Geol Soc. vol. iii. p. 28, &c.

in which little more than fucoidal remains have been found, or bodies which Pander has termed Platydolenites.* This shale, in which green grains and a few thin sandy courses appear, is so soft and incoherent, that it is even used by sculptors for modelling, although it underlies the great mass of fossil-bearing Silurian rocks, and is, therefore, of the same age as the lower crystalline and hard slates of North Wales; so entirely have most of these oldest rocks in Russia been exempted from the influence of change, throughout those enormous periods which have passed away since their accumulation!

The shale is followed by a sandstone, in parts coherent, which in other tracts is a green-grained calcareous grit (as under the Castle of Narva), containing the little horny Obolus Apollinis (the ‘Ungulite’ of Pander), one or two species of Siphonotreta, and the Acrotreta of Kutörga. These strata of green sand are covered by and sometimes interlaced with bituminous schist†; whilst the limestone which succeeds is identical, through many of its fossils, with the chief Lower Silurian limestone of Sweden; some of the

* Bullet. de la Soc. Geol. de Fr. t. viii. p. 252. In some greenish, sandy beds subordinate to the lowest Silurian strata of the Baltic provinces of Russia, M. Pander also recently detected very minute bodies, not larger than pins' heads, which he supposed might be the teeth of fishes; but the naturalists to whom they have been submitted, both in France and England, including M. Barrande and Dr. Carpenter, can see in them, when placed under powerful microscopes, no evidence whatever of *bone*. They are probably, it is suggested, fragments of the hard, crustaceous ends of the segments of some trilobite.

† In addition to various able Memoirs which he has written on the crystalline and palæozoic rocks of Russia, and to the geological maps which he has prepared, my associate in the Imperial Academy of St. Petersburg, Colonel Helmersen, has published an account of the nature and position of this schist, which in some places contains about 25 per cent. of bitumen and a considerable portion of inflammable gas. It is certainly one of the oldest deposits which have combustible properties, though the anthracitic layers of the South of Scotland seem to lie still lower in the series. It is probable that most of the bituminous and anthracitic beds of these early periods were formed out of the marine vegetables, or fucoids, of which the ancient rocks offer impressions and casts, occasionally of large size. — (See Helmersen, Sur le Schiste argileux-bitumineux d'Esthonie, Journal des Mines de Russie, 1838, p. 97.)

species being also well-known fossils of the British Isles. This limestone, for the most part thin-bedded and marly, occupies various low
hills south of St. Petersburg, and thence ranges into the cliffs of the
Gulf of Bothnia, where it thus appears in relation to the inferior strata.

LOWER SILURIAN ROCKS.—CLIFFS NEAR WAIWARA.

a. Lower Shale, obscured by fallen blocks.   *b.* Ungulite grit.   *c.* Bituminous schist, with
graptolites.   *d.* Orthoceratite Limestone (covered by granite blocks which were transported
from Sweden in the glacial periods, 'Russia in Europe,' vol. i. pp. 34. 511., *et seq.*).

These lower limestones, which constitute the chief members of
the Lower Silurian of Russia, are filled, as in Scandinavia, with a

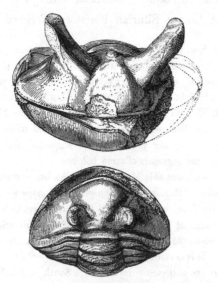

ASAPHUS EXPANSUS, var. CORNUTUS, from the St. Petersburg Limestone.
(From "Russia and the Ural Mountains," vol. i. p. 87.)

profusion of trilobites, large orthoceratites, univalve shells, and
cystideæ, characteristic of this age. One of the most remarkable of
the trilobites, the Asaphus cornutus, is here represented.

Orthis calligramma and the great Orthoceras duplex among the shells, and Spheronites aurantium among the Cystideæ, may be considered quite characteristic of this formation.   But we must refer to the works of Pander and Eichwald, and of my colleagues and self, in which these rocks and their contents are described, for a full description of the fossils.

Upper Silurian. — Upper Silurian deposits, like those of Gothland, were only spoken of by my companions and self as existing in the Isles of Oesel and Dago. With the little time at our command, and in a flat country much obscured by drift, we could not pretend to draw any well defined line between the Lower and Upper Silurian. Seeing, however, that Pentamerus oblongus occurred in a calcareous band ranging between the Lower Silurian cliffs of Esthonia and the Devonian rocks of Livonia, it was natural in our rapid survey to refer that stratum, as I had done in England, rather to the top of the lower than to the base of the upper group.

Subsequent researches, however, and the preponderance of certain fossils, have led the Russian geologists to class this Pentamerus band with the Upper Silurian.*   In truth, the whole series in Russia, as

* M. Ozersky, who translated the work of 'Russia in Europe' into the national language, was the first person to call attention to the succession of those upper calcareous Silurian strata, though, when he wrote, the fossil distinctions above alluded to were not elaborated. Afterwards, in a short article (Erman's Archiv für Russland, 1848, p. 236.), M. Pander noticed the superposition of a rock charged with Pentamerus oblongus and many corals, to the well-known lower limestone of St. Petersburg. The most important, however, of these contributions which clear up the subdivisions of the Russo-Baltic Silurians, is by Professor Kutorga, as given in a recent map of the government of St. Petersburg, the result of ten years of labour (" Compte rendu de la Société Minéralogique de St. Petersbourg, 27 Janvier, 1852"). Representing a continuous band of Upper Silurian between the Devonian and the Lower Silurian to the *west* of St. Petersburg (a tract which I never examined), he admits that, to the south and east of the metropolis, the succession is what my friends and self represented it to be ; viz., Devonian rocks at once resting on Lower Silurian, with one or two minute traces only of the lower band of the Upper Silurian, or the limestone with Pentamerus oblongus.

in Scandinavia, is of such very small vertical dimensions—not, perhaps, a fortieth part of the magnitude of the grand British deposits—that in a region where all these strata are conformable, it is only by close examination of organic remains that such distinctions can be drawn. The lower limestone, with Orthoceras duplex (Foss. 59.),

FOSS. 59. ORTHOCERAS DUPLEX. WAHLENBERG.

A characteristic Russian and Scandinavian Fossil.

Orthis biforatus, many Cystideæ, Trilobites of the genera Asaphus, Ampyx, Cybele, Illænus, &c., and many other true Lower Silurian fossils, is conformably overlaid by another limestone, in which certain Wenlock forms, particularly a multitude of corals, predominate, but in which the Pentamerus oblongus also occurs. Both these calcareous masses (Lower and Upper), which are perfectly united and conform-

The Russian geologists seem, however, to have to learn, that in Britain this species of Pentamerus ranges from the top of the Caradoc or bottom of the Wenlock shale down into the middle part of the Llandeilo formation; and they will therefore understand the nature of my reasoning pp. 86. 98. *antè.* Wherever, as in Russia and America, this species *appears for the first time, in ascending order,* in strata otherwise exclusively charged with Wenlock fossils, the geologists of such countries have a perfect right to class the rock as Upper Silurian. According to Kutorga, the band of true Lower Silurian limestone above noted is first followed upwards by a course of limestone without fossils, which he leaves in the lower division ; next by a magnesian limestone with Pentamerus, and this again by beds containing fucoids, encrinites, and corals. According to this author, these beds are surmounted, in the islands of Öesel, Dago, and Möen, by still higher Silurian strata. These Russian isles seem, in fact, to be equivalents of the southern end of Gothland (Hoburg), where I have described certain strata which probably represent the Ludlow Rocks. (Quart. Journ. Geol. Soc. Lond. vol. iii. p. 1.) In another work, " Uëbersicht des Silur-Schichten Systems Lieflands und Esthlands, 1852," M. Schrenk, the well known explorer of the botanical and mineral condition of the north Ural Mountains, has, in conjunction with Professor Schmidt, extended considerably the range of the Upper Silurian. He speaks of a mixture of the fossils of the Lower and Upper Silurian rocks.

able, constitute only what would be called one formation, not distinguishable except by fossils. The upper division occupies the whole of the isles of Öesel, Dago, Wörmes, and a narrow zone on the mainland, as indicated on the recent map published by Professor Kutorga. The chief mineral features of these overlying rocks, in ascending order, consist of fine-grained limestone, passing in some beds almost into marble, and thus presenting a contrast to the incoherent Lower Silurian limestones of St. Petersburg; next, of magnesian beds (dolomite) occasionally somewhat crystalline and cavernous, here and there charged with white silicious concretions. In other parts, they are friable and of grey and yellow colours.

Rocks having the same mineral aspect, and containing the same fossils, appear, as I am informed by Count A. von Keyserling, along the edge of the Timan Hills, in the distant region of the Petchora, at the mouth of the river Tatchina, and near the shore of the Icy Sea. It is further to be remarked, that in no part of Russia has there been detected the slightest unconformity between the Lower and Upper Silurian; and that with the exception of some of their probable equivalents in the isles of Dago and Öesel, the Ludlow rocks have no true representatives.

The inferior part of the so-called Upper Silurian of the Russian geologists is chiefly characterized by the following fossils: — Pentamerus oblongus, P. borealis, Leptæna depressa, Atrypa reticularis, Bumastus Barriensis? Orthoceratites, some Trilobites, Leperditia marginata, and L. Baltica; with a profusion of corals of the genera Stromatopora, Halysites, Cyathophyllum, Favosites, &c., chiefly known elsewhere in the Wenlock formation.

Devonian Rocks of Russia. — This formation constitutes one of the most widely spread deposits of the Empire, and extends over a region more spacious than the British Isles. This great extension is not, however, due to the thickness, but the very slight inclination of the beds, and to their having been but rarely disturbed. Though subjected to broad and very slight bends only, they rise into the Valdai Hills, and also into a broad dome near Orel; thus constituting the highest eminences of this comparatively low region.

In addition to many shells, some of them peculiar to the region, but others well known in strata of this age in Devonshire, the Rhenish provinces, and elsewhere, fossil fishes occur in numerous localities, and strikingly characterize this deposit in Russia. Several of them are identical with species of the Scottish Old Red Sandstone, as shown by Agassiz; whilst, from a microscopic examination of their teeth, Owen has proved that others belong to Dendrodus, a genus* of sauroid fishes, equally characteristic of the formation in North Britain. The forms of Coccosteus, Cephalaspis, Acanthodes, Diplacanthus, Cheiracanthus, and Cheirolepis, have not, it is true, been found; but other Scottish species not less remarkable have their exact representatives in Russia. Such are the great Asterolepis (Chelonicthys) Asmusii, Ag.; and the smaller species, A. minor, Ag.; Glyptosteus favosus, Ag.; Bothriolepis ornata, Eichw. (Glyptosteus reticulatus, Ag.); Holoptychius Nobilissimus, Ag.; Dendrodus strigatus, Owen; D. biporcatus, Owen; Cricodus incurvus, Ag. (Dendrodus, Owen); Ptericthys major, Ag., &c.†

RAVINE OF THE BELAIA IN THE VALDAI HILLS.
(From p. 46. Russia in Europe.)

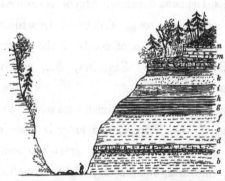

The lower beds a, b, c, d, e, f, consist of various sands and marls, in which ichthyolites are disseminated, but of which the bed d is a complete congeries of fish-bones surmounted by a copious mass of red, white, and green argillaceous marls. Then follow bituminous schists (g, h, i) with courses of bad coal, constituting the bottom beds of the carboniferous deposits, which, after other alternations of sands and marls (k, l), are followed by the carboniferous limestone m, n, with many characteristic fossils, including even species which are well known in Britain, such as the *Productus hemisphæricus, P. punctatus,* and *P. semireticulatus.*

* Russia in Europe, &c., vol. i. p. 635.   † Ibid. p. 40., and fig. p. 636.

Reposing on Silurian rocks, these Devonian strata are in numberless places overlaid by carboniferous limestone. In some tracts the bottom beds are flag-like shelly limestones and marls, in others, sandstones both soft and hard, which range over extensive northern districts to the banks of the lake of Onega and the shores of the White Sea.

In certain parts, as in the Valdai Hills, their upper beds, represented in the opposite woodcut, are red and green marls, containing several of the above-named fossil fishes of the Old Red Sandstone of Scotland; the whole mass being covered by carboniferous limestone, with its usual fossils, including large Producti.

When my colleagues and myself explored Russia, the connexion, between the character of the fossils and the nature of the matrix in which they are imbedded, was more pointedly brought before us in our range over that vast Empire, than in any other region with which we were acquainted. In Courland, Livonia, and the Baltic governments, as well as in the great central dome-like region of Orel, to which the system extends, and where it has since been ably examined and described by Helmersen—finely laminated limestones alternate with, and are subordinate to, great masses of sand, marl, and flagstone; and whilst in the calcareous courses mollusks prevail, with occasionally some remains of fishes, the latter are found, almost exclusively, in sandy and marly beds. In tracing these rocks from the Baltic provinces to the White Sea and Archangel, the limestones thin out, and the group is there represented only by sands and marls. Accordingly we find that where the rocks have the sandy character of the chief masses of Old Red Sandstone in the British Isles,—there they are also tenanted by fishes;—a remarkable geological phenomenon, showing an accordance between the lithological character and organic contents of rocks of the same age in very distant countries.

Another and not less remarkable fact obtained by the survey of

Russia is, that fossil fishes, of species well known in the Old
Red of Scotland, have in several places been found to be intermixed, in the same bed, with those shells that characterize the
system in its slaty and calcareous form, as exhibited in Devonshire and
many parts of the continent.   This phenomenon, first brought to
light by my colleagues and myself, demonstrates more than any
other the identity of deposits of this age, so different in lithological
aspect, in Devonshire on the one hand, and central England and
Scotland on the other.   The same admixture has since been recognized to some extent in the Devonian rocks of the Eifel and the
Hartz, and also in North America, as will be stated in the next
chapter.

DEVONIAN AND CARBONIFEROUS ROCKS IN THE GORGE OF THE TCHUSSOVAYA RIVER.
(From Russia in Europe and the Ural Mountains, vol. i. p. 386.)

In speaking, however, of identical formations having a very dissimilar aspect, both mineralogically and pictorially, we can find no greater contrast than exists in the different parts of the vast Russian Empire itself. Thus the soft red and green Devonian marls of the Valdai Hills are represented on the western flank of the Ural Mountains by hard, contorted, and fractured limestones, as in the preceding sketch, made in the gorge of the river Tchussovaya.

With the progress of research, M. Kutorga has divided the Devonian rocks of Russia into the following three stages, in ascending order: 1, clay and marl, with Lingula bicarinata; 2, compact sandstone, with many ichthyolites; and 3, argillaceous limestone, with many brachiopods.

M. Pander, in the notice previously referred to, has even shown that the highest of these beds are covered by other sandy strata, in which fossil fishes equally abound, whilst in the more southern tracts near Voroneje*, it was shown, by my associates and myself, that many species of brachiopods and fishes were absolutely commingled in one and the same marly stratum. It follows, therefore, that the minute distinctions traceable in one district are not applicable to another; and that the limestones are mere mineral accidents in a great arenaceous and marly series of strata; whilst, on the whole, the identification of the group with its equivalents in other parts of the world has been established in the clearest manner.

In the distant region of the Petchora, Count Keyserling assured himself that certain beds charged with Goniatites are inferior to all the other Devonian strata of Russia, whilst, as we shall presently see, the chief goniatite deposit on the banks of the Rhine is underlaid by fossiliferous limestone and sandstones of great thickness, which there constitute the central and lower members of the Devonian

* Russia in Europe, &c., p. 60.

series of the Rhine. It would, indeed, appear that there is no known equivalent in Russia of the lower shelly greywacke, or great ' Spirifer-Sandstein,' which forms the base of the Devonian rocks of the Rhine.*

That the Devonian is completely independent of the Silurian in Russia, is not only demonstrated by the marked difference in its fauna, but also from its reposing in one tract on the Upper, in another on the Lower Silurian, as demonstrated by Helmersen†, Kutorga, and others.

Of the Carboniferous system, as exhibited in Russia in Europe, it is enough to say, that its lowest members only are there well developed. In the small section above given (p. 328.), we see in the uppermost strata the representatives of the carboniferous limestone, with coal bands inferior to it — immediately lying on the red and green rocks with Devonian fish. To the north-east, or on the Andoma River in the government of Olonetz, this division swells out in its calcareous members, but loses even the few courses of poor coal above mentioned. In its range by Moscow to Tula and Kaluga, the limestone, which is in parts yellow and a true earthy dolomite, contains also thin subordinate seams of coal. Again, in the southern region of the Donetz, between the Don and the Dnieper, where these lower coal strata reappear, they are of much larger dimensions, with several beds of coal of better quality, which, as in other places (Ch. XI.), are interstratified with beds of true carboniferous limestone.

In this last tract, however, the prevailing horizontality of the palæozoic strata of this vast country no longer exists. The limestone, sandstone, and shale, with coal, have there been violently

* Goniatites occur in Upper Silurian in Bohemia, according to M. Barrande.

† In addition to Colonel Helmersen's survey of the Devonian rocks, which extend so largely throughout Central Russia, M. Raimund Pacht has also described in some detail the nature of these rocks, and some of their fossils, in Livonia. (Devonische Kalk in Livland. Dörpat, 1849.)

extruded to the surface through a cover of surrounding jurassic and even cretaceous deposits, and are very highly dislocated.

But if, as it has been suggested, Russia* should open out deep shafts *around* the field of the Donetz, the beds of lower coal which she possesses may be found to lie in valuable and regular masses beneath the unbroken secondary rocks.

Some of the grandest examples of the carboniferous limestones of Russia are to be seen in the banks of the Volga, near Syzran, where thick beds of the rock, containing fossil species common to this division in many quarters of the globe, are loaded with small Fusulinæ, which were first noticed by Pallas, and are remarkable in being the oldest of this class of foraminifers; the Fusulinæ being in truth the prototype of the Nummulite, which becomes so abundant in the lowest tertiary deposits.

Other and not less gorgeous masses of the same limestone, but wherein no coal exists, are displayed on the flanks of the Ural Mountains, where they undulate conformably with underlying limestones which are charged with Devonian fossils.

Without repeating sketches which are given in another work, the reader may be referred to the preceding woodcut, p. 330.; for several of the distant peaks there represented are carboniferous limestones overlying rocks of Devonian age. The clearest proofs are thus exhibited (the two formations being conformable in all their undulations on the western flanks of the Ural), that the portion of the series, which is slightly coal-bearing in central Russia, and much more so on the Donetz, is here entirely wanting.

These carboniferous limestones, every where more or less characterized by their Producti, are overlaid along the western edge of the

---

* In the work on Russia I suggested, that whenever the Imperial Government might wish to make trials for *what may very probably prove to be a valuable and unbroken coal-field*, sinkings should be made through some of those surrounding horizontal secondary formations not far distant from the edge of the upcast and dislocated coal-field. (See Russia in Europe, &c., p. 118. *et seq.*)

Ural Mountains by sandstones and grits, which occupy much the same place in the general series as the millstone grit of England.

In Russia these strata are, however, more calcareous, and contain goniatites. They constitute the uppermost courses of the carboniferous system, properly so called, in the vast territory between the Volga and the Ural Mountains, where they are seen in many localities to be at once succeeded by the Permian deposits before described (see diagram, p. 296.). As they only contain at intervals very rare and thin traces of coal, it is manifest that the place of the great upper coal-fields of England is unoccupied by any due representative in the Russian Empire.

The deposits of Silurian, Devonian, and Carboniferous age, which spread in such wide, horizontal, and slightly broken masses over Russia in Europe, are thrown up in the Ural Mountains as metamorphosed and highly mineralized rocks, some idea of which is afforded in the previous sketch (p. 330.) of the Devonian and carboniferous rocks in the gorges of the Tchussovaya river. Other views given in a subsequent chapter on the auriferous rocks, will realize to the eye the condition of the more central and completely altered rocks of that chain. In the mean time it may be stated, that hard and crystalline palæozoic strata, associated with numerous eruptive rocks, form the nuclei of all the mountainous ridges in Siberia, whether in the lofty Altai on the south, where they have been well described by M. Pierre de Tchihatcheff*, or in those hills which act as separating barriers between the great rivers Ob, Jenissei, and Lena, and extend to the Sea of Ochotsk and Behring's Straits on the north-east. Grand, therefore, as is the spread of soft and slightly coherent primeval rocks over Russia in Europe, the hard stony tracts of this age which extend over the Asiatic regions of the same great empire are probably much more extensive.

* See the magnificent work, L'Altai Orientale, of M. Pierre de Tchihatcheff. Some Devonian fossils have been found near Lake Baikal.

Restricting our view to Russia in Europe, we see the Silurian and Devonian rocks arise partially to the surface in the southern provinces of Podolia, on the banks of the Dnieper; and, although no rocks of such high antiquity have yet been detected in the Caucasian mountains, they reoccur in the same latitude near Constantinople*, have been recognized in several parts of Asia Minor to the south of the Black Sea†, and also traced (most of them belonging by their fossils to the Devonian type) to even the southern flanks of Mount Ararat, and thence eastwards into Persia.‡

Looking, therefore, to Scandinavia as a great mass of ancient crystalline rocks, whose origin is buried in obscurity, we know that it has been successively overlapped by those Silurian, Devonian §, Carboniferous, and Permian strata, which occupy such an enormous region in Russia. The discovery in Spitzbergen of certain palæozoic fossils, one of which Leopold von Buch considered to be of carboniferous age, and others of which have been clearly identified by De Koninck‖ with Permian forms, at once shows how widely spread were these primeval deposits, and how they were successively accumulated around a large crystalline nucleus of preexisting rocks. In North America, we shall presently see another grand example of this phenomenon, and in which there is a much more complete development of Silurian rocks.

---

* Proc. Geol. Soc. Lond., vol. ii. p. 437. The Turkish fossils are of Lower Devonian age.

† Tchihatcheff, Bull. Geol. Soc. Pr., vol. vii. p. 389.

‡ Abich, Quart. Journ. Geol. Soc. Lond. 1846, vol. ii. 2nd pt. p. 96.; and Russia and the Ural Mountains, vol. i. p. 652. Prof. Abich also found carboniferous limestones everywhere overlying these Devonian rocks, and characterized, too, by a small *Fusulina*, of a species distinct from that found in Russia. In extending his observations to the eastward, M. Hommaire de Hell discovered Devonian rocks in Persia, where they form the axis of the Elburz chain and the southern slopes of the mountains of Tchihenneme, and are highly fossiliferous in the valley of Touwa. (Bull. Soc. Geol. Fr., vol. vii. p. 500.)

§ See p. 319, *antè*.   ‖ Bull. Acad. Roy. Belg., vol. xiii. and xvi.

CHAPTER XIV.

PRIMEVAL SUCCESSION IN GERMANY AND BELGIUM.

GENERAL SKETCH OF THE CHARACTER OF THE OLDER ROCKS EXTENDING WESTWARDS
FROM POLAND AND TURKEY IN EUROPE INTO THE ALPS AND CARPATHIANS.
DEVONIAN, CARBONIFEROUS, AND PERMIAN ROCKS OF SILESIA AND MORAVIA.
SILURIAN ROCKS (LOWER AND UPPER) OF BOHEMIA. LOWER SILURIAN, DEVONIAN,
CARBONIFEROUS, AND PERMIAN ROCKS OF SAXONY, THE THÜRINGERWALD, ETC.
GENERAL VIEW OF THE SUCCESSION OF DEVONIAN AND CARBONIFEROUS ROCKS
IN THE RHENISH PROVINCES AND BELGIUM.

IF the general order of the older strata be so clearly exposed in that
vast portion of Russia in Europe, which has from the remotest antiquity been exempted from all intrusion of every description of
plutonic or volcanic rocks, we no sooner pass to the south-west and
enter the Danubian and Turkish provinces, than we meet with rocks
more or less crystalline, which have been penetrated at numerous intervals by ancient igneous matter.   These extend from the Balkan
into the ranges of Thrace and Transylvania on the one hand, and into
the Carpathians on the other, but have as yet been so little examined
in detail, as to leave us in ignorance of the extent to which a palæozoic
classification can be applied to them.   We know, however, from the
works of Boué, Smyth, and others, that these rocks are usually so
crystalline, that few spots remain, like those near Constantinople,
whence the palæontologist can expect to derive fossils.

When, however, we reach the Eastern or Austrian Alps, we
find that, metamorphosed as most of the strata have been in that
mighty pile of mountains, there are nevertheless certain " oases "
at wide intervals, which indicate a succession very similar to that
which we have been following through other countries. Thus, in
the ridge south of Werfen in the Salzburg tract, Orthoceratites,

Orthidæ, Cardiolæ, and other fossils, have been detected, which mark a remnant of a true Silurian zone, the chief mass of which is in a crystalline state.   In the Alps near Gratz in Styria, certain grey schists and calcareous flagstones contain a good many Devonian fossils, and similar rocks have there been traced through a district of some extent.

Near Bleiberg, in the Carinthian portion of the Eastern Alps, is a limestone with large Producti, which is universally recognized to be of carboniferous age ; whilst in various parts of the Western Alps, the rocks contain courses of anthracite, associated with plants of the same era.   On the whole, however, it may be said, that in nearly all the countries extending over the southern regions of Germany, the clear separation of the palæozoic rocks, which can be easily effected in many other parts of Europe, is impracticable; and this is doubtless owing, in great measure, to their numerous mutations of structure, as well as to the dislocations and inversions which all the rocks there have undergone, from the Silurian to the Eocene Tertiary inclusive.

When, however, we travel westward in a more northern parallel, and pass from the central provinces of Russia into Poland, Prussia, and the northern states of Austria, we find, that although the palæozoic rocks of those regions have no longer the wholly unaltered facies which they presented to us on the banks of the Neva, the Volga, and the Dwina, and though they have, unlike their Russian equivalents, been penetrated by many eruptive rocks, yet they still contain numerous examples wherein order and the normal relations may be clearly observed.

In Poland, and in many low districts of Northern Germany, so widely spread are the younger secondary and tertiary deposits, that the rocks under consideration rise only in small patches to the surface.   Thus, around Kielce*, to the south of Warsaw, a nucleus

* See Russia in Europe, vol. i. p. 39.

of Devonian rocks, with much limestone, and charged with many
characteristic fossils, is followed by carboniferous limestone and by
thick-bedded coal-seams, over a small district that ranges from
Russian Poland into the Silesian coal tracts of Prussia. Thence
to the westward, rocks, which have been or may be classified as
Silurian, Devonian, Carboniferous, and Permian, occupy detached
districts in the south-east of Prussia, in Saxony, and the smaller
states of Germany, and spread over large tracts in the northern territories of Austria (particularly in Bohemia), and in parts of Moravia.

In traversing the Riesen-Gebirge to the south of Breslau, the
geologist acquainted with the full development of the palæozoic rocks
of other regions, is struck with the comparative tenuity of their
chief upper members — the Devonian and Carboniferous deposits.
In large portions of these mountains, whose mineral characters have
been recently examined in great detail by M. Gustav Rose, and
whose geological distinctions will soon be laid down on a map by M.
Beyrich, the strata, which are probably of Silurian age, have
undergone a complete metamorphosis. They have indeed passed so
entirely into a state of crystalline schist and marble, that no fossils
of the earlier era have rewarded the labours of the eminent men
above mentioned, or of my old friends Dechen and Öynhausen,
who first explored these ranges. That these older crystalline strata
are, however, the equivalents of the Silurian rocks, is to be surmised,
because they are clearly surmounted and followed by greywacke
schist and limestone, charged with many Devonian forms. In
the district of Waldenburg, south of Breslau, and at Silberberg,
to the north of Glatz, I have seen the Devonian limestone, charged
with Clymeniæ and many other typical fossils: it has a thickness of
50 to 60 feet, and is covered by an overlying limestone, of similar
thickness, laden with great Producti and several other shells well
known in the mountain limestone of Britain*, especially those of

---

* At one spot, Falkenberg, I collected in the carboniferous limestone the fol-

the lower carboniferous zone. The two beds are separated by a very thin mass of schist. This carboniferous limestone is overlaid by rocks occupying the place of the millstone grit of English geologists, and which, assuming very much the same mineral characters as the rocks whereon they repose (and being perfectly conformable with them), were in old times confounded with the so-called and undefined ' grauwacke.' It is in this portion of the series, *i. e.* in the schists and sandstones associated with the carboniferous limestone, that most of those plants have been found, which Professor Göppert has described as transition plants (*Calamites transitionis, &c.*). I am assured by M. Beyrich, that nearly all the plants so described (for others will hereafter be mentioned in true Devonian rocks, at Saalfeld, p. 358.) have been taken from such carboniferous deposits. Above these come the small productive coal-fields of Waldenburg, succeeded by a copious development, particularly along the south flank of the Riesen-Gebirge, of red rocks, which, many years ago, I grouped with the Permian deposits.* At Ruppersdorf, for example, these red rocks are micaceous flagstones with dark, bituminous schist, containing fishes of the genus Palæoniscus, the Holacanthodes gracilis, Beyr., and the Xenacanthus Decheni.

Resting upon conglomerates, the whole of these rocks are viewed by Beyrich as an equivalent of the Rothe-todte-liegende. Even taken in this sense, they constitute the lower half of the Permian rocks; for their ichthyolites are closely allied to those of the Zechstein, whilst their plants can only be viewed as a special flora, which, though approaching to the carboniferous type, is marked by peculiar

lowing fossils, viz. Productus gigas, P. antiquatus, Leptæna analoga, Orthis crenistria (var.), Spirifer lineatus, S. pinguis, S. glaber, with a minute Phillipsia, a small Orthoceras, and several species of corals.

* See Russia in Europe and the Ural Mountains, vol. i. p. 200. I examined this tract when alone, in the year 1844.

forms. No Stigmarian or Lepidodendron, such as are found in the true coal, have been found in this zone, but it exhibits Coniferæ approaching to the Voltzia*, with those specific forms of Neuropteris and Odontopteris, which characterize the Permian flora.

Again, in Moravia, from all that I can learn, whether from examining the collections of Professor Glocker of Breslau, who has so long explored that province, or from consulting a recent publication of Professor von Hingenau†, or from a partial, personal survey in 1847, it would appear, that the order of the palæozoic rocks is very analogous to that which prevails in Silesia. There I have observed crystalline schists and limestone, which may represent Silurian rocks, followed by masses of limestone, which at Rittberg and other places around Olmütz are charged with Devonian fossils. Some of the strata, in tracts that I have not seen, are referred to the carboniferous era. The crystalline rocks which range thence, or from the Sudeten mountains into the Böhmerwald-gebirge, and have been generally described by Partsch‡, Heinrichs, and others, consist of gneiss, mica schist, chlorite schist, and hornblendic slates, with masses of saccharoid limestone; the whole penetrated by granite, syenite, greenstone, serpentine, &c. Although portions of these vast masses may represent the Silurian rocks, the greater part of the schistose rocks, including limestones in which fossils have been detected, are clearly of Devonian age, as I have elsewhere shown.§

Silurian Rocks of Bohemia. — If the traveller has worked his

---

* This inference respecting the character of the Silesian plants is derived from an examination of a collection shown to Professor Morris and myself by M. Beyrich, during a recent tour in Germany.

† Uebersicht der Geologischen von Mähren und Osterr. Schlesien. Wien. 1852 (with a Map).

‡ Jahrbuch der K. K. Geol. Reich. 3 Heft. Sept. 1851.

§ See Edinb. Phil. Mag. 1847, p. 66., and 'Devonische Formation in Mähren,' Leonh. N. Jahrb. I., 1841. The observations were made conjointly with my old colleagues, Von Keyserling and De Verneuil, in the year 1847.

way round from Britain through Scandinavia and Russia into Germany, and has there lost the clue whereby he could have identified the lower crystalline schists and limestones of Silesia and Moravia, with their representatives in the western, northern, and eastern regions of Europe, he will, indeed, be rejoiced when he enters upon the territory around Prague, and there sees, under the guidance of M. Barrande, a most complete and symmetrical exposure of the whole Silurian System, whether as respects the clear order of the strata, or the vast abundance of organic remains.

The geologist who would form an adequate idea of the relations of this ' Bassin Silurien,' so admirably developed by that author, must traverse the tract as I have done, from Przibram and Ginetz on the south-east, to Skrey on the north-west, or visit the environs of Beraun. He will then at once recognize the truthfulness and accuracy of the divisions, which have been established after a most indefatigable and persevering study of the rocks and their organic remains during the last twenty years. The simple and regular form of this basin, and the order in which its concentric deposits are arranged, have enabled its explorer to ascertained the existence of three characteristic stages, into which he divides what he has there called the '' Silurian System.'' (See the annexed diagram.)*

---

* Though the conclusions are stated briefly in the text, I must be allowed to say that they could only have been arrived at through intense labour and unwearied research, as well as by the liberal employment of Bohemian workmen. The last twenty years have been occupied in unravelling this grand and rich Silurian fauna. To most persons the acquisition of the Bohemian or ' Czeck' language would have been an insurmountable barrier; but finding that without this key he could not direct his workmen, M. Barrande made himself a Bohemian scholar in order to become a thorough geological expounder of these rocks. His Silurian collections are truly more interesting than those ever made by any one individual in my day; and the evidences of faithful scrutiny and of a philosophical spirit pervade all the pages of a work which will, I am sure, be duly appreciated by every geologist and naturalist who consults it.

GENERALIZED SECTION ACROSS THE SILURIAN BASIN OF CENTRAL BOHEMIA, BY M. JOACHIM BARRANDE. (From Barrande's Systéme Silurien de la Bohême, vol. i. p. 56 b. See also Leonhard's N. Jahrb. 1847, tab. i.).

Just as in the British Isles, the lowest fossil-bearing strata, C of the section, repose on a vast mass of azoic schists, conglomerates, and greywacke, A, B. These are in parts semi-crystalline, and rest upon gneiss and granite. The argillaceous schists, C, which are the lowest in which fossils occur, he terms "Zone Primordiale." This primordial fauna is specially characterized by trilobites of the genera Paradoxides, Conocephalus, Ellipsocephalus, Sao, and Agnostus; and it also contains a rare Orthis and some Cystideæ. According to the opinion of M. Barrande, in which I entirely agree, this zone is represented in Britain (as before said, p. 43.) by the rocks containing Lingula, Agnostus[*] and Paradoxides in North Wales, and, possibly, also, by those which in the south-west flank of the Malvern Hills contain the Olenus; whilst, as has already been noted, Olenus, together with Paradoxides and Agnostus, are typical of the lowest band containing animal life in Scandinavia, (Regiones A, B, of Angelin). The large species of Paradoxides in Bohemia, also belong to the same primordial group. And to the same epoch must be referred the curious forms which have recently

---

[*] Mr. Salter has added Agnostus and Conocephalus to the list. They were lately detected by him in the "Lingula Flags" of North Wales.

been described by Dr. Dale Owen, as characterizing the lowest Silurian rocks of vast tracts in North America.

The next zone or the stage D, which is one of schists and quartziferous sandstones (quartzites), and represents the great and superior mass of the Lower Silurian (Llandeilo and Caradoc), is charged with abundant casts of Trinucleus, Ogygia, Asaphus, Illænus, Remopleurides, and many other trilobites. In it also has been observed the same development of the genus Orthis among the brachiopods, and of Cystideæ among the echinoderms, which stamps the character of the great mass of the Lower Silurian rocks of Scandinavia, Russia, Britain, and North America.

Prague and its Imperial palace, the Hradschin, stand upon these Lower Silurian rocks, which, dipping under Upper Silurian schists and limestones to the south of the city, rise up again to the surface on the banks of the Moldau, a few miles further south, and thus exhibit the superior members enclosed within a trough.

M. Barrande places the base of his Upper Silurian, or bottom of his stage E, where the quartzites with Trinuclei &c. are overlaid by black schists containing many graptolites. Several bands of greenstone and trap, which the author considers to be of contemporaneous age, are interpolated in this portion of the series, and the shales are followed by grey, argillaceous limestone replete with a very rich fauna, unquestionably of an Upper Silurian type. For, although it may be contended, that the schists which contain both the double and simply serrated graptolitic bodies, ought rather to be classed with the lower division, no one can doubt that the limestones into which those schists graduate upwards are of that true Upper Silurian type, so well marked in England and Gothland. Thus, we here meet with the well-known fossils Cardiola interrupta, C. fibrosa, Phragmoceras ventricosum, P. aurantium (Cyrtoceras, Barr.), Orthoceras annulatum, O. nummularius, Rhynconella (Terebratula) navicula, R. obovata, R. Wilsoni, R. marginalis, Nerita haliotis, Pentamerus Knightii, P. galeatus, Spirifer trapezoidalis, and forms of Leptæna, all of them published in my original work as Upper Silurian *species*. Again, most of the zoophytes which so characterize the Wenlock and Gothland limestone are here found in profusion, including the widely known corals, Halysites (Catenipora) escharoides, Favosites Gothlandica, &c. Even that fossil, which has not otherwise, I believe, been detected out of England, the Ischadites Königi, Sil. Syst., has been here collected. It is also in this rich stage, that the Bohemian trilobites have their maximum of specific development: M. Barrande having described no less than seventy-eight species from this one limestone; the chief genera being Acidaspis, Calymene, Cheirurus, Cyphaspis, Lichas, Phacops, Harpes, Bronteus, and Proetus.

Ascending from this limestone (E of the section), the remaining and superior stages of calcareous rock (F and G) are so entirely conformable, are so often knit together in parallel folds in the same hill with the stage E, and are so connected zoologically as well as stratigraphically, that M. Barrande has united them all in one natural group, or his third fauna. The band F is a thin-bedded, hard limestone, occasionally containing huge concretions like the ballstones of the Wenlock limestone of England: it is usually of a light colour, and somewhat crystalline, and therefore offers a strong contrast to the dark and earthy rock of the stage beneath it. All intervening bands of shale and schist disappearing, the band F is at once surmounted by the highest limestone G, a rock usually of small concretionary structure, but which when opened, as in large quarries on the banks of the Moldau (Branik, Wiscocilka, &c.), exhibits extensive, flattish and slightly undulating surfaces, the upper layers of which are covered by grey shale, H of the diagram.

M. Barrande refers all these three limestones and their shales, inferior and superior, to the Upper Silurian group, since he has found (just indeed as I long ago anticipated *), that the British divi-

---

* See Silurian System, pp 196. 301., *et passim.*

sion into Wenlock and Ludlow Rocks, was not to be looked for in distant lands. In truth, he has detected abundant Ludlow rock fossils in the lower limestone, E; and whilst the upper stages, F, G, H, contain some of the same species as that rock, they are also distinguished by many which are locally peculiar to each of them. In this way F is marked by its profusion of brachiopods, whilst the cephalopods, so abundant in the lower limestone, are reduced to a few. Of trilobites, though there are seventy-five species, they all belong to the same genera as those of stage E.

In stage G, or the highest limestone, there would seem at first sight to be some reason for rather classifying this mass as Devonian, seeing that, as well as the rock beneath, it contains Goniatites, a genus of cephalopod unknown in the Silurian of Britain, but abundant in the Devonian and carboniferous rocks of many countries. This one feature, however, (and the Bohemian species are peculiar and not known Devonian types,) cannot prevail, in the opinion of the author, in the face of the much more decisive fact, that many true Silurian types pervade the three limestones, and unite them in one natural group. Thus, we here meet with forty species of trilobites of the genera mentioned in the opposite page, with the addition of Dalmanites (Phacops). Now, although all these genera, with the exception of Calymene, are known in the Devonian rocks, no one of the Bohemian species has been found in them *; and the resemblance is therefore confined to the brachiopod shells, of which a few species are certainly found in Devonian strata.†

In an invaluable table appended to his work (plate 51. vol. i.), which contains the result of as much philosophical thought and profound research as I ever saw embodied in a single page of natural history,

---

* I have the authority of M. Fridolin Sandberger to state, that he knows of no species of a trilobite of the Upper Silurian of Bohemia in the Devonian rocks of the Rhine.

† Barrande, in Haidinger's Naturwissensch. Abhandl. 1847 vol. i.

M. Barrande brings at once before the eye the proof, that every one of the few trilobites which occur in the Devonian rocks belongs to a genus which took its rise and had its maximum development in the Silurian period. In other words, the Devonian trilobites are only the expiring remnants of the crustaceans of the first great natural epoch in which those animals flourished. M. Barrande has also defined the upper limit of Silurian life by showing that his highest limestone contains three species of Calymene, a genus never yet found in the Devonian rocks; the form which was originally taken for a Calymene in the lowest Rhenish 'grauwacke' being now recognized as a Phacops.

In short, more than 1200 species of fossils have already been obtained through the labours of this naturalist from his rich Silurian basin; and, just as in Britain and other parts of the world, so in Bohemia, the most assiduous, nay, microscopic, researches, continued during many years, have failed to *detect the trace of a vertebrated animal* in any of the Silurian rocks, except their *uppermost* strata. After all his laborious researches, M. Barrande has collected two fragments only of fish-bones; and it is remarkable that in this same upper stratum he has also found the peculiar crustacean named Pterygotus, and has thus ascertained the same union of inhabitants in the uppermost Silurian rock of Bohemia, which I had observed in the Ludlow rocks of England (see Sil. Syst., tab. 4 ; and pp. 143. 205. *antè*).

Notwithstanding their strict conformity, the Lower and Upper Silurian rocks of Bohemia are supposed to be more separated from each other by their fossils than in their wide geographical distribution in the British Isles. This fact may to a great extent be explained, by an appeal to the usual phenomena of clear separation *wherever a limited tract only is surveyed*, as before adverted to.

But here we have not even occasion to search for any other cause than the obvious one on which M. Barrande dwells, to account satisfactorily for the specific separation, in this tract, of the animals of his three Silurian zones. He shows, in short, that after the accumulation of what he calls the primordial zone, the sea bottom of this region was powerfully disturbed by the eruption of porphyries; and that again, after the completion of his second zone or at the close of the older Silurian period, other igneous rocks (greenstones, trap, &c.) were so copiously and repeatedly evolved, as to account for the destruction of nearly all the animals then living in these environs; graptolites especially being the creatures which survived such turbulent conditions.

Yet even under this point of view, the author of the ʻ Bassin Silurien de la Bohême ʼ has proved, that not less than fifty-seven species of fossils are common to his stages D and E. Thus, when the observer is working his way up through the Lower Silurian schists and quartzites of the stage D, with their Orthidæ and Trinuclei, he meets unexpectedly in them with intercalated bands of dark shale, mineralogically similar to those of the superior stage, and in which, out of sixty-seven species which have been discovered, fifty-seven are forms that characterize the Upper Silurian band E. This is the more remarkable, too, as the fossils otherwise known in the stage D are by no means similar to those of the overlying rocks.

M. Barrande, broaching a theory at once original and ingenious, has suggested, that these schists so insulated among the older strata, may be regarded as having been the seats of colonies of animals which, existing in some other part of the world whilst the Lower Silurian fauna prevailed in Bohemia, were carried to these spots by currents, and thus for a time inhabited the region, before the more general introduction of exotic species at a subsequent period.

For my own part, I must say, that I rather prefer to explain the fact by referring to what may almost be called a geological law; viz., that fossil animals are frequently associated with particular conditions, disappearing and reappearing with changes of the sea bottom.

Thus, with the distribution of mud and silt, accompanied by the evolution of calcareous springs, it is to be presumed, that here the graptolites with certain mollusca of the Wenlock date began to abound. A return however, of the sandy condition of the sea bottom, occasioned by new currents, brought back Trinuclei, Orthidæ, and other previous tenants of this bay of the sea, to be eventually succeeded by a still more prolific spread of the younger races. But whether the geologist should adopt the ingenious speculation of M. Barrande, or rest satisfied with my simple view of the case, the alternation is undeniable, and thus *the Lower and Upper Silurian rocks are irrevocably linked together through this physical interchange of a number of their respective fossils.*

Before I take leave of this rich continental centre of Silurian life, which my eminent friend has rendered so classical, let me invite the attention of the reader to one salient proof of his acumen and sagacity as a naturalist. Every one knows that living crustaceans, from the king crab (Limulus) and lobster downwards, proceed from eggs; and in reaching maturity, many of them, even of the higher grades, are known to pass through a metamorphosis. Now, M. Barrande has discovered, after examining myriads of fossils, that the trilobite or earliest crustacean underwent a similar metamorphosis from the embryo to the adult state, and passed through many changes of form, as observed in living crustacea. In the genus Sao, for example, he has distinguished no less than twenty stages of development of the same species, each stage being marked by an addition to the thoracic ribs of the animal; and he has thus taught us, by true natural proofs, that several so called genera and many species named by contemporary authors really belong to this one creature.* The same analysis of forms has, indeed, been extended by the author to other and higher deposits of the Silurian system; for he has indi-

* See particularly Corda, Prodrom. einer Monog. der Böhmische Trilobiten, in which the species *Sao hirsuta* (Barr.) has been divided into ten genera and eighteen species!

cated twenty-two changes in his beautiful new species Arethusina Konincki of the stage E; and altogether he has found the phenomenon to hold good in twenty-six species of the Lower and Upper Silurian strata. He has even followed out his subject into other families, and has traced the fossil Nautilus from the egg, through twenty variations of form, to its completion with a perfect mouth.

In a word, the work of M. Barrande will convince every one who studies it, that we have now obtained quite as clear an insight into the first recognized stages of life upon the globe, as into that of younger deposits, whose secrets are so much more easily wrung from the stony records of nature.

Palæozoic Rocks (Silurian, Devonian, Carboniferous, and Permian) of Thuringia, Franconia, Saxony, and the adjacent Principalities. — The older sedimentary formations occupy a considerable region on the north-western flank of that devious chain of granitic, gneissose, and other crystalline rocks, which, ranging from north-east to south-west, divides Saxony from Austria, and trends into the Fichtel-Gebirge of Bavaria. From that chain, the deposits in question descend into and spread over a broad and comparatively low, undulating tract, which in its central part is cut through transversely by the river Saal, as it flows from Hof on the south-east to Saalfeld on the north-west. On their north-western boundary, these sedimentary rocks again rise into the lofty eminences of the Thüringerwald; the whole succession under consideration having a dominant strike from south-west to north-east. In this way the younger strata may be said to occupy a broad trough, ranging lengthwise from Ronneburg and Gera on the north-east, by Schleitz, Plauen, and Hof, to Upper Franconia on the south-west; whilst the older rocks of the series, rising up on either side, are often in a highly metamorphic state, but chiefly on the south-eastern flank of the depression.

The first effort to co-ordinate the various sedimentary rocks of this region with their equivalents in the Rhenish provinces and

the British Isles, was made in the year 1839 by Professor Sedgwick and myself.* We commenced our survey by examining the
mountainous elevations or greywacke of the Thüringerwald, and
thence extended our observations to the shelly limestones of Upper
Franconia, wherein Count Münster had collected his transition
fossils. As the result, we indicated that which has proved to be
the true general succession. We spoke, for example, of a slaty
greywacke with greenish slate, and quartzose flagstones, and grits
like those of the Longmynd; of quartzites in roofing slates with
a greenish tinge (Schwartzburg, &c.), that reminded us of the
lower slates of Cumberland and Westmoreland. The whole of them
were said to exhibit great undulations, to be powerfully affected by
a slaty cleavage, but on the whole dipping to the south-east, and
surmounted by subcrystalline limestones with alum slate (pyritous
shale of Doschnitz, &c.). These, again, were described as covered
by other grey schists and greywacke, containing also courses of limestone; and all such beds disappeared, we said, under more earthy
and arenaceous strata, resembling the Devonian rocks of the Rhenish
provinces we were then describing. Finding, however, no fossils
except encrinites in the lower part of these rocks, we could assign to
them no more definite place than that they were probably of the age
of the slates of the Ardennes. On the other hand, we considered
the fossiliferous limestones of Franconia (Elbersreuth, Hof, Gattendorf, &c.), which are replete with Clymeniæ, Goniatites, and
Orthoceratites, as being unequivocally of Devonian age; and, lastly,
by actual sections and physical proofs, we confirmed the view of
Count Münster derived from fossils only, and showed that the limestones and associated schists of Trogenau, Regnitz Losau, and Würlitz
near Hof, were truly of the age of the carboniferous or mountain
limestone.†

* Trans. Geol. Soc. Lond., 2nd ser. vol. vi. p. 296.
† The fossils collected in these limestones were, Productus punctatus, P. anti-

In calling attention to this view of succession, which, in a recent visit with Professor Morris, I found to be substantially correct, it is by no means contended that Professor Sedgwick and myself had done more than throw out a general suggestion; for at that time, not only was there no true geological sketch of this complicated country, but not even any ordinary geographical map which could be depended on. In the fourteen years which have since elapsed, good trigonometrical surveys have been made of large portions of the region, and able geologists of Saxony and Meiningen have constructed geological maps, and have described many fossils entirely unknown in former days.

Thus, Professors Naumann, Cotta, and Geinitz, in Saxony, and M. Richter of Saalfeld in Meiningen have elaborately worked out the demarcation of the mineral masses of which these old rocks are composed, and some of their lower members having been found to contain graptolites, trilobites, and other fossils, such inferior greywacke has very properly been referred by those authorities to the Lower Silurian.* With this view I quite agree, guarding the definition of the word 'Lower Silurian' by explaining, that some of the black schists of this region which contain graptolites, may be of the age of that

quatus, P. sublævis, P. pustulosus, a large Orthis near to O. crenistria, Phill., Chonetes papilionacea, Euomphalus pentangulatus, and several corals of the genera Syringopora, Cyathophyllum, &c.;—a perfectly carboniferous assemblage.

* I must not omit to add to this list the name of M. Engelhardt of Ober-Steinach, who has made a large collection of fossils, some of which are manifestly of Lower Silurian, and others of Devonian age. The strata near that place are so dislocated and in great part inverted, that I can well understand how that gentleman, from whom I recently experienced every courtesy when I visited Steinach, should have been led to indicate an order of strata whereby the rocks which are really Devonian were placed under the Lower Silurian greywacke. The formations referred by him to the Upper Silurian (under the names of Wenlock, Aymestry, Ludlow, &c.; Zeit. Deutsch. Geol. Gesellsch; Berlin, vol. iv. p. 234.) are for the most part Devonian, as proved by fossils, which, not relying on my own judgment only, I referred to M. Barrandc.

zone in Bohemia, which forms the base of the Upper Silurian of M. Barrande.

The following, therefore, may be taken as an improved view of the succession, completing the first sketch by Sedgwick and myself.

1. Ancient rocks of the Thüringerwald, consisting of greenish, talcose schists with white quartz veins, which in former times afforded some gold. They range from Steinheide by Igleshieb, or from south-west to north-east, in association with certain bands of ferruginous and purple-coloured grey-wacke, not unlike the bottom rocks of my British sections. These masses occupy the highest grounds of the Oberland of Meiningen, attaining an elevation of near 3000 English feet above the sea. 2. The beds of this lofty plateau fold over both to the north-west and south-east, and are throughout affected by a slaty cleavage, which extends to all the overlying formations. As the planes of this cleavage, usually plunging to the north-west at a high angle, are dominant, and often obliterate the lines of true bedding, they have misled some observers with respect to the physical succession of the strata. Attentive observation, however, shows that, rolling over in undulations to the south-east, the above rocks ('Grüne Grau-wacke' of Richter), containing fucoids, are overlaid by the group which that author terms 'Graue Grauwacke.' The latter is composed of lime-stone and aluminiferous schists, subordinate to much greywacke, slate, and sandstone, and occasionally contains in its lower beds fine examples of fucoids. Among these are the Phycodes circinnatum, and the Fucoides Allegha-nensis, together with Myrianites, and the (graptolites?) Cladograpsus nereitarum, Nereograpsus * Sedgwicki, and N. cambrensis, (Nereites, Sil. Syst.). The upper beds specially contain graptolites, orthoceratites, and some forms of trilobites. Of the truth of this succession I convinced myself in examining, with M. Engelhardt and Professor Morris, that portion of the section of Ober-Steinach which pertains to the Silurian rocks. The same rocks occupy a broader surface between Schwartzburg and Saalfeld; one of their members being extensively quarried for the manufacture of slate pencils (Griffelstein). A species of trilobite found by M. Engelhardt approaches near to, if it be not identical with, Ogygia (Asaphus) Buchii; and as some of the worm-like bodies (Nereites), and several of the grapto-lites, both double and single, are not distinguishable from forms known in the British Isles, there can be no doubt, that we have in this slaty group a true equivalent of the Lower Silurian formation.

* M. Geinitz regards this as belonging to the Graptolite family.

In fact, several of its numerous graptolites are species identical with those described by Portlock, Salter, M'Coy, Harkness, Nicol, and myself, from the Lower Silurian of Britain; and among these, as seen in the work of Geinitz, or identified by M. Richter, are the following: Diplograpsus folium, His.; D. palmeus and D. ovatus, Barr.; D. teretiusculus, His.; Graptolithus priodon (Ludensis, Sil. Syst.); G. Sedgwicki, Portlock; G. Becki, Barr.; G. latus, M'Coy (?); G. spina, Richt.; G. turriculatus, Barr.; G. Nilssoni, Barr.; G. sagittarius, His.; G. colonus, G. Proteus, and Rastrites peregrinus, Barr. * ; with some other species common to this region and Bohemia, including the remarkable graptolite, Retiolites Geinitzianus, Barr. Several of the species occur also here, and in the Silurian rocks of Scandinavia, North America, and other countries.

The Lower Silurian rocks of the Thüringerwald and of the Saalfeld tract, which are penetrated at intervals by porphyries and greenstones, are irregularly overlapped, towards their south-eastern flanks, by masses of Devonian age (the 'rothe grauwacke,' Richter), which will presently be considered. Silurian rocks, with schists containing similar graptolites, again emerge from beneath younger strata in a series of low undulations far to the east of the lofty, wooded Thüringerwald. The base rocks, in fact, of the low ridges extending from Ronneburg on the north-east, through Schleitz to Lobenstein on the south-west, are composed of graptolite schists and roofing slates. Similar rocks reappear in parallel undulations in the environs of Plauen and Hof; and it is probable that considerable portions of the Thon-Schiefer of the general map of Saxony will be afterwards found to belong to the Lower Silurian division.

The exact definition of the outline of these Silurian rocks, and their separation from all the older unfossiliferous and crystalline strata on the one hand, and from the Devonian rocks on the other, will be a work of no small labour. Their regular order has been repeatedly interfered with by pseudo-volcanic or eruptive masses,

* See some of these forms of Graptolite, p. 46, *antè*, and pl. 12. f. 1. of this work.

which, whether they constitute stratified layers or appear as intrusive
bosses, have singularly affected the ordinary sediments.   These igneous rocks are, indeed, most accurately laid down on the map by
Professor Naumann and his associates. So rapidly does the geologist
here proceed from one formation to another, that one small hill is
often seen to be composed of Lower Silurian schists with graptolites,
and another, within gunshot, of true Devonian limestone.   These
changes are well seen near Schleitz, where the low ridges of black
brittle schist, which form the pleasure-grounds of the reigning Prince
(Heinrich's Ruhe), are laden with Graptolites, Orthidæ and other
fossils *; whilst, on the western side of an intervening small hill of
eruptive rock, the geologist has before him a limestone charged with
Clymeniæ, Cypridinæ, and true Devonian fossils!   On the whole,
however, it may be said, that although they occupy so broad an area in
the lofty tract of the Thüringerwald, the Lower Silurian rocks only
make their appearance partially, and within comparatively narrow
bounds, in the central portion of that region which has here been
designated a vast *undulating trough*.   Even then, their superior
members are alone visible, one characteristic band of which is composed of the black schists and slates, extending from Ronneburg by
Schleitz, to Lehestein, and Lobenstein.

The Devonian rocks of this portion of Germany, previously known
to us through the palæontological illustrations of Count Münster,
have recently been much better developed by the labours of Professor Geinitz and M. Richter.

From the densely wooded cover of the lands, it is exceedingly
difficult to trace any absolute junction of these Devonian rocks with

* All the Silurian fossils figured from this spot in the work of Prof. Geinitz
were previously found by M. Berner, an intelligent apothecary of Schleitz, who
had, it appears, detected Graptolites in these strata, many years before the attention of geologists was called to them. M. Berner has extensive fossil collections
on sale.

the subjacent Lower Silurian of the Thüringerwald. That they lie, however, irregularly upon their older neighbours, is manifest from the faithful delineation of their outlines, as laid down in the geological map of Richter, which shows the Devonian, or the 'Rothe Grauwacke' of that author, in patches or hummocks only, in relation to a great spread of the more ancient rocks.* That this is their true position, I can now affirm.

The extent to which this Devonian group of Thuringia, Franconia, and Voigtland, can be paralleled in its details with that of the Rhenish provinces of Prussia, is not yet completely ascertained; for beginning with what is now usually recognized as the lowest fossiliferous rock of the Rhine, or the 'Spirifer Sandstein,' I have nowhere been able to detect a section where such strata are infraposed to the fossiliferous limestones of the region.

The next question then is, can we even divide the Devonian of this region in Central Germany into lower and upper limestones, as was first done in Devonshire, and afterwards in the Rhenish provinces?

This separation has, indeed, been partially made by Geinitz, who places at the bottom of this group certain schists near Ronneburg, which Naumann has identified in several other localities. These schists contain Tentaculites

---

* See Beitrag zur Paläontologie der Thüringer Waldes von Reinhard Richter, Dresden and Leipsig, 1848; and Die Versteinerungen der Grauwacken Formation in Sachsen, &c., von Hans Bruno Geinitz, Leipsig, 1853. In the first part of his able work, Prof. Geinitz describes and figures many species of graptolites, before referred to, p. 353; and in the second he adds to these the following Silurian fossils:—Nereograpsus tenuissimus, Emmons; Orthoceras Brongniarti, Troost; O. tenue, Wahlenberg; Leptocheles (Dithyrocaris) Murchisoni, Agass.; Tentaculites tenuis, Sow.; Pterinæa Sowerbyi, M'Coy; Nucula levata, Hall; Cytherina subrecta, Portlock (?); and Orthis callactis, Dalm. These fossils are, however, very obscure, and their identification therefore doubtful. In the Devonian rocks he describes one genus of trilobite only, the Phacops; of chambered shells, 24 species, including Orthoceratites, Gomphoceras, and Cyrtoceras, 11 species; Clymenia, 6; Goniatites, 4; Gasteropods, 8 species; Lamellibranchs, 24 species; Brachiopods, 11 — including the universal Atrypa reticularis; Crinoids, 9 or 10 species; and corals about 11 species.

lævigatus, Roemer; T. subconicus, Gein.; with Phacops. Then follow certain limestones near Plauen, Wildenfels, and other localities, including the well-known Elbersreuth, which are also classed as older Devonian. These are characterized by Orthoceratites, O. interruptum and O. dimidiatum, Münst., Clymenia lævigata, Münst., and peculiar corals and crinoids. Then succeed stratified, igneous rocks, some of which are greenstones, others coarse trappean breccias, and others again finely levigated ash-beds, occasionally calcareous and undistinguishable from the 'Schaalstein' of the Rhine (Plauschwitzer-schichten of Naumann). Above these is said to be placed the Clymenia limestone, properly so called, and in which many fossils abound, including Goniatites and several species of Posidonomya (P. inversa and P. regularis, Goldf., &c.), together with a vast profusion of Cypridina serrato-striata, Roemer. Not professing to have worked out the proofs of these subdivisions, I must say that in the tracts I have examined there are no evidences of two Devonian limestones, separated from each other by slaty rocks, as in England and on the Rhine.

In short, I have nowhere seen in Saxony, or the adjacent tracts, *the true representative of the Eifel limestone, as characterized by its Stringocephali, Calceolæ*, &c., overlaid by another which represents the Clymenia rock of England (or the Kramenzel-stein of the Rhine). But, as towards the end of this chapter we shall indicate that those two calcareous masses, which are so clearly separated in some districts, are nearly brought together in others, so is it possible that the divisions partially indicated through their fossils, by Geinitz, may be established in Voigtland, and that the trappean tuff (schaalstein) of Plauschwitz and other places, may, as he believes, separate these two rocks.

However this may be, the ascending order, in several places, is clear, from a highly fossiliferous and nodular limestone, laden with Clymeniæ, Goniatites, and Orthoceratites, and which is everywhere characterized by Cypridinæ, into overlying strata in which land plants begin to appear. The Devonian limestone is surmounted by a copious accumulation of sandstones and schists, occasionally siliceous,—the 'jüngste grauwacke' of Geinitz,— charged with land plants. These are sometimes followed by the carboniferous limestone or 'kohlen-kalk' with its large Producti, and by other strata with which a different series of land plants is associated.

Such an ascending order is seen at Hof, on the right bank of the Saal, between the town quarries replete with Cypridinæ and an overlying coralline limestone. This succession is more clearly viewed between Gattendorf and Troguenau; the intervening space between the Devonian

or Clymenia limestone of the former, and the Carboniferous or Productus limestone of the latter, being occupied by ferruginous greywacke
with traces of plants, and by a mass of 'kiesel schiefer,' which there
occupies the same place as the rock so called by the geologists of the Rhine
country.*

Again, in the gorge of the Saal near Saalfeld, the cliffs, which
M. Richter has described in a section†, expose a magnificent mass
of limestones, which are throughout characterized by Cypridina
serrato-striata and C. calcarata, Roem., with many other upper
Devonian types, but which offer no evidence whatever of a second
limestone between them and the Lower Silurian rocks. Here
M. Richter has detected Phacops latifrons, Bronn, and P. granulatus, Posidonomya minuta and P. intercostalis, Roem., with
trails of worm-like animals. Here, also (no igneous rocks appearing), the ascending order seemed to me to be clear and unequivocal.
In spite of the dominant slaty cleavage, the planes of which dip to
the north-west, the Cypridina limestones, after those fine convolutions which render the gorge so picturesque, are seen to pass under
the mass of red rocks, or 'rothe grauwacke' of Richter ‡; the whole
being covered unconformably by terraces of 'Zechstein.'

The lower part of this reddish coloured sandy and schistose greywacke is interlaminated with the Cypridina-schiefer, the mass of
which immediately succeeds to the limestone; and from this point, the

---

* Much confusion may arise, in comparing the local descriptions of German
geologists, from the use of mineral terms applied to rocks. Thus, in this region
of Thuringia and Saxony, 'Kiesel Schiefer,' which here usually designates some
of the older Silurian schists and also some Devonian beds, might be strictly
applied to the flinty, schistose strata overlying the Cypridina limestone and its greywacke near Hof, and which are unquestionably on the parallel of the so-called
'Kiesel Schiefer,'—i. e. lower carboniferous—of the Prussian and Hessian geologists.

† Beit. Paläont. des Thüringer-Waldes. Dresden, 1848. Tab. i.

‡ I convinced M. Richter of this fact, by closely scrutinizing this section in company with that gentleman.

beds begin to contain land plants, which augment in quantity and size in the overlying or younger strata. These consist, first of reddish and grey, and then of greenish shale, of considerable thickness, which on the right bank of the Saal extend from the cliffs of Bohlen by the Pfaffenberg to the Kleitsch Hill, and finally, at the foot of the Rothenberg, are surmounted by brownish, red, micaceous flagstones, containing a great quantity of fossil plants. I direct particular notice to this section, because it exhibits, more clearly than any other with which I am acquainted, the extent to which the land vegetation augments as we ascend in the Devonian rocks.

The lowest plants, as discovered by M. Richter in the Cypridina schists, consist, according to Professor Unger, to whom they were referred, of many species of Ferns, of new or undescribed genera, and even of new families. Some of them are considered to be intermediate between Ferns and Equisetaceæ,—others seem to be primitive forms of Cycads and Conifers, possessing characters of which (says the Professor) no one has as yet had an idea; and some present such a singular organization, that he terms them—the "prototype of the gymnosperms!"*

This section is still further interesting in demonstrating a passage upwards into other and overlying beds under the Rothenberg,—viz. into the micaceous sandstones and flagstones, which, from the plants they contain, must be classed with the lower carboniferous rocks. Such, for example, are Calamites transitionis, Göpp.; Megaphytum (Rothenbergia) Hollebenii, Cotta; with Knorria, &c.; plants which are well known in the Rhenish provinces of Prussia, where they invariably occupy the lower carboniferous rocks, and never descend, like the group above-mentioned, into the Devonian, properly so called. Here, then, on the edge of the Thüringerwald, M. Richter

---

* Letter from Professor Unger to M. Richter. In some of the Conifers of these upper Devonian plants, the resin even of the wood has been preserved.

has collected data to prove, by plants alone, a succession from the Devonian to the Carboniferous period.

Having called the attention of this author, on the spot, to the importance of applying these facts respecting the distribution of the fossil plants to the tracts between Saalfeld and Schleitz, in parts of which such remains were for the first time observed in a recent excursion*, I now learn from him, that he has considerably extended the area which is occupied by the Lower Carboniferous rocks in that country, as defined by its plants.†

These plant-bearing rocks also occupy a considerable fringe of elevated country at the southern extremity of the Thüringerwald between Sonneberg and Teushnitz; and they occur in abundance in the gorge of the Steinach river, north of Koppelsdorf. As they offer a good line of geological demarcation, I also strongly urged M. Credner to distinguish this 'Jungste Grauwacke' of my German friends from all its older associates which bear that unmeaning family name. It will, of course, require much accuracy of local observation to draw the line between the plant-bearing rocks of Devonian, and those of Carboniferous age; or rather, it will best become the progressive state of science, not to attempt to mark any hard and rigorous line between them, but to shade off on a map the colour of one rock into that of another; thus imitating the succession of nature, in which there is no mistake.‡

---

* The observation was made by M. Richter and Baron von Baumbach, in company with Professor Morris, and myself.

† The ground between Saalfeld and Schleitz is all laid down in the geological maps of Saxony under one colour, or as older 'grauwacke;' whereas a very large portion of it must now be assigned to the *lower carboniferous formation*. When will my valued friends, the mineralogists and geologists of Germany, abandon a word which has led to such endless confusion? I cannot but regret that a work of such ability as that recently issued by Geinitz should bear the title of the "Fossils of *the* 'Grauwacke Formation,'" under which name he treats of Silurian, Devonian, and Carboniferous rocks.

‡ At the southern end of the Thüringerwald, Professor Morris and myself

Such is a meagre outline of the palæozoic succession of this diversified region of Central Germany, the complete local elaboration of which calls for the full employment of all the able men who are occupied in working out its highly interesting features.

A slight allusion only has been made to the south-eastern flank of the great undulating trough of Plauen, Schleitz, and Hof; for,. although Naumann and his associates have shown in their maps, that older and more crystalline rocks appear, we have yet to learn how much of their primary clayslate, or 'thon-schiefer,' is to be abstracted from the unfossiliferous rocks, and grouped with the lower members of the series we have been considering.*   Other enquirers

partially examined, in company with M. Büttner of Kronach, a zone of coal measures containing some irregular seams of coal, and exhibiting a rich fossil flora, which, flanking the loftier so-called 'grauwacke' chain of the Thüringerwald (Köppelsdorf, &c.), is interposed between it and the rothe-todte-liegende. This coal formation rises out conformably from beneath a vast mass of inclined beds of rothe-todte-liegende, not less than from 1500 to 2000 feet thick (Stockheim, Neuhaus, &c.). As the coal and rothe-todte-liegende are there, as in Gotha, followed by copper slate, zechstein, and a very full development of the Bunter Sandstein (which at Kronach occupies noble escarpments, and is manifestly divided into a whitish lower, and a red upper sandstone), this district is well worthy of a monograph to develop the features of the Permian group with its relations to the coal beneath, and to the lower Bunter Sandstein above it.

The general relations of these overlying deposits to the so-called 'grauwacke' rocks of the Thuringian mountains, will be seen in the forthcoming southern sheet of a geological Map of the Thüringerwald by M. Credner, a copy of which he presented to me after I had examined the above-mentioned coal tract.   In classifying the igneous rocks of the Thüringerwald, M. Credner believes that the granites, with hypersthene rocks, &c., have caused the great metamorphism of those strata, which when they range into southern Thuringia, I have described as Lower Silurian, &c., but which, in the northern parts of the forest where granite abounds, become mica schists, &c. The greenstones and amygdaloids of the tract are associated, according to Credner, with carboniferous rocks; and the two porphyries of the region, the one containing quartz, and the other without it, are connected by that good practical observer with the period of the rothe-todte-liegende; the former, which is the oldest, being traversed by the '*quartz freier*' porphyry.

* In travelling from Plauen to Franzenbad it seemed to me impracticable to

may seek to ascertain to what extent many of these ancient schists and slates, evidently of sedimentary origin, have been converted into mica schists, and even into the metalliferous so-called 'gneiss' amid which the illustrious Werner himself taught his lessons at Freiberg.

It would seem presumptuous that a passing geologist should hazard any opinion on such a point. Still I venture to state, that much of the so-called gneiss in the plateaux around Freiberg, is a rock very different in age from the antique and crystalline gneiss of Scotland and Scandinavia. I would indeed suggest, that those portions of it which are separated by way-boards, and exhibit several of the features of bedding and jointing of aqueous deposits, may prove to be of no higher antiquity than the lower members of the sedimentary formations under review.

In the mean time, if we reason upon the fact, that this region contains no Upper Silurian rocks, we may surmise that it was raised above the waters, and constituted dry land during a long period, and was afterwards depressed to great depths, to receive accumulations of the Devonian era, at a period when the bottom of the sea in these localities was powerfully agitated by volcanic action.

This same country also contains clear evidences of a phenomenon to which Professor Sedgwick and myself called attention in 1835, viz., — that whilst the older 'grauwacke,' now known to be Lower Silurian, with the Devonian and Lower Carboniferous rocks, have partaken of the same movements of elevation, contortion, and dislocation, all these rocks have been abruptly separated from the upper coal-fields and the deposits which have since been termed Permian.

separate geologically and physically the 'thon-schiefer' or glossy shillat of that district from the contiguous graptolite schist of Plauen. In fact, the rocks on either bank of the Elster are mineralogically the same, and have similar bedding, joints, and cleavage; the faces of the last plunging to the north-west. The discovery of any Silurian fossils in the greywacke, which rises into the hilly tracts on the flanks of the Erz-Gebirge, would remove these rocks from the 'thon-schiefer.'

This feature of a great physical rupture, which seems to have pervaded Germany and France, and has been duly noticed by M. Elie de Beaumont, has, however, its well-defined limits even in Europe; for, grand as it may be, the phenomenon is local only, and has not extended to Russia on the east, or Britain on the west.

Palæozoic Rocks of the Hartz, the Rhenish Provinces, and Belgium. —In advancing westwards from central Germany to the Hartz *, and the Rhenish provinces of Prussia, the geologist loses nearly all traces of the Silurian rocks of Bohemia, Saxony, and the Thüringerwald, whilst deposits of the Devonian and Carboniferous ages become vastly more expanded.

Yet, with this absence of the oldest fossiliferous deposits, the regions under consideration present quite as venerable an exterior, and contain rocks possessing a structure quite as crystalline, as those of which we have just taken leave. For, if we first glance at the range of the Hartz — that shrine at which so many poets have wor-

---

* Not having visited the Hartz since my second survey of that tract, with Prof. Sedgwick, in 1839, I am unable to do justice to so highly interesting a subject, and can only say that I believe our general order as given to the public was substantially correct, — subject, always, to that *reformed* interpretation of the exact relative value of the various sedimentary rocks which is now applied to the Rhenish provinces, and also to the lights which M. Adolf Roemer has very recently shed upon the tract in which he resides. In referring to that instructive work, the Palæontographica of Dunker and Hermann von Meyer (Band 2. lief 2.), the reader will perceive, at p. 71, how inverted and confused the order of succession appears in a section which is there offered. The feature, however, of this work which most concerns the observations in the text, is the publication (1852) of a group of fossils which, according to Adolf Roemer, lie in "Upper Silurian limestone." I must leave the arbitration of this point in the hands of my palæontological friends. Unquestionably, the great mass of the greywacke of the Hartz is Devonian, a point further corroborated by M. A. Roemer's discovery, in the heart of the chain, of the *Coccosteus*, a fish of the Old Red Sandstone. But whether there be really any fossiliferous rocks of older date in the Hartz than in the Rhenish provinces of Prussia, must yet remain in doubt; seeing that some of the forms published by Adolf Roemer in his plate of the so-called Silurian fossils, are well-known Devonian species.

shipped Nature in her wondrous freaks, and where the German geologist long regarded his old 'Grauwacke' as a mass whose order could
never be defined, and whose age was fathomless — its chief portions
are now ascertained to be of no more remote antiquity than the Devonian era. The Brocken itself, the giant of the chain, so sanc-
tified in many a legend, was preceded in this very district by submarine volcanic eruptions, which disturbed the bottom of a primeval
sea. It is composed of a comparatively modern granite, which, having
burst forth long after the slaty 'Grauwacke' had been accumulated,
has through ages of decomposition arranged itself into those chaotic
piles or 'felsen meer,' so graphically described by Leopold von
Buch. Again, subsequent outbursts of porphyry were also manifestations of the subterranean forces which produced the last great
elevation of the Hartz, and gave to the surrounding masses of
the secondary formations their present outline. For, unlike the
prevalent north-east and south-west alinement of the older rocks in
Britain, Scandinavia, Russia, and Germany, the geographical direction of the Hartz is from north-west to south-east; and thus its
appearance upon a map is not derived from its more ancient mineral
nucleus, but from the unconformable and enveloping younger strata.

As, then, the great masses of the slaty rocks of the Hartz which
contain fossils, are known to be of the same age as those of the
Rhine, and like them are also flanked by lower carboniferous strata,
we may now pass on, referring the reader by the way to the works
of Hausmann, Adolf Roemer, and others; merely adding, that in
mineral composition, association with igneous rocks, and the inversion
of great masses, the two regions are also alike. In truth, through
the combination of many disturbing causes in this one district, the
chain of the Hartz has literally been riven into detached fragments,
the relative age of which cannot be proved by order of superposition,
and can only be interpreted through an examination of the organic
remains of each mass. In the Rhenish provinces, on the contrary,

though large portions of their strata are infinitely contorted and broken, and sometimes also inverted, the Northern or Westphalian frontier exhibits a perfect and complete succession of formations in their normal ascending order.   The reader's attention will therefore be now directed to that region.

The convoluted and broken rocks presenting such an antique slaty aspect, and which, crowned with castles, form the chief features of the gorges of the Rhine and the Moselle, exhibit nowhere any fossiliferous band so old as the Upper Silurian. This remark applies to all the territory on the right bank of the Rhine, from the Taunus mountains on the south-east, to the coalfields east of Dusseldorf on the north-west.   This vast and formerly undivided ' Grauwacke ' tract, including the Duchy of Nassau, and having its northern frontier in Westphalia, is bounded on the east by the secondary rocks of Hessia, which range southwards by Marburg and Giessen to Frankfort.   On the left bank of the Rhine, the same upward succession occurs between the Lower Devonian rocks of the Hundsrück on the south-east, and the coal tracts of Aix-la-Chapelle and Belgium on the north-west.   It is only by deflecting westward into the mountainous tract of the Ardennes, that we meet with older slaty rocks, rising from beneath all the other deposits.

Although this view of the age of the Rhenish strata has for some years prevailed among scientific men, it is right to explain how it has happened, that English geologists*, who fourteen years ago applied the classification they had worked out in their own country to these Rhenish rocks, and modified the views of their precursors, should in their turn have seen reason to admit the value of certain important corrections made by their successors.

The clear general views of that Nestor of geologists, D'Omalius d'Halloy, the remarkable work and map of M. Dumont, as well as

---

* See Sedgwick and Murchison, Trans. Geol. Soc. Lond. vol. vi. p. 211.

the previous labours of Prussian geologists, including the maps of Leopold von Buch, Hoffman, von Dechen, and Öynhausen, unques- tionably led the way in the succession of efforts, through which our present knowledge has been obtained. After the publication of the above works, Professor Sedgwick and myself endeavoured to show (1839) that, like Devonshire and Cornwall, the Rhenish provinces contained a great mass of those strata, intermediate between the Silurian and Carboniferous deposits, which we had called Devonian; the equivalent, in our belief, of the Old Red Sandstone of Scotland and Herefordshire. Our contemporaries have admitted that, in our excursion of one long summer in Germany, we succeeded in proving the existence of such an intermediate series both in Prussia and Belgium, and also in showing how, on the right bank of the Rhine, the uppermost 'grauwacke' was divisible into lower Carbonife- rous and upper Devonian rocks. Misled, however, by an erro- neous interpretation of some of the fossils (for at that time the Lower Devonian forms had been little developed), we adopted the belief, that the inferior 'fossiliferous grauwacke,' or that which has since been called the 'Spirifer Sandstein' of the Rhine, was an equivalent of the Upper Silurian. I have been convinced, through the palæontological labours of Ferdinand Roemer and the brothers Sandberger, that the types of that lower Rhenish subdivision are distinct from the Upper Silurian, and in harmony with the lowest Devonian group of other countries. And for some years I have been aware that, whilst our sections representing the succession of the mineral masses were correct, the interpretation or synonymy to be attached to the lower division was erroneous. Again, in the superior portions of the group now recognized as Devonian, the geo- logists of Prussia and Nassau have made subdivisions, both minera- logical and zoological, which it is essential to notice.

It is, however, satisfactory to have ascertained in a recent visit to my old ground, that all the knowledge acquired in the fourteen

years which have elapsed since our survey was made, has but led to a
much more complete identification of the Rhenish provinces with
Devonshire, than that which was proposed by my colleague and self.
In short, it now appears that not some only, as we thought, but
all the palæozoic strata of Devon have their equivalents on the banks
of the Rhine. So, that, starting from the North Foreland of the
Bristol Channel, and ascending into the heart of the culm-fields, as
described p. 256., the geologist has before him the successive representatives of the Rhenish deposits.

Those persons who may refer back to the sixth volume of the
Geological Transactions of London, will, therefore, understand that all
the Rhenish ground which is described or coloured in the map and
sections as Upper Silurian, is now embodied in the Devonian rocks;
whilst to their admirable description of the fossils, MM. d'Archiac
and de Verneuil have but to add the one plate of the few so-called
Silurian fossils (executed in England), to their thirteen plates of
true Devonian types, and all the general features of our labours
will be in harmony with subsequent observations. The absence of
Upper Silurian rock in this region will modify to a very small
extent the reasoning of our distinguished French colleagues, as to
the relative distribution of the different classes of fossil animals, — a
point already well known to those who follow the progress of palæontology. In truth, if the field geologist makes his survey faithfully, and establishes a correct order of superposition, his facts
will always eventually be found to be coincident with the truths
of natural history. Of this the Rhenish provinces and Belgium have
afforded the best illustrations; for notwithstanding the opinion of a
distinguished contemporary M. Dumont, who has, I think, too much
undervalued the weight of fossil evidence, it is essentially the study
of organic remains which has led to the clear subdivision of the vast
mass of older rocks, which were there formerly merged under the
unmeaning term ' Grauwacke.'

Ascending Series in the Rhenish Provinces. — The slaty masses of the Ardennes, or the oldest rocks of the region, may by some persons be considered Silurian, from certain rare trilobites, orthoceratites, and shells, which have been detected in them.   Forms of Homalonotus, from the upper part of the slates, which I examined many years ago *, seemed to me to be of that age ; and many better specimens of the genus having been found, it is suggested, that the species in the Ardennes slates are different from those of the Devonian rocks of the Rhine.   It has even been surmised, that the true Upper Silurian is here omitted, as in Saxony and the Thüringerwald, and that some of the oldest slaty rocks of this region are Lower Silurian.

However this may be, neither in the gorges of the Rhine between Bingen and the mouth of the Lahn, where the rocks have been so contorted and so much subjected to a crystalline slaty cleavage, nor in the quartzose ranges of the Taunus and the Hundsrück, has any one been able to detect typical Silurian fossils.   On the contrary, the recent identification of the limestone of Stromberg on the south edge of the Hundsrück with the Eifel limestone†, and the discovery of certain fossils by Fridolin Sandberger (by no means lowest Devonian species) along the edges of the Taunus, have substantiated the views originally embraced by Professor Sedgwick and myself, that those mountains were simply metamorphosed masses of the same age as the chief fossiliferous 'Grauwacke' of the Rhine. Passing, however, from these defaced leaves in the book of geological succession, and commencing inquiry where the order of superposition, aided by palæontology, diffuses a clear light, we find in the Rhenish countries the following ascending order of the Devonian rocks properly so called : —

Lower Devonian or ' Spirifer Sandstein,' Wissenbach Slates, &c. — Slaty schists (those of Caub, for example), with greywacke sandstones

---

* Trans. Geol. Soc. vol. vi. p. 275.   They are not, however, the *H. Knightii.*

† MM. F. Roemer and Fridolin Sandberger agree in this opinion.

and quartzose rocks, and a rare trace of impure limestone; the whole
being exhibited between Coblentz and Caub. This is the 'Aeltere Rhei-
nische Grauwacke' of F. Roemer, or the 'Spirifer Sandstein' of Sandberger. It contains a good many remarkable fossils ; among which may be
noticed the large and broad-winged Spirifers—S. macropterus and S. speciosus, Terebratula Archiaci, many species of Pterinea, some Orthidæ,
especially O. circularis, Leptæna plicata, Chonetes semiradiatus, Sow. (C.
sarcinulata Schloth.?), Pleurodictyum problematicum, Phacops (Cryphœus)
laciniatus, Phacops latifrons, and both smooth and spinose species of Homalonotus,—H. Ahrendi, and H. armatus*, &c.

Looking to this list, and the plate representing some of these oldest
Rhenish fossils, as given in the Transactions of the Geological Society†, it
will be seen that, according to the palæontological knowledge of the year
1839, it was very natural to group this member of the series as Upper
Silurian. In fact, it was not then known, that Homalonoti were ever
found in Devonian rocks ; and the few imperfect casts of trilobites (since
ascertained to be of the genus Phacops) were then supposed to belong to
Calymene, a genus still unknown in Devonian rocks. Again, a large Pentamerus from Greifenstein was believed to be the Silurian form P. Knightii;
and though not identical, the distinction is but slight.‡

According to the sections made by my colleague and self, the beds which
next follow in this disturbed region, are the slates of Wissenbach, with their
numerous small pyritized orthoceratites and goniatites. This order is also

---

* The Homalonotus Herschelii, 'Sil. Syst.,' a fossil formerly sent to me from the
Cape of Good Hope by Sir John Herschel, and all the spinose forms of this
genus, are now believed to belong to Lower Devonian, and not to Upper Silurian
rocks. (M. Sandberger and Mr. Salter are of the same opinion.)

† Trans. Geol. Soc. 2nd series, vol. vi. pl. 38.

‡ The insulated shelly conglomerate of Greifenstein, with its large Pentameri,
may perhaps still be held in doubt, as being possibly the representative of an
Upper Silurian rock? The Pentameri occur in a highly dislocated mass, the
relations of which to other strata are not seen; and besides these fossils,
Chonetes lata (?) is said to have been recently found in it. At this spot, Greifenstein, the undulations and breaks are very numerous, so as to destroy all sure
indication of the dip of the quartzose grits in which the fossil shells occur. These
sandy strata have been, in great part, converted into quartz-rock, and are penetrated by basaltic eruptions. See Sedgwick and Murchison, Trans. Geol. Soc.
vol. vi 2nd series, p. 256.

sustained by the brothers Sandberger, who, admitting that the fauna of this band is rather peculiar, affirm that it contains from eight to ten species like those of the Spirifer Sandstone : among which are Nucula solenoides, Goldf. ; Phacops laciniatus, Roem. ; and P. brevicauda, Sandb. Thus, through their imbedded animal remains, a natural union subsists between these two rocks ; which, in the Rhine country, constitute the Lower Devonian. *

This Lower Devonian mass — at least, all that portion of it which is characterized by broad-winged Spirifers — was long ago recognized † as extending largely over the Prussian provinces on the left bank of the Rhine, which are watered by the Moselle ‡, the Lieser, and the Ahr ; and thence by the flanks of the Ardennes into Belgium, where its position was clearly marked by M. Dumont.

Middle Devonian of the Rhine. — ('Agger and Lenne' group of Dechen, or Calceola Schiefer, followed by Stringocephalus or Eifel limestone.) The strata which succeed to the Lower Devonian on the right bank of the Rhine, consist, like the previous beds, of schists and sandstones, but in which lenticular-shaped masses of limestone prevail at intervals.   They

---

* In the Caub slates, which are, indeed, quite intercalated in the lower division, the brothers Sandberger have found Phacops laciniatus, with P. latifrons, Homalonotus planus, and some Orthoceratites.

† Sedgwick and Murchison, Trans. Geol. Soc. vol. vi. p. 280.

‡ A work has recently appeared (1853), entitled ' Geognostische Beschreibung der Eifel,' with a map and plates of fossils, by Professor Steininger, in which that author endeavours to show, that the whole of the Eifel region, ranging from south-west to north-east, is a geological trough, supported on the one side by the clay slate and quartzose rocks of the Ardennes, and on the other by rocks on the banks of the Moselle, which he conceives to be of similar age. This latter inference (my respected friend M. Steininger must forgive me) is at variance with the belief of other geologists who have explored this region, and who, from organic remains, class the Moselle slates with the rocks of Coblentz and the mouth of the Lahn, *i. e.* the Lower Devonian of the present day.   Though it may seem ungrateful in me not to support that view of M. Steininger, which refers the ' grauwacke' under the Eifel limestone to the Silurian era, I am bound to say that a careful revision of all the fossil evidences collected by palæontologists, where fossils are more numerous and perfect than those to which the learned professor of Trèves appeals, compels me to adhere to the opinion expressed in the text.   A reference to the plates of his work will, indeed, satisfy naturalists that most of the figures given by M. Steininger belong, not to the Silurian, but to the Devonian rocks.

there occupy a wide tract which is watered by the rivers Agger, Vohne, and Lenne. On the left bank, this group constitutes the greywacke of the banks of the Ahr or Système Ahrien of Dumont. After various undulations and the protrusion of many bosses of eruptive igneous rocks, and the diffusion of much iron ore, these strata on the right bank of the Rhine, plunge under the great band of limestone which, ranging by Elberfeldt and Schwelm to Iserlohn, and appearing also at Paffrath and Refrath near Cologne, is the well-known equivalent of the Eifel limestone. Although my colleague and self thus described the stratigraphical position of these rocks, we did not attempt to separate them from the underlying Spirifer Sandstone, with which they are connected closely by mineral characters, and to which they are everywhere conformable. A laborious examination of the country, such as native geologists alone could execute, has, however, led to the separation of these strata from the lower division or the 'Spirifer Sandstein,' and has shown, that whilst they are united with the Stringocephalus or Eifel limestone which covers them, they are separated from the subjacent shelly 'grauwacke.' It has, in short, been found that, whilst they contain few species in common with the underlying rocks, they are laden with fossils that occur in the superior limestone. Hence, the Prussian geologists, viz., Von Dechen, F. Roemer, and Girard, have for some time been delineating these strata separately on an exquisitely finished geological map of the Rhenish provinces of Prussia, the result of infinite research, and they apply to these rocks the name of 'Agger and Lenne' group.* This subdivision is another of the many proofs which patient inquiry has brought forth to demonstrate, that it is not possible to form correct geological groups by appealing to mineral characters and superposition only; for these were precisely the grounds on which the strata in question were closely united by my colleague and self, with the greywacke of Ems and Coblentz.

It is now believed that the 'Agger and Lenne' group is identical with those more schistose courses on the left bank of the Rhine, which also underlying the Stringocephalus limestone are known as the 'Calceola Schiefer.' There they are characterized by the presence of the Calceola sandalina, Dalmania (Cryphœus) punctata, C. stellifer, Spirifer cultrijugatus, S. speciosus, S. heteroclytus, and many other fossils which are

---

* This beautiful Map occupying 30 sheets, "Geognost. Karte der Rhein-provinz und Westphalen," will be one of the noblest memorials of the school of Leopold von Buch, the great geological cartographer of the period.

common both in the Eifel country, and in the limestones of Devonshire. In following these formations into Belgium, the same physical order is known to prevail; and I learn from M. de Koninck, that a band containing the same lower Eifelian group of fossils alluded to, is clearly exposed between Couvin and Chimay.

Stringocephalus or Eifel Limestone.—This is the great central, calcareous mass, which gives to the Devonian rocks their dominant and independent characters; for if the Lower Devonian of the Rhine exhibits, in some of its forms, analogies to the Upper Silurian type, the limestone to which we have now ascended, and the approach to which is clearly indicated by its fossils in the strata that preceded it, contains a fauna which is unmistakeably peculiar, and wholly unlike that of the Silurian below or the carboniferous above it. Among the most striking Rhenish types of this rock are Stringocephalus Burtini, Uncites gryphus, Davidsonia Verneuilii, Spirifer undiferus, and S. lævicosta, Megalodon cucullatus, Lucina proavia, Murchisonia bilineata, and the corals, Cyathophyllum cæspitosum, Favosites polymorpha, Heliolites pyriformis, &c.

Laden with a profusion of corals, crinoids, and other fossils, many of which have been made known through the beautiful works of Goldfuss, it is unnecessary here to dilate further on the numerous organic remains of this rock, very many of which are common, as before said, in Devonshire. It may, however, be stated, that in the Eifel country, and also in the Hartz, the Coccosteus and other ichthyolites have been detected, which, as in Russia (see p. 328.), identify the rock as a member of the Old Red Sandstone of Britain. Among the organic remains are also several types, which connect this band with the subjacent strata, such as Phacops latifrons, Bronn, and Pleuracanthus punctatus, Steininger; with Spirifer speciosus, Schloth., Spirifer cultrijugatus, Roem. &c.[*]

Upper Devonian of the Rhine and Belgium.—(Clymenia and Goniatite limestone, Cypridina-Schiefer, Sandb.; Kramenzel-stein, F. Roem., with schists (Flint-schiefer, and sandstone, &c.).) Although we observed and noted, that the chief masses of the Devonian limestone in the Rhenish provinces were frequently followed by other courses of impure limestone, my colleague and self did not propose that separation which is seen in

---

[*] See the comparative table of Fred. Roemer, Geol. Kenntn. Nordw. Harzgebirges; Dunker and von Meyer's Palæontographica, 3 Band., p. 1., 1850; also the memoirs of the brothers Sandberger, Versteinerung, Rhein. Schichtensyst, Nassau, 1850; and Frid. Sandberger, Geol. Verhaltn. Nassau. (with map) Wiesbaden, 1847, &c.

Devonshire, and has already been adverted to as there existing between the lower or Stringocephalus limestone and the superior or Clymenia limestone of Petherwin. The researches of the able palæontologists of the Rhine, to whom allusion has already been made, and those of M. de Koninck in Belgium, have, however, led to a separation which more clearly identifies the succession of these foreign tracts with that of Devonshire. M. Ferdinand Röemer, for example, divides this upper group into the following ascending series: 1. Receptaculite schists, so called because they are charged with the Receptaculites Neptuni, Defrance; 2. limestone characterized by Goniatites auris, and other species; 3. schists with many Goniatites, Clymeniæ, and Cypridinæ; 4. and lastly, schists containing Rhynchonella cuboides and Productus subaculeatus,— beds which are paralleled by that author with the Upper Devonian strata of North Devon. Looking, however, at this Upper Devonian division in a broad point of view, as it generally appears on the Continent, it seems to me to be more frequently characterized by the small crustacean Cypridina serrato-striata, than by any other fossil. Where the calcareous courses thin out, and Clymeniæ and Goniatites, or other characteristic shells, are not frequent, the minute crustacean is almost everywhere present, often ranging through a considerable succession of beds, and giving to them their prevailing zoological character. In each country, however, through which this division ranges, it exhibits some peculiar features, though in most tracts it is chiefly marked by containing Cypridinæ, Clymeniæ, and certain Goniatites. The name, therefore, of 'Cypridina Schiefer,' adopted by the brothers Sandberger, who have described so many of the organic remains of this remarkable band of rocks, is, I repeat, highly characteristic of it as a whole.

In Nassau, where the upper limestone is in some places very little removed from the lower or massive limestone (as near Weilburg), it has only to be followed a short distance eastwards, to be seen divided from its neighbour by copious strata of the plutonic rock called " Schaalstein," which was formed of submarine volcanic ejections.

But, it is along the northern frontier of the Devonian rocks in Westphalia, where they all subside conformably beneath the overlying carboniferous deposits, that they assume an importance, which can only be well understood when the remarkable map, before spoken of, shall be published. There, the group above the Stringocephalus limestone exhibits, first, slaty schists with some thin layers of grey and black limestone containing Goniatites retrorsus; strata which at Nuttlar attain the great thickness of 1000 feet (Flint-schiefer of the Prussian geologists). Next follow micaceous sandstones often running into concretionary forms, as seen at the 'Rauhe Hardt,' near Iserlohn.   Then appear the Cypridina schists and a nodular limestone, which, from the cavities it weathers into, has been called 'Kramenzel-stein' or ant-stone*, the greatest thickness of which is about 200 feet.

It is this nodular Kramenzel-stein, and the associated schists and sandstone, frequently of a reddish colour, which are most charged with Cypridinæ and Clymeniæ, and these give to the upper group its chief character.   When most extended, including the schists called 'Flint,' the group has the dimensions of upwards of 1300 feet. This Upper Devonian or Cypridina Schiefer, as before explained, is much developed in Saxony and in the adjacent tracts of Thuringia and Franconia.   (See last Chapter.)

In terminating this sketch of the Devonian rocks of the Rhenish provinces, it may be observed, that notwithstanding a certain general aspect, and the occurrence of a few similar species, even the lowest Devonian here indicated is very different from the Upper Silurian type of any part of Europe; whilst the middle or typical Devonian is completely distinct. Neither of these rocks contain, for example, any Silurian species of crustacean or cephalopod; and graptolites, which occur from the bottom to the top of the Silurian rocks, are unknown in them.   The Devonian and Upper Silurian are, therefore, infinitely more distinguished from each other, as natural-history groups, than the Lower and Upper Silurian, which, on the contrary, are intimately linked together, as proved in the earlier chapters, by a very considerable number of the same species of crustaceans, cephalopodous and brachiopod shells, corals, and graptolites.

In following these Devonian rocks into Belgium, it is found that, though their main divisions are persistent, their details differ considerably from those of Westphalia.   At Aix la Chapelle, where the Belgian type may be said to begin, all the members of the series are much attenuated. We there find a very meagre equivalent only of the lower shelly greywacke; and the calcareous representative of

* The cavities are frequently filled with ants' nests; hence this name, given by the workmen.

the Eifel limestone is but a poor coralline rock, in parts dolomitic, followed only by certain nodular schists charged with the Receptaculites Neptuni and Spirifer Verneuilii, which alone represent the copious masses of Flint-schiefer, Kramenzel-stein, &c., of the previously mentioned tracts. Again, the Belgian succession, properly so called, seems to be deficient in the equivalents of those lower carboniferous strata, which in Prussia are intercalated between the Upper Devonian and the Carboniferous limestone, or its equivalent Kiesel Schiefer, &c.; for the highest Devonian nodular schists, with Spirifer Verneuilii, are at once surmounted by the limestone with great Producti (Carboniferous Limestone).* The mineral character of these various rocks, and the changes they undergo, must be studied in the works of Dumont; and their zoological distinctions, in those of de Koninck, &c.†

According to de Verneuil, the Lower Devonian of the Rhine is the equivalent of the Oriskany sandstone of the United States. Other analogies have been indicated, through their fossils, between these Continental deposits and the Marcellus slates and Hamilton groups of North America. But, for the present, let us simply bear in mind that, whether we regard the physical order of the masses, or their imbedded remains, the Rhenish rocks, which have afforded more than 450 species of fossil animals, are not only really remarkable counterparts to the succession in Devonshire, but, through the vigour of their explorers, have already much outnumbered in fossils their British equivalents.

Carboniferous Rocks of the Rhenish Provinces.—The upward succession

---

* I owe the knowledge of these facts to M. Ferdinand Roemer. M. de Koninck, however, tells me that between Couvin and Philippesville he has observed an order like that of the Rhine.

† See particularly the memoir of Mr. D. Sharpe, comparing the palæozoic formations of America with those of Europe. Quart. Journ. Geol. Soc. vol. iv. p. 145.

from the Devonian rocks into the Carboniferous deposits is clear on both banks of the Rhine and in Belgium; and if in the latter country certain lower carboniferous schists are attenuated, this thinning out of one portion is more than compensated by the presence of noble masses of carboniferous limestone.   The carboniferous limestone of Belgium has been distinguished zoologically by de Koninck into two stages; the inferior of which, as in Russia and many other countries, is marked by the presence of the large Productus giganteus and several other types of that genus.   The upper stage contains the Spirifer Mosquensis, Fischer; Terebratula Royssii, L'eveillé; and Actinocrinus stellaris, Kon.*

The splendid rocks of this formation, which the traveller admires as he passes along the railroad from Namur to Liége, gradually thin out as they range towards the Rhine.   They make, in short, their last appearance as a solid limestone at Cromford near Ratingen, on the right bank of the Rhine; for thence, in their extension to the east and north, they lose their calcareous character.   Instead of two or more massive divisions, they dwindle into one or two thin, flat beds of black limestone, very much like the black culm-stone of Devonshire, which there in like manner is the feeble representative of the great Carboniferous or Mountain Limestone of other parts of England.   Like the Devonshire rocks, the Westphalian strata are also characterized by the very same species, the Posidonomya Becheri; which shell, even where the calcareous matter is entirely absent, as in the schists of Herborn and other places, distinctly marks the *centre* of the Lower Carboniferous rocks.

In the German Rhenish provinces, the lower member, or the schists

---

* The most characteristic fossils of the two stages of the Carboniferous Limestone of Belgium are as follows; the list having been furnished, at my request, by M. de Koninck.   See also his "Description des An. Foss. Carb. de Belgique." Liége, 1842.

Inferior or Visé Limestone.—Cardiomorpha oblonga, Sow., sp.; Arca obtusa, Phill.; Pinna flabelliformis, Martin; Avicula Dumontiana, De Kon.; Productus giganteus, Martin, sp.; P. striatus, Fischer, sp.; P. sublævis, De Kon.; P. undatus, Defr.; P. plicatilis, Sow.; P. proboscideus, Vern.; Chonetes comoïdes, Sow., sp.; C. papilionacea, Phill., sp.; Spirifer bisulcatus, Sow.; S. striatus, Martin, sp.; S. crassus, De Kon.; S. convolutus, Phill.; Rhynconella acuminata, Martin, sp.; R. angulata, Linn., sp.; Bellerophon costatus, Sow.; Euomphalus pugilis, Phill.; Chemnitzia constricta, Martin, sp.; Orthoceras giganteum, Sow.; Nautilus cyclostomus, Phill.; N. globatus, Sow.; Goniatites sphæricus, Sow.; Psammodus porosus, Agass.

Superior or Tournay Limestone.—Pentremites caryophyllatus, De Kon. and Le Hon.; Actinocrinus polydactylus, Miller; A. stellaris, De Kon. and Le Hon.; Spirifer Mosquensis, Fischer; S. tricornis, De Kon.; Terebratula Royssii, Leveillé; Chiton priscus, Münster; Euomphalus tabulatus, Phill.; E. tuberculatus, De Kon.; Gyroceras aigoceros, Münster, sp.; Goniatites rotatorius, De Kon.

below the limestone (Lower Sandstone shale), is usually very silicious, and is, in that tract, the 'Kiesel Schiefer' of the Prussian geologists, which occasionally expands into a deposit of considerable dimensions. In some districts it has been named 'Pön Sandstein.' The limestone which reposes on these schists and sandstones is again surmounted by a copious succession of sandstone, mineralogically not very unlike many of the inferior divisions of the so-called 'grauwacke,' and which was also formerly known under the name of 'Fletz-lehrer Sandstein.' Professor Sedgwick and myself showed, that while the Posidonomya beds represented the 'culm,' or true Carboniferous limestone, this 'Fletz-lehrer Sandstein' or 'jungste grauwacke' was simply the equivalent of our British millstone grit, and that, as in England, it lay immediately beneath the most productive coalfields.

This identification was, in truth, one of the most satisfactory assurances, that the principle derived from the collation of organic remains with physical order, had led us to a correct view in placing as a member of the carboniferous deposits, strata which had been previously connected with much more ancient greywacke rocks.

In the extensive districts of Westphalia which he has surveyed and mapped, Dr. Girard has still more clearly delineated the distinctions between those bands of rock, which formerly were all merged under the unmeaning term 'grauwacke.' Even above the 'Kramenzel-stein,' or Clymenia limestone, there are, in that region, two superior sandstone formations, both of which, as well as the inferior rocks and Spirifer Sandstein, were included in the Rhenish 'grauwacke' of old authors. The sandstone covering the 'Kramenzel-stein,' locally called 'Pön sandstein,' is a fine-grained, micaceous rock, occasionally silicious, in which no fossils, except fragments of plants, have been detected. This sandstone seems to occupy the place of the lower shale of the South of England, of the yellow sandstone of Ireland, and of the lower coaly sandstone of Scotland; for, like them, it is immediately covered by the carboniferous limestone or its equivalent, often in a state of chert or petro-silex, with Posidonomya Becheri and Goniatites crenistria. The rock which next succeeds is, as before said, the distinct equivalent of the British millstone grit, like which it underlies the great productive coalfields.

Whilst the Devonian rocks of the Rhine, unlike those of Saalfeld (p. 358.), have afforded few traces of land plants, it is important to observe, that it is specially in this Lower Carboniferous group, that nearly all the so-called transition and 'grauwacke' fossil vegetables of various German authors have been collected. Seeing that this series, in the Rhenish provinces, is so very analogous to that of Britain, it will indeed be interesting to compare the plants described by Göppert, Unger, and others, with the rich Flora of the same age found in Northumberland*, Scotland, and parts of Ireland.

Inversions. — Inadequate as this brief sketch may be to convey any just idea of the nature and contents of the older deposits of this interesting region, it would be still more imperfect if a reference, however cursory, were not made to the phenomenon of the inversion of some of its stratified rocks. Adopting the explanation of M. Dumont, of such reversals of the normal order in Belgium, Professor Sedgwick and myself applied it to account for the position of certain masses of the Eifel country. We pointed out how, near Münster Eifel, the overturning of the beds had been produced by movements which acted throughout masses of sediment several thousand feet in thickness; the phenomenon being illustrated by this diagram.†

INVERSION IN THE EIFEL EXPLAINED.

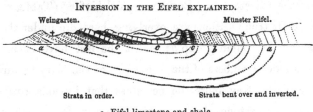

Weingarten.   Münster Eifel.

Strata in order.   Strata bent over and inverted.

c. Eifel limestone and shale.
a, b. Shelly Devonian greywacke.

* See remarks, by Mr. G. Tate, on the fossil flora of the coal formation, in the ' Natural History of the Eastern Borders,' by Dr. Johnston, 1853.

† From Trans. Geol. Soc. Lond. vol. vi. p. 277.

In this case, whether by following the strata upon their strike, or by making a short traverse across them, as in the preceding section, the true order was soon detected.

Again, on the northern or outer frontier of the Devonian rocks of Westphalia, where the order is very regular, all such strata being seen to plunge under the Carboniferous, as before explained, the attention of the observer is roused when, exploring north-westwards to Brilon, he finds a large tract of country (much penetrated by very ancient igneous rocks) which, extending towards Berleburg, exhibits everywhere the older beds overlying the younger. This phenomenon, long ago known to the Prussian geologists von Dechen and Erbreich, is expressed in this woodcut, taken from part of a larger section published by my colleague and self in the Geological Transactions. *

INVERTED STRATA SOUTH OF BRILON.

N.N.W.   S.S.E.

Brilon.

c. Lower Carboniferous (Kiesel and Posidonomya Schiefer).
b. Upper Devonian schist, lime, and iron ore } Devonian.
a. Slaty Devonian -   -   -   - }
(* Trap and igneous rocks.)

The reader will see that the lower carboniferous rocks (c), consisting of the 'Kiesel' or 'Posidonomya Schiefer' (Table, p. 382.), which in their true positions are overlying, here dip under the upper or calcareous Devonian group b; whilst the last have been carried under still older members of that series (a), or the slaty 'grauwacke' of this region. Nay more, he perceives that igneous rocks, porphyries, and greenstones, &c., some of which were formed successively on the bottom of the same seas, in which the Devonian strata (b)

* See Sedgwick and Murchison, Trans. Geol. Soc. Lond. vol. vi. p. 239. The woodcut in the Geological Transactions shows the coal strata to the north covered by cretaceous rocks.

to which they are parallel, were accumulated, have also been over-thrown *en masse* with all the other rocks. It follows, therefore, that the regular order in which these formations are traceable along so wide an adjacent region is here completely deranged, and that all the rocks are inverted, the older being incumbent on the younger! Now, this very same phenomenon of the overturning of huge masses of the solid crust of the earth has not had its limit in the Rhenish provinces, but has extended into Belgium, where the same strata, as described by M. Dumont, are affected along a band quite parallel to those great folds in Westphalia and the Eifel.* They are, therefore, to be considered as great parallel, wave-like undulations, similar to those which have been so ably explained by the brothers Rogers of the United States, and which will be adverted to hereafter.

In looking back to p. 144. we see, indeed, that in Cornwall, this great European inversion has proceeded still further to the west, and that masses of an antiquity unknown in the Rhenish provinces and Belgium (*i.e.*, the Lower Silurian) overlie Devonian strata. In that example, both the rocks being fossiliferous, no doubt as to the intensity of the overthrow can exist.

In the earlier part of this Chapter it was also shown that, in one district of Thuringia, certain plant beds, which belong to the lower carboniferous division, plunge under upper Devonian beds charged with Cypridinæ and Clymeniæ, and that near Ober Steinach, in Meiningen, these Devonian strata are apparently surmounted by true Lower Silurian!

If a young geologist had detected, for the first time, such abnormal relations in a country like the Eifel, where, besides igneous erup-tions of remote periods like those in the last diagram, the strata have been pierced by volcanos which have certainly been in activity under the atmosphere, he might for a moment suppose, that such modern

* See the very remarkable geological map of all the volcanic tracts in the Rhenish Provinces, by M. v. Öynhausen.

eruptions, exhibiting, as they do, many craters and coulées, must have exercised an influence analogous to the causes that produced the above-mentioned inversions. But even a cursory examination would dispel such an hypothesis; for we see that these sub-aerial volcanos, with their scoriæ, pumice, lava, and ashes, are nothing more than mere superficial pustules, which in bursting forth have produced no sort of alteration in the position or nature of the rock masses, but have merely vomited their contents into depressions already prepared to receive them; thus showing us that when these volcanos at length appeared, the present system of hills and valleys had been completed.

Infinitely more intense, therefore, must have been that much deeper seated agency, which giving rise to such internal writhings of nature, inverted thousands of feet of strata over vast areas, and twisting them like ropes, coiled them over the rocks which lay above; thus placing them over deposits formed at immeasurably subsequent periods!

But, striking as are the above-mentioned features of inversion, still grander examples will be adverted to in the sequel. The reader will, however, recollect that, after all, such phenomena are local and abnormal only, as respects the surface of the globe, on which, fortunately for science, the original impress of order is visible over by far the greater portion.

In reverting then to such order, I annex the following table of the divisions of the older Rhenish rocks, as now understood by the geologists and palæontologists who have been cited in the preceding pages, and whose discoveries, coupled with my recent observations, have led me to divide the Devonian rocks of the region into three parts. By having alongside of this table the general succession of the same rocks as classified by M. Dumont*, the

---

* See Dumont, Mémoire sur les Terrains Ardennais et Rhénan, in Mém. Acad. Bruxelles, vol. xx. In his Introduction, in which he treats of the invalidity of

reader will at once see, that although the mineral order which that
author indicates is doubtless correct, no one of his three 'terrains' is
in unison with a classification based upon the distribution of organic
remains.

fossil evidences, the author is mistaken in attributing to me alone the establishment
of the 'Devonian System;' and he also has misunderstood the process of inquiry by
which the older rocks were classified in England.   I have only referred slightly in
the text, to the terms used by M. Dumont in his more recent classification of Belgian
rocks, illustrative of his general map, because his distinctions and divisions, as
founded on the mineral features of the successive masses of rock, do not coincide, as
above stated, with what geologists know to be their natural history characters. Admiring the method by which, after a laborious survey, this able geologist and sound
mineralogist reduced to order the complicated and often inverted ancient strata
of his native country, I am compelled to say that the divisions in the 'Tableau des
Terrains,' which he has published, to accompany his Map of Belgium, is, to a great
extent, at variance with the opinions of numerous contemporaries, as well as myself,
and is so much in contradiction to the classification recorded in this volume, that
I beg to call attention to the annexed table, in which M. Dumont's chief divisions
are paralleled with those which I have adopted.   Setting aside the whole of his
'Terrain Ardennais,' or lower great group, in which few or no fossils are known,
it will be seen that his Terrain Rhénan embraces not only all that is here shown to
be Devonian, i. e. Spirifer Sandstein, etc., (his Système Coblentzien,) but includes
his Système Ahrien, or just that member of the Devonian which the geologists of
Prussia and Nassau have agreed to separate from the rocks of Coblentz, and to
group with the Eifel limestone, but which, judging from mineral characters only,
M. Dumont has left in his Terrain Rhénan.   Again, his Système Condrusien
unites what is here termed Upper Devonian and Lower Carboniferous, including
the Clymenia, and carboniferous, limestones ; whilst he separates the latter from
the coalfields, of which, according to all geologists, it forms the natural base.
Lastly, if the Système Gedinnien (Dumont) really contains some Upper Silurian
forms, as suggested by Forbes and Sharpe, then, indeed, each of the great Terrains
of M. Dumont is still more antagonistic to a classification by organic remains.
See Mr. D. Sharpe's Explanation of the equivalents of M. Dumont's classification,
Quart. Journ. Geol. Soc. Lond. vol. ix. p. 18.; also Mr. Austen's view, ib. vol. ix.
p. 244.; and compare their tables with that which is here given.

ORDER OF THE OLDER ROCKS OF THE RHENISH PROVINCES AND BELGIUM.

CLASSIFICATION OF M. DUMONT.			CLASSIFICATION ADOPTED IN THIS WORK.					
Terrains.	Systèmes.	Lithological Divisions.		Fossiliferous Divisions.	Localities in Prussia.	German Names.	Geological Groups.	Classes.
ANTHRAXIFÈRE.	Houiller.	Ampélite, Psammite, Schiste, Houille.	Upper Carb.	Coal Measures, i. e. Shale, Sandstone and Coal.	Werden, Herdecke, &c. (Westphalia).	Stein-Kohl Formation.	Carboniferous.	UPPER PALÆOZOIC.
			Lower Carboniferous.	Millstone Grit, &c.	Schwerte, Menden (base of Coalfields).	'Fletz-lehrer Sandstein,' or 'Jüngste-Grauwacke.' (Jungere Grauwacke.)		
	Condrusien.	Calcareux—Calc. à Crinoides et à Productus, Dolomie, Silex, Anthracite.		Carboniferous, Mountain, or Productus Limestone. (Culm Limestone with Posidonomya Becheri.)	Ratingen (Cromford), Arnsberg, Herborn.	'Berg-Kalk' and 'Posidonomya Schiefer.'		
				Limestone, Shale, and Sandstone.	Schelke, Langescheid.	'Kiesel-Schiefer,' 'Pön Sandstein,' &c.		
		Quartzo-Schisteux (supérieur.) { Psammite grisâtre, Macigno, Anthracite. Schiste grisâtre, Calc.-Schiste, Calcaire, Oligiste oolitique.	Upper Devonian.	Schists and nodular Limestone with Clymeniæ, Cypridinæ, &c.	Mecklinghausen, Laasphe, (Westphalia). Weilburg and Selters, (Nassau).	'Kramenzel-stein,' or 'Cypridina-Schiefer.'	Devonian. (Old Red Sandstone.)	
	Eifelien.	Calcareux. Calcaire et Dolomie.		Schists with Goniatites retrorsus.	Meschede, Nuttlar (Westphalia).	'Flint-schiefer.'		
		Quartzo-Schisteux (inférieur). { Schiste gris fossilifère, Calc. Schiste et calcaire-argileux, Oligiste oolitique. Poudingue, Psammite et Schiste rouge. (Pepinster).	Middle Devonian.	Great central Coralline Limestone with Stringocephalus Burtini, Megalodon cucullatus, and the chief Devonian types.	Eifel, and Paffrath, Brilon, Limburg, Iserlohn, &c. (Westphalia). Cubach (Nassau).	'Eifel-Kalkstein,' or 'Stringocephalus-Kalk.'		
RHÉNAN.	Ahrien.	Grès, Psammites et Schistes, gris-bleuâtres.		Schists and 'Grauwacke' Sandstone with Calceola sandalina.	Escarpments of the Eifel Limestone and banks of the Ahr, Left bank of Rhine. Agger-thal and banks of River Lenne, Right bank of Rhine.	'Calceola Schiefer,' or 'Agger and Lenne Gruppe' (von Dechen).		
	Coblentzien	Grès et Phyllades, gris-bleuâtres.	Lower Devonian.	Wissenbach Slates, Caub Slates, &c.	Wissenbach near Dillenburg, and South flank of the Taunus (Nassau).	'Wissenbach-Schiefer,'		
				Coblentz Grauwacke, (Spirifer Sandstone,) with Spirifer macropterus, Pleurodictyum problematicum, &c.	Banks of the Rhine (*passim*) from Bingen, by Caub to Coblentz. (Shelly, lower Grauwacke of the Eifel).	'Spirifer Sandstein' (Sandb.), or 'Aëltere Grauwacke' (F. Roem.).		
	Gedinnien.	Poudingues, Grès verts et Phyllades rouges, verts, ou aimantifères.					ilurian ?	
ARDENNA	Salmien.	Quartzo-Phyllades. Ottrélitifères ou oligistifères.						
	Revinien.	Quartzites et Phyllades gris-bleuâtres.						
	Devillien.	Quartzites blancs ou verts, et Phyllades						

N. B. The few fossils yet found in the strata called 'Gedinnien' by Dumont, and which overlie the slates of the Ardennes, are too imperfect to authorize a comparison with any known Silurian types. Some geologists believe that the opposite subdivisions of Dumont may merely represent the bottom of the Rhenish series; others suggest, in the absence of fossils, that the lowest rocks of this region, particularly some of those of the Ardennes, may even be Lower Silurian, the Upper Silurian being unrepresented, as in Saxony.—

CHAPTER XV.

SILURIAN, DEVONIAN, AND CARBONIFEROUS ROCKS OF FRANCE, SPAIN,
PORTUGAL, AND SARDINIA.

ALTHOUGH the palæozoic rocks* occupy considerable spaces in
France, it is not possible on this occasion to offer more than a slight
sketch of their chief features in those tracts where they are clearly
exhibited. In Brittany, where they appear in force, the authors of
the geological map of France † divide them into two principal masses.
The inferior is composed of, 1. Glossy schists (schistes satinés luisans)
of great thickness, in which a few thin courses of grit and shaly
limestone occur. In their mineral aspect, *and in their entire want of
fossils,* these strata remind the geologist of the rocks which underlie
the lowest fossiliferous deposits of Bohemia, and they may not unaptly
be compared with some of the crystalline and subcrystalline rocks of
Anglesea and the Longmynd, or hardest bottom rocks of the Welsh
and English series. Their mean direction in Brittany is, like that of
the Longmynd, from east 20° north, to west 20° south, and they were
termed Cambrian by E. de Beaumont and Dufrenoy. 2. Under the
name of Silurian, these authors include a series of thick and complex
fossiliferous strata, which they have again divided into two groups. The
lowest of these they thus arrange : — 1st. Conglomerates and silicious
sandstones ; 2nd. Bluish schists, which at Angers, Poligny, &c., furnish good slates, and correspond through their fossils ‡ to the Llandeilo
formation of Britain. So far, the succession in Brittany, as dependent
on organic remains, is in unison with that of Britain ; but the chief
portion of the next division, consisting of compact limestones and

---

* Often, however, in a metamorphic state.

† MM. Elie de Beaumont and Dufrenoy.

‡ Trinucleus and Ogygia, with Illænus giganteus, &c., are abundant.

schists, has been abstracted by de Verneuil from the Silurian, and
shown to belong to the Devonian system. In this way the order in
Brittany is analogous to that of Cornwall on the one hand, and many
parts of Germany on the other, in both of which tracts, as already
shown, there is an equally sudden transition from Lower Silurian to
Devonian.

The second group of the authors of that magnificent work the geological map of France, is made up, first, of silicious conglomerates, coarse
grits, and argillaceous schists, which many persons would term greywacke, then of beds with coal, and, lastly, of a limestone specially characterized by the presence of Amplexus and Productus. Now d'Archiac
and de Verneuil have demonstrated, that this last and highest group
is neither Silurian nor Devonian, but a true member of the Carboniferous rocks. It is, in fact, identical, through its fossils, with the
carboniferous limestones of many regions.

All these fossiliferous deposits, whether Lower Silurian, Devonian,
or Carboniferous, are conformably inclined, whatever may be the
inclination of the beds; and striking from east 15° south, to west 15°
north, they are transgressive to the older unfossiliferous rocks on
which they repose. It was the concordance in their direction and
inclination which naturally led Elie de Beaumont and Dufrenoy to
embrace, in the first instance, all these fossiliferous deposits in the
Silurian period. But in the interval which has elapsed since that
classification was proposed, Brittany has been examined by many
geologists, including Frappoli, de Fourcy, Durocher, Blavier, Rouault,
as well as by de Verneuil and d'Archiac, all of whom agree in
adopting the order and nomenclature here announced. The eastern
limits of the province, towards the departments of La Mayenne and La
Sarthe, have indeed been studied in the greatest detail by M. Triger,
to whom the execution of the geological map of the last-mentioned
district was confided. The collection of fossils made in the palæozoic
rocks around Cherbourg, by my late old and indefatigable corre-

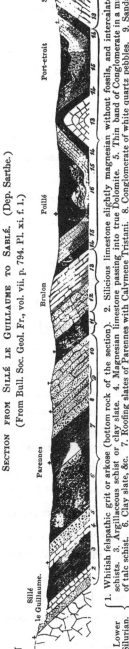

Section from Sillé le Guillaume to Sablé. (Dep. Sarthe.)
(From Bull. Soc. Geol. Fr., vol. vii. p. 794. Pl. xi. f. 1.)

Lower Silurian. { 1. Whitish felspathic grit or arkose (bottom rock of the section). 2. Silicious limestone slightly magnesian without fossils, and intercalated in schists. 3. Argillaceous schist or clay slate. 4. Magnesian limestone passing into true Dolomite. 5. Thin band of Conglomerate in a matrix of talc schist. 6. Clay slate, &c. 7. Roofing slates of Parennes with Calymene Tristani. 8. Conglomerate of white quartz pebbles. 9. Sandstone and schist with several courses of Ampelite charged with Graptolites. 10. Ferriferous grit and sandstone like that of May in Normandy. 11. Schists or Ampelite shale, with Cardiola interrupta, Orthoceras pelagium, and Graptolites, forming a passage into Upper Silurian.

Devonian. { 12. Quartzose whitish sandstone with Dalmannia calliteles, Homalonotus Gervillei, and other Devonian fossils. 13. Schists and limestones with Devonian fossils.

Carboniferous. { 14. Sandstone, schist and Anthracite, worked at Sablé;—base of the Carboniferous group. 15. Carboniferous limestone with Productus gigas, Chonetes, comoides, &c. 16. Anthracitic schists, &c. &c. Poillé, Fercé, &c.

spondent M. de Gerville, has also contributed essentially to clear away the difficulties attending a right classification ; and lastly, the Geological Society of France (during its meeting at le Mans, 1850) under the leadership of M. Triger, made an admirable section across the older rocks to the north of Angers, the *resumé* of which was published by de Verneuil and de Lorière.

This diagram, reduced from their publication, exhibits the principal geological features of the country between Sillé le Guillaume and Sablé, a distance of about twenty-five English miles; and it clearly develops a succession of the palæozoic formations, from the Lower Silurian to the Carboniferous rocks inclusive.

In this section, the lowest stratum is a white sandstone (1) resting on porphyry. The still more ancient crystalline and glossy schists of Brittany are here wanting, and the lowest beds which are visible

(1 to 6), and in which some limestone appears, have not yet afforded any fossils, to serve as equivalents for those of the lower zone of Bohemia and Scandinavia, or the Lingula flags of Britain.

The great schistose and slaty mass which follows (Ardoises de Parennes, or 7 of the diagram), forms the inferior limit of Silurian life, as known in France. From its wide extension, and its fossils, this formation merits great attention; for it is doubtless, as M. de Verneuil states, the representative of the Llandeilo and Bala rocks of Britain, of the second zone of M. Barrande in Bohemia, and of the Orthoceratite limestones of Scandinavia, Russia, and North America. It is, in fact, the dominant member of the Lower Silurian rocks, so termed by all the authors who have described them in those countries.

These slaty masses of Brittany are characterized, like their equivalents elsewhere, by a profusion of Trilobites. Unfortunately these crustaceans are, for the most part, much flattened and distorted; but the following species have, notwithstanding, been recognized, some of which were long ago figured by Guettard and Brongniart: —

Calymene Tristani, Brong.; C. Arago, Rouault; C. Verneuili, Rouault; Asaphus Guettardi, Brong.; Ogygia Desmaresti, Brong.; Illænus giganteus, Burm. (the two last are gigantic species); Ill. Beaumonti, Rouault; Acidaspis Buchii, Barr.; Phacops (Dalmannia) socialis, Barr.; P. Dujardini, Rouault; Placoparia Tournemini, Rouault; Cheirurus claviger, Beyr; Lichas Heberti, Rouault; Trinucleus Pongerardi, Rouault,* &c.

Mollusks are scarcer than trilobites; the well-known British species Bellerophon bilobatus (Sil. Syst.) is, however, not unfrequent. But, as a rule, neither the species of crustacea nor of shells are the same with those of Britain, while there are many identical with the fossils of Spain and of Bohemia.

These Lower Silurian schists traverse an extensive tract, as may be seen by referring to the geological map of France, and crop out near

* De Verneuil and Rouault, Bull. Soc. Geol. Fr., 2d ser. vol. iv. p. 320.

the harbour of Brest and Bay of Crozon, where De Verneuil has detected in them Calymene Tristani. The spot where the deposit affords the best roofing slates is Angers, where the quarries are 300 to 400 feet deep. There the lines of original deposit and the laminæ of cleavage coincide. In fact, the remains of the included trilobites occur along the divisions by which the highly inclined slates are cleaved; a phenomenon sometimes observed in Britain, where the cleavage is, however, usually transverse to the bedding. (See p. 34.)

The peninsula of the Cotentin also contains these trilobite schists, near Siouxville, where M. de Gerville collected the Calymene Tristani.*

The lower slaty rocks of Llandeilo age are succeeded in La Sarthe by an arenaceous group, represented by the 8 and 9 of the previous woodcut. In general, this division exhibits, in ascending order, conglomerates, white and reddish silicious sandstones, with some subordinate beds of 'ampelite' schists containing graptolites, chiefly the G. colonus and the G. testis of Barrande. The position of these beds is probably the same as that of the graptolite schists of Dalecarlia and parts of Sweden. They agree also, according to de Verneuil, with the great schistose member of the American Silurians called the 'Hudson river group,' which is equally characterized by graptolites, and which overlies the Trenton limestone or Llandeilo flags, the equivalents, as above said, of the slates of Angers. As to the silicious sandstones in which the schists are enveloped, they are identical with the rocks of Jurques, Gahard, and May in Normandy, which have been placed by French geologists without hesitation on the parallel of the Caradoc sandstone of Britain.

In some districts the rock is so ferruginous as to be the seat of iron mines, and the ore in parts is of an oolitic structure : trilobites are found in it.

The sandstones are not rich in fossils, but where I have examined them in company with M. de Verneuil, they contain the British species Bellero-

---

* Crustacés Fossiles.   Paris, 1822.

phon bilobatus, with Conularia pyramidata, some species of Dalmannia, Homalonotus Brongniarti, and H. (Plesiocoma) rara ; the last-mentioned fossil occurs also in the Lower Silurian of Bohemia and Spain.

Like the lower and larger portion of my Caradoc formation in Britain, this group of sandstones and schists is distinctly connected by its fossils with the inferior slates.

Above all the strata, however, which have been mentioned, there is in France, an upper course of 'ampelite' schist, which, though often confounded with the lower, must be separated from it. These upper ampelitic schists (11 of the section), which are black and occasionally bituminous, have, as usual, given rise to wasteful searches after *coal*. They are distinguishable from the inferior masses by containing concretions of black limestone with a brilliant fracture, and are characterized by certain imbedded fossils. The most characteristic are Graptolithus priodon, Orthoceratites and Cardiola interrupta.* As this last-mentioned species is chiefly an Upper Silurian fossil in Britain, and occurs in Bohemia at the base of that division, it would seem fair, in the first instance, to consider this French deposit as of like age. At the same time we must observe, that Cardiola interrupta, on whose presence the comparison was drawn, is occasionally, though rarely, found in the Lower Silurian both of Britain and Bohemia. Whilst, therefore, this zone is clearly recognized, through extensive districts of France, as the highest of the rocks which are referable to the Silurian epoch, and whilst it is clearly at the base of the Upper Silurian, it is by no means an equivalent of the great British formation of Wenlock, and still less has it any reference to the Ludlow Rocks. We must, indeed, recollect that in Britain, as in Bohemia and Sweden, such schists, with Cardiola interrupta and Graptolites, underlie limestones with the well-known fossils of Wenlock, Dudley, and Gothland.

* See an account of the extension of this zone of ampelite schists and its intercalated minerals, by M. de Boblaye, Bull. Soc. Geol. France, 1 ser., vol. x. p. 227.

In France, therefore, the ascending series of strata is very different from that of Britain. The Upper Silurian is essentially wanting; the succession being analogous, I repeat, to parts of Germany where the Lower Silurian is succeeded by Devonian rocks, with only graptolitic schists between them. (See p. 354.)

France is by no means poor in Devonian equivalents. Numerous species of fossils derived from the strata (12 and 13) of the above diagram have been compared by de Verneuil with those of the lower shelly greywacke of the Rhine and the limestone of the Eifel, and he has found that many of them are identical. Among these, the broad-winged Spirifer macropterus, the Terebratula Archiaci, Grammysia Hamiltonensis, and Pleurodictyum pro-blematicum, are striking Devonian forms, unknown in any true Upper Silurian rocks. A still greater number of species in the Lower Devonian limestones of La Sarthe*, and Nehou† in Normandy, are identical with those of the Eifel.

Having established the identity between the Devonian rocks of the West of France and the lower portions of the system in the Rhenish provinces, de Verneuil observes, that in such a consecutive development we might expect to meet with some species, which on the continent of Europe and the British Isles are considered to be Silurian. And such is the case. Thus, Bronteus Brongniarti, Capulus robustus, Terebratula eucharis, T. Haidin-geri, Atrypa reticularis, Orthis Gervillei, Leptæna Bouei, L. Bohemica, and L. Phillipsi, are found in the Devonian of France, and also in the uppermost Silurian of Bohemia. And Pentamerus galeatus, Atrypa reticu-laris, with the corals Heliolites interstincta, H. Murchisoni, Stenopora fibrosa, and Chonophyllum perfoliatum, are common to the Upper Silurian of England or Sweden, and the Lower Devonian of France.

The Devonian group of Brittany and the West of France is also usually curtailed, like the Silurian, of its superior members,—those which, in the Rhenish provinces, are so fully developed between

---

* De Verneuil and Rouault, *antè*, p. 386.

† At this locality abundance of Devonian fossils were obtained by M. de Ger-ville, and distributed among his friends. Both trilobites and shells are found there; and among the latter are the common Devonian species, Spirifer heteroclytus, Athyris concentrica, and Calceola sandalina. More than 45 species of Brachio-poda were lately found here by Mr. Davidson.

the above-mentioned rocks and the base of the Carboniferous system.

In the 'Bas Boulonnais,' however, we meet with a fair representative of the Upper Devonian, rich in fossils, and on which, in 1840*, I published a short memoir. This member of the system is there immediately overlaid by the carboniferous limestone, and both of the formations extend into Belgium and the Rhenish provinces, occupying the same relative position.† This Upper Devonian is every where

---

* Bull. Soc. Geol. France, vol. xi. p. 229. At that time the full relations of the strata had not been determined, and I was deceived as to the nature of a small imperfect body, probably a coralline, which I termed a graptolite, a fossil unknown in Devonian rocks. M. Delanoue, who has recently published a small geological map on this country, has ascertained by sinkings, that the Devonian deposits range underground towards Bethune and Arras. Bull. Soc. Geol. France, 2nd series, vol. ix. p. 399.

† Mr. Austen, whose former labours threw so much light on the real nature of the limestones of Devonshire, has recently surveyed this district of the Boulonnais in detail, and has endeavoured to show that its lowest strata (limestones, dolomite &c.) represent the rocks of the Eifel, and are charged with the same species of corals, and many of the same mollusks, which are so well known in Devonshire and on the Continent. The overlying calcareous courses with shale he considers to be equivalents of the Upper Devonian as exhibited at Petherwin (see p. 256. *et seq.*); the comparison being further sustained by a description of certain flag-like, light-coloured sandstones, charged with plants, and the same species of Cucullæa — C. trapezium and C. Hardingii, Sow.—as the yellow sandstones of Marwood and Baggy Point in North Devon, which sandstones constitute one of the upper members of the Devonian Rocks. Mr. Austen's section does not descend so far in the series as to exhibit any representative of the Spirifer Sandstein or bottom fossil rock of the Rhine. But the rest of the Devonian rocks of the Rhine and Belgium have, according to his view, their equivalents in the north of France. (Quart. Journ. Geol. Soc. vol. ix. p. 231.) It must, however, be stated, that de Verneuil, and others, do not adopt this opinion, and continue to view the chief masses of the Boulonnais as Upper Devonian; while Sharpe would class them as lower carboniferous.

Mr. Austen has successfully dwelt upon the complete conformity and passage upwards of these Devonian rocks of the North of France into the lower carboniferous strata with their two bands of limestone and interstratified seams of coal, and has made some valuable suggestions respecting the probable extension westwards, even perhaps to England, of the beds of coal which have already been opened out near St. Omer beneath the cretaceous deposits.

recognized by the abundance of its smaller Spirifers, among which Spirifer disjunctus, Sow., and its numerous varieties, are the most characteristic. The Productus subaculeatus and Athyris concentrica are also frequent fossils. The limestone recurs in different bands with dolomites, sandstone, and schist, composing a group which has been supposed to correspond to the ' *Etage quartzo-schisteux supé-rieur du terrain anthraxifére,*' of the classification of Dumont. Like the Lower Devonian of Brittany, the rocks in the ' Bas Boulonnais ' have a strike from the west-north-west to east-south-east, but are soon overlapped, first by the carboniferous, and then by the cretaceous and younger deposits.

Though the Devonian system of France is of no great vertical dimensions, and not comparable in thickness to the deposits of like age in Scotland, England, or the Rhenish provinces, it is of considerable economic importance, particularly in certain schistose and arenaceous districts which it traverses, and where its limestone is much used in agriculture. At the extremity of the peninsula of Brittany these rocks appear in the Bay of Brest.

In this work it is impossible for me to convey an adequate idea of the Carboniferous system, particularly that portion of it which is developed in the extensive upper coal-fields of France. The following few observations relate, therefore, only to the lower and calcareous members of the system, which immediately succeed to the subjacent rocks; the great coal-fields of Valenciennes in the north, and those of Lyons and Auzun in the south, are therefore unnoticed. In the Bas Boulonnais there is no gap nor omission, as above stated, in this portion of the series; for the Upper Devonian is at once followed by the Carboniferous Limestone, with many of its large Producti, and other typical fossils. But in Brittany and the West of France, the lower carboniferous deposits rest at once upon the Lower Devonian; as is seen between Sablé and Juigné (p. 385.), and at the coal pits of Fercé. The anthracitic group of the department of the Sarthe (the

14, 15, 16, of the preceding section) is composed of conglomerates, sandstones, shale, coal, and limestone, all which unquestionably belong to the Carboniferous system, of which they constitute a lower member,—a rock which in Central and Southern France *is always unconformable to the coal-fields above them.*

A like physical arrangement has been alluded to as occurring near Hof, in Bavaria, where true carboniferous limestone, with large Producti, was described by Professor Sedgwick and myself, in the year 1839[*], and where that rock is inclined conformably with the Devonian, on which it rests, and is unconformable, as before stated, to the horizontal coal-fields of Bohemia. It was this unconformity which led E. de Beaumont to group the inferior carboniferous member with the lower palæozoic rock to which it is conformable, though their respective organic remains refer them to different formations. The geologist, who combines stratigraphical with zoological evidences, is therefore bound to state, that the dislocations of the older rocks on the Continent have occurred at periods different from those of the same age in Britain and America, in both of which countries the great superjacent coal-fields *have* partaken of the movements which affected the limestones on which they rest. And thus we are more than ever confirmed in the belief, that all fractures of the crust of the earth are local phenomena only.

As the strata containing large Producti (Sablé, Fercé, Juigné, La Bazouge, &c.), were shown to be truly of carboniferous age, their fossils being absolutely those of the mountain limestone of England, it followed that many other rocks in the east of France, which had been considerably metamorphosed, but in which such fossils were detected, must also be referred to the same era of formation. Thus de Verneuil and Jourdan have proved, that the schists, slates, and quartz rocks of Central and Eastern France, particularly in the department of the Haute Loire, (Roanne to Lyon) which, from their crystalline aspect, had been grouped with those of a much older epoch, are clearly to be classed with the Carboniferous system. Though

* Trans. Geol. Soc. Lond., 2nd series, vol. vi. p. 298. Pl. 23. f. 15.

in parts highly altered, and having a very antique and crystalline aspect. they have been found to contain well known fossils of the Coal period, such as Productus Cora, Chonetes papilionacea, Spirifer bisulcatus, Orthis crenistria, Goniatites diadema, &c.

It was in the northern extension of this chain of hills from Thiers to Cusset near Vichy, where the schists have a still more ancient character and are pierced by numerous porphyries and syenitic rocks, that by the help of a few fossils, I was enabled to satisfy myself, that the slaty rocks on the banks of the Sichon belonged to the Carboniferous era.*

The Permian Rocks of France may be disposed of in a phrase ; for, in truth, they are very inadequate representatives of the diversified and complex assemblage of strata which constitute the Permian of Russia, Germany, and Britain. The lower portion of a series of red rocks in the Vosges Mountains, usually known as the ' Gres des Vosges,' is considered by E. de Beaumont and myself to be of this age, and it has also been shown how other red sandstones, near Lodêve in the south of France, may from their plants be classed as Permian. But nowhere within the limits of the country is there any limestone to represent the great calcareous centre or Zechstein, and the group is wanting in those animal remains which constitute its essential distinctions.

In the hard and subcrystalline rocks of the south of France, which lie to the south of the great granitic plateau of that region, (Lodêve, Pezenas, Beziers,) no clear succession of the older rocks has been determined, and until very recently they were considered, to a great extent, unfossiliferous.

Within these few years, however, MM. Fournet and Graff discovered at Neffiez near Pezenas † a very singular small *oasis*, not an English mile in length, in which some beds with several characteristic Lower Silurian fossils, and others containing Cardiola interrupta and graptolites occur, also red limestones charged with unequivocal Upper Devonian types, as recognized by de Verneuil. These are Goniatites amblylobus and Car-

---

* Quart. Journ. Geol. Soc. Lond., vol. vii. p. 13.

† Bull. Soc. Geol. France, 2nd series, vol. viii. p. 44.

dium palmatum of Westphalia, Nassau, and the Eifel. This rock is also identical with the 'marbre griotte' of the Pyrenees, which the lamented Leopold von Buch classed as Devonian. Again, we have also in this one little spot, true carboniferous deposits; for though there are only a few beds of their lower members, their fossils are unequivocal; viz. Productus gigas, P. semi-reticulatus, Euomphalus acutus, Caninia gigantea, &c. The series is even terminated by some workable beds of coal.

Such, says de Verneuil, is the palæontological order of the strata, if we look only to the fossils, and compare them with those of all other known countries in the world. But whilst this is the universal and normal order in America, Asia, and Europe, these members of the series, so distinct elsewhere, seem to be physically intermixed at this little anomalous spot of Neffiez. In short, the schists, with Graptolites and Cardiola interrupta, appear to be interposed in the middle of the carboniferous group, or placed between the Productus limestone and the coal!

The tract of Neffiez has, however, manifestly been subjected to violent dislocations, and I therefore believe, with my associate de Verneuil, that the apparent anomaly is due to one of those inversions or reversals known in the Alps and in America, and to which I have called attention in my remarks on Germany (p. 377.).

The palæozoic formations again protrude to the surface in the Corbiéres to the south of the plains of Languedoc, and rising from beneath the younger deposits they also, as has been long known, form a central portion of the Pyrenees. In that chain, the altered limestones, or 'Marbres Griottes,'* which occur as concretionary masses in the schists, and are charged with Goniatites, are unquestionably Devonian. The schists of Gedre, between St. Sauveur and Gavarnie, contain other fossils of the same age.

Just as in the Eastern Alps, the Ural, and many other mountain chains which have undergone much metamorphism, the observer fails to detect here the equivalents of the Lower Silurian Rocks.† But

* This 'Marbre griotte' of the valley of Campan is well described by M. Dufrenoy.

† Bull. Soc. Geol. France, 2nd series, vol. i. p. 137.

in certain spots, though at very rare intervals, the Pyrenees afford traces of one member of the system, as characterized by Graptolites, Cardiola interrupta, and the Orthoceratites, O. Bohemicum and O. gregarium;—fossils which have already been spoken of as probably representing the base of the Upper Silurian division.

Palæozoic Rocks in Spain, Portugal, and Sardinia.—To the south of the great chain of the Pyrenees, where several of their members occur in a more or less crystalline state, the Lower Palæozoic rocks have much the same development as in France. Our acquaintance with the order of these rocks in Spain, is chiefly due to the researches, during the last five years, of my colleague de Verneuil. For, whilst it is true that living Spanish and French geologists and mineralogists of ability, such as Casiano de Prado, Schultz, and Paillette, had made themselves well acquainted with the physical and lithological features of many tracts in which the older rocks prevail, and that an eminent French mining engineer, Le Play, had published a mineral sketch of the Sierra Morena, it was only by an examination of the imbedded fossils of the strata in each chain, that the relative age of the rocks was at length fixed, and correct comparisons established.*

* This observation has reference only to the order and classification of the stratified rocks; for in reference to mineralogy and other departments of geological science, a vast number of works and memoirs have been written in the last century, by native and foreign authors. See de Verneuil and Collomb, Coup d'Œil sur la Constitution Géologique de quelques Provinces de l'Espagne. Appendice Bibliographique. Bull. Soc. Geol. Fr., vol. x. p. 138., where a very copious list of authors is given, comprising upwards of 160 publications in the last hundred years At the meeting of the British Association for the advancement of Science in 1850, I presented, on the part of M. de Verneuil, what I believed to be the first geological map of the whole peninsula, accompanied by a short notice. Since that time my colleague has twice revisited Spain. The high antiquity of the crystalline rocks of Gallicia, is affirmed on the authority of M. Schultz, who has made a geological map of that province, and is completing a beautiful map of the Asturias. The tracing of Silurian rocks into the Toledo mountains, is due to the zealous researches of M. Casiano de Prado, to whom we also owe the proofs

A glance at the map shows that the Peninsula is marked by dominant chains of mountains more or less parallel, which, trending from west-south-west to east-north-east, are separated from each other chiefly by enormous basins of tertiary age that lie at much lower levels. Let us first say a few words on the Guadarama, the chief central ridge of the kingdom, which passes to the north of Madrid. Casiano de Prado, the zealous Spanish geologist, who is charged with the execution of the geological map of the province of Madrid, has shown that the gneiss and other crystalline schists with subordinate limestones, are pierced by granites, and being much altered, rise to heights exceeding 7000 feet. These are flanked by schists and silicious sandstones, which this author refers to a Lower Silurian age, because they contain a fucoid, Cruziana or Bilobites, like that which occurs in rocks of this age in France and Britain.

True Silurian rocks of the lower division, and more clearly determinable as such by their fossils, occur in the Sierra Morena, and partially in the mountains of Toledo and those of Arragon. In crossing the Sierra Morena from Almaden on the south to Cordova on the north, M. de Verneuil recognized an ascending order. The inferior strata consist of schists and some intercalated dark limestones, with quartzose sandstones; the latter not unlike the British Stiper Stones. (See p. 36.) These, being the hardest and least decomposable portion of the strata, form the summits or peaks of the low ridges, and clearly exhibit the strike or range of the masses, which is from east by north, and to west by south. The cele-

that the quicksilver mines of Almaden are in Lower Silurian rocks. Many of the fossils (chiefly Devonian) from the North of Spain were first found by M. Paillette, whose description of the rocks is given in the Bullet. Geol. Soc. France, vol. ii. p. 439. A sketch of a geological map of Spain was also published by M. Esquerra del Bayo in Leonhard and Bronn's Jahrbuch, 1851. In his last excursion M. de Verneuil was accompanied by M. Collomb, who has also given a sketch of the general views they then acquired of the structure of the peninsula, in the Bibliothêque Universelle de Geneve.

brated mines of quicksilver of Almaden occur at the foot of one of the quartzose ridges of this age, as determined by M. Casiano de Prado.* The Lower Silurian rocks of the Sierra Morena range eastwards into Murcia, where they occupy the largest portion of the mountainous tracts of that province. They have recently been described by M. Pellico, who states that they are much metamorphosed, and are only recognizable through their large ill-preserved Orthoceratites and certain corals. There, the Lower Silurian limestones are the sites of some of the richest mines, particularly near Carthagena, and also in the Sierra Almagrera, so famous for its ores of lead and silver.†

The fossils found in the Sierra Morena and the Toledo Mountains occur usually in black shivery schist, with occasional graptolites of species well known in Brittany, Normandy, and Bohemia. The most prevalent fossil is perhaps the Calymene Tristani; but recently many other well-known French forms of trilobites have been detected‡, including Illænus Lusitanicus? or I. Salteri; Asaphus nobilis; Calymene Tristani; C. Arago; Dalmannia socialis, and D. Phillipsi, Barr.; Trinucleus Goldfussi, Barr.; Homalonotus (Plesiocoma) rara; Placoparia Tournemini; Orthoceras duplex; Bellerophon bilobatus; Redonia Deshayesiana; &c.

Now, whilst nearly all these species are also found, as has been stated, in the slaty schists of Angers and Vitré, in France, which the French geologists have named and mapped as Lower Silurian,

---

* The mercury of Almaden is said not to occur in veins, but to have impregnated the vertical strata of quartzose sandstone associated with carbonaceous slates. The association of mercury with such rocks is still more remarkable in the Asturias, where mines of mercury are worked in coal strata.

† It is most probable, that the still more metamorphosed rocks of the Sierra di Gador and Sierra Nevada, with their limestones and rich lead mines, are of the same age.

‡ In an examination by Casiano de Prado, Eusebio Sanchez, and E. de Verneuil.

several are identical with Bohemian species, published as of the same age by Barrande. It is also well worthy of notice, since the general type of the fossils is so much that of central Europe, that the very common British species Bellerophon bilobatus, and the remarkable cephalopod Orthoceras duplex (so characteristic of the Lower Silurian fauna of Scandinavia and the North of Europe) should also be found, as M. de Verneuil remarks, in this southern parallel. Some of these Silurian fossils were first discovered by M. Casiano de Prado, near Molina, in Arragon. They have subsequently been there traced by M. de Verneuil, and also along the eastern borders of New Castile, where they occur in black graptolite schists, which jut out from beneath the secondary deposits. These Silurian schists have further been followed by the same author in the mountains of Checa, Horca, Origuela, and Monterde.

If any Upper Silurian rocks can be said to exist in Spain, they are only, as in France, thin courses of black schist, with spheroidal, calcareous concretions in which Graptolites occur, together with the Orthoceras Bohemicum and O. styloideum of Barrande, and the British species Cardiola interrupta. Such rocks, which are feebly exhibited in parts of the Sierra Morena, are also seen on the south flank of the Pyrenees, near Ogasa and St. Juan de las Abadesas, where they are recognized* by the presence of Cardiola interrupta and Orthoceratites. It is also supposed that a band of silicious limestone pierced by *elvans,* or granitic dykes, and which extends along the south side of the chain from Gerona, by Hostalrich, to Barcelona, may be referred to the same age, Orthoceratites having been found in it near the last-mentioned city. †

---

* By M. de Verneuil and M. de Loriére. The fossils near San Juan de las Abadesas were discovered by M. Amalio Maestre; they have been quoted by Prof. Leymerie, from near St. Béat on the north side of the Pyrenees, and exist also in Sardinia.

† By Mr. Pratt, F.R.S., who has successfully explored large tracts in Spain,

Devonian Rocks in Spain.—Devonian rocks are developed in the Sierra Morena north and south of Almaden, occurring in several repetitions troughed by Lower Silurian rocks, the fossils lying generally in sandstones, or in small bands of impure limestone.   The most characteristic species are, Productus subaculeatus; Leptæna Dutertrii; Spirifer Verneuili; S. Archiaci; S. Bouchardi; Orthis striatula; Terebratula reticularis; T. Orbignyana; T. concentrica; Phacops latifrons; &c.   All but one (Terebrat. Orbignyana) of these are common fossils at Boulogne; and there are many others which are well known in French and German localities, besides some species peculiar to Spain.   In the southern parts of Cuenca, near Hinarejos, de Verneuil and Collomb also detected deposits containing Devonian fossils, including the broad-winged Spirifer of the lower beds of the Rhine[*]: and, last summer, these authors discovered the same deposits at the eastern extremity of the Guadarama range between Senianza and Atienza.

Whilst the eastern part of the Sierra Cantabrica, extending westwards from the high road from Leon to Oviedo, is composed of quartzose and schistose rocks supposed to be of Silurian age (though no fossils have been detected), the medial part of this chain presents a very rich development of Devonian deposits.   These rocks consist of red sandstone, shale, and grey limestones, which, owing to powerful dislocations, assume bold and peaked forms, which are visible

chiefly in the Asturias, the Pyrenees, and Catalonia.   For his notice of the Orthoceratites, see Quart. Jour. Geol. Soc. Lond. vol. viii. p. 270.

[*] M. de Verneuil has identified nearly 80 Devonian species, from Sabero in Leon, and from Ferrones, and Aviles, in the Asturias, which ought, he says, to be places of pilgrimage for all collectors.   Of the works on this region, see, 1st, " Reseña geognostica de principado del Asturias, vistazo geolog. sobre Cantabria," per G. Schulze, Insp. Gen. de Minas.   2nd. "Recherches sur quelqu'unes des rôches des Asturies, et les Fossiles qu'elles contiennent," par Paillette, De Vern. et D'Archiac.   Bull. S. G. Fr., 2nd series, vol. ii. p. 439.   " Sur les Env. de Sabero (Leon)," par Casiano de Prado et de Verneuil.   Bull. S. G. Fr., 2nd ser. vol. vii. p. 137.

at great distances from the plains of Castile. They are as prolific in mineral wealth as in organic remains; the iron of Mieres and Sabero being extracted from them.* Thanks to the labours of Schultz, Casiano de Prado, Paillette, de Verneuil, and d'Archiac, the respective formations of these countries are now becoming much better known.

Carboniferous Rocks of Spain. — Proceeding to the east, the Devonian rocks of the Sierra Cantabrica are succeeded in the Asturias by the richest coal-field in Spain. Its lower beds consist of massive limestones, so resembling the Devonian rocks on which they lie, that except for their respective fossils they could with difficulty be separated. The fossils are, however, decisive of the superior rock, for among them are several well-known British species of Productus, such as P. semi-reticulatus, together with Spirifer Mosquensis, and even Fusulina cylindrica, the foraminiferous shell so characteristic of the carboniferous limestone of Russia. Above these masses, beds of coal first begin to alternate with other and smaller courses of limestone; so that here, again, as in Scotland and Russia, the carbonaceous portion is subordinate to the lower members of the group. Then follow conglomerates and sandstone with fossil plants, which are said to have a thickness of ten thousand feet, and which, probably representing the millstone grit of Britain, are here copiously charged with coal. About eighty beds of coal have, in fact, been recognised in strata for the most part vertical.

The limestones that form the inferior limit of these carboniferous deposits, truly vindicate the propriety of the old English name Mountain Limestone; for they rise to the highest points of the Cantabrian chain, and constitute the mountains of Cabrales and Cobadonga, as well as the peaks of Europa. They

---

* Though the coal of Sabero is *apparently* included in Devonian rocks, M. Casiano de Prado thinks that this appearance may be due to inverted folds of the strata; so much has this tract been convulsed.

also advance to the sea near Ribadesella, and penetrate on the east into the provinces of Santander and Palencia.*

The carboniferous deposits of the Sierra Morena range along the southern part of that chain. Like similar formations in the north of Spain, their lower beds generally contain limestones, in which occur the same species of Productus and other marine fossils that are known in many other regions. The coal is chiefly associated with overlying conglomerates and sandstones: some of it, however, as in the Asturias, is in the calcareous series. The best coal-fields of this southern region are near Seville, and in the neighbourhood of Belmez, between Almaden and Cordova: the strata are highly inclined.

On the flanks of the crystalline schists, probably metamorphic, which form the Sierras east of Burgos, towards Escaray and the Moncayo, there are also masses of sandstone and shale, with impressions of coal plants and traces of coal, associated, as in the other Spanish coal-fields, with a few marine fossils. These rocks are only mentioned to show, how generally the same palæozoic succession has prevailed all over the peninsula, before the country was thrown up into those ridges which now form the lines of separation between its different provinces.

Whatever be their direction or inclination, the Silurian, Devonian, and Lower Carboniferous rocks of Spain have all been conformably, and apparently simultaneously elevated, as in France.

The existence of Permian rocks in Spain is very doubtful; no fossils of that age having yet been found. Led, however, by the analogy of rocks and stratigraphical indications, Professor Naranjo y Garza has referred to that formation the red magnesian limestone and the gypsiferous marls of Montiel and of the lakes of Ruidera, the

---

* During the summer of 1852, MM. de Verneuil, Casiano de Prado, and Loriére ascended one of the highest of these peaks of Mountain Limestone, and found, by barometrical observation, that it was 2500 metres, or about 7750 English feet, above the sea. It was covered with snow on the 1st August.

sources of the river Guadiana. In the same group this author includes the famous cave of Montesinos in La Mancha, which, in
Cervantes' immortal work, Don Quixote is made to explore.

Silurian and Carboniferous Deposits in Portugal. — Geologists are
indebted to Mr. Daniel Sharpe for nearly all that they know of the
real structure and succession of the sedimentary rocks of the kingdom
of Portugal, whether of palæozoic or secondary age.*

Of the former, the only examples pointed out in previous years,
—though probably many more will be detected, in addition to another
case to be presently noticed,—occur at Vallongo, in the immediate
environs of Oporto. There, the Lower Silurian rocks, precisely of
the same mineral type as those of France and Spain, and including
the same fossils, together with two or three British forms and some
new species, rise up in highly inclined and vertical strata; and, as at
Angers and other places, are quarried for roofing slates.

An anomalous arrangement of the lower and higher deposits, is,
however, there apparent, which is at variance with the facts observed
in other parts of the world. The coal-field of Vallongo, which has
supplied a considerable portion of the city of Oporto with coal, and
in which are certain plants not distinguishable from those of the
Carboniferous era, dips under the Lower Silurian schists with their
characteristic trilobites! Now, if this had been really the normal
position of the plant-bearing strata, we should have to believe in the
existence of a terrestrial flora before the Lower Silurian formations
were accumulated,—a flora composed of the same vegetation as that
which, in other regions, we find only when we reach the Upper
Devonian deposits, and approach to the horizon of the great coal-
fields. Believing fully in the accuracy of Mr. Sharpe's sections, I
think that, however, from his own good description of this district
and the adjacent tract, we may without difficulty surmise how this
apparent anomaly has been brought about. The Lower Silurian
rocks and the contiguous coal strata (for the coal is not found within

* See Quart. Journ. Geol. Soc. London, vol. v. p. 142.

the body of the lower slates) are both situated between two ranges of eruptive rocks; the one on which Oporto stands being of granite, and the other, to the east, of syenite.   On the flanks of the granite of Oporto, micaceous schists abound, which, if metamorphic, may be of any age.   But, even if these be of ante-Silurian date, we have simply to imagine a trough of coal, of the true carboniferous date, placed between these schists, on the one hand, and the clay-slates with Lower Silurian fossils, on the other, and then by a movement, of which we have many well authenticated examples both in Europe and America, this trough has been placed in an inverted and dislocated position.

The enormous length of time which must have elapsed between the accumulation of the Lower Silurian and the formation of the Devonian rocks, and during which interval we have no trace of land plants having appeared, forbids us, indeed, to adopt the view of an infra-Silurian coal-field, until we have exhausted every other means of explaining the anomaly.

Thus, the plications of the strata in Belgium, as delineated by M. Dumont, those in Westphalia mentioned in this volume, or those in North America, described by the Professors Rogers, explain how strata really inverted in one place, may be followed until they resume their regular order.   Wedged in as these Silurian and Carboniferous masses of Portugal are, between two flanking parallel ridges of eruptive rock, it was no doubt difficult to detect their regular order; though, even in describing a transverse section from Oporto to Aveiro, Mr. Sharpe himself states, that clay slates lying on gneissose and micaceous schists, are surmounted by carbonaceous shale and red sandstone.   This is, I conceive, the natural order; because even in the adjacent districts of the north of Spain, to which the author first pointed attention, as being likely to contain Silurian rocks, there are no traces whatever of anthracitic coal-fields beneath or associated with such ancient rocks; all the carboniferous deposits of that region being, as previously stated, in their usual and normal position.

Again, after these pages were written, Mr. Sharpe himself communicated to the Geological Society of London* the discovery of fossils and other important observations made by M. Carlos Ribeiro, which show the prolongation of the same axis of Silurian rocks from north-north-west to south-south-east, far beyond the Douro. The rocks in question form, in fact, the crest of the Serra de Busaco, on which Wellington and Massena first tried their strength, and thence extend to the south-south-east beyond the river Mondego † into the Serra de Mucella. During the greater part of their course, these Silurian rocks are surrounded or flanked by older unfossiliferous masses, consisting of mica and chlorite schists with clay slate, &c. These crystalline rocks are at once followed by dark brown indurated shales, somewhat slaty, which abound in Lower Silurian fossils, among which Mr. Sharpe has recognized numerous species of Leptæna and Orthis, with trilobites, many of them previously undescribed. This author has, however, identified the most characteristic of the trilobites with species well known elsewhere; Trinucleus Pongerardi, Rou., Calymene Tristani, and C. Arago of Brittany, and Phacops socialis of Bohemia; so that this lower fossil rock of Busaco is, obviously, in Portugal, as it is in France and Spain, the oldest zone with fossil remains, and the same which Mr. Sharpe described at Vallongo near Oporto. It is surmounted, and chiefly along the middle of the ridge of Busaco, by a band of hard ochreous shale, frequently altered by eruptive masses of greenstone trap, and generally breaking up into prisms, the bedding being scarcely distinguishable. It is probably equivalent to the sandstone

* Quart. Geol. Journ. vol. ix. p. 135.

† At the mouth of this river, and at the age of sixteen, the author of this work, then an Ensign in the 36th regiment, disembarked (August 1st, 1808) from a boat alongside of that which conveyed Sir Arthur Wellesley, the future illustrious Wellington.

of May in Normandy. This rock is full of small corals, such as Favosites fibrosa, besides species of Reteporæ, also many species of the simple plaited Orthides, and some species of trilobites common to the subjacent deposit; the whole of the evidence proving clearly, that it is simply a superior member of the Lower Silurian group. Again, the uppermost of these Silurian rocks in Portugal is precisely what we have been considering in Spain and France; viz. a blueish shale or argillaceous schist, containing the well-known British species Cardiola interrupta and Graptolites Ludensis, with numerous crushed specimens of orthoceratites and other mollusks; fossils which, according to Mr. Sharpe, are referable to the Wenlock shale or inferior portion of the Upper Silurians.

Now, all these Silurian rocks are overlaid unconformably by a deposit of true carboniferous age, the shales and sandstones of which are full of ferns and other plants, of species common in the coal deposits of France and Germany. Thus, when traced out, the true order of the north of Portugal is the same as that of other regions.

Sardinia.—Extending our inquiry eastwards from Northern Portugal and Central Spain, to like parallels of latitude on the other side of the Mediterranean, we again meet in Sardinia with a similar succession of palæozoic deposits. Geologists owe this determination to the researches of General Albert della Marmora, so long favourably known to geographers for his beautiful topographical map of that island. Judging from certain Orthidæ and Orthoceratites, which this author sent to me in the year 1848, I have for some years had no doubt that Silurian rocks existed in Sardinia, as laid down by Collegno, in his geological map of Italy. I have since been informed by M. Barrande, who has personally inspected the rocks and fossils in the southern part of the island, near Flumini Maggiore, that in his opinion both Lower and Upper Silurian are there present.

In anticipation of a work by General della Marmora which is about to appear, accompanied by an elaborate map, it may be stated, that some of the fossils collected by him and his assistant M. Vecchi, which were submitted to Professor Meneghini of Pisa, have been identified with well-known Lower Silurian forms; such as Orthis Lusitanica, Sharpe; O. testudinaria and O. vespertilio, Sil. Syst. &c. Other fossils are, on the contrary, Upper Silurian types, such as Orthoceratites ibex, O. gregarium, Sil. Syst. (?), and several others, besides Cardiola and Avicula.*

General della Marmora has informed me, that all these rocks and the metamorphic masses with which they are associated, are distinctly surmounted by strata containing anthracitic coal, charged with many of the same species of fossil plants which prevail in the old coal deposits of other regions. Thus, among the plants from Seni and Sculo, Professor Meneghini has recognized Pecopteris arborescens; P. dentata; P. unita; P. polymorpha; P. hemiteloides, and several others; with species of the genera Odontopteris, Neuropteris, Sphenophyllum, Annularia, Asterophyllia, Sigillaria, Syringodendron, and an abundance of the well-known Calamites Suckowii.

In fact, the strata in which these coal plants occur, and which terminate downwards in conglomerates†, rest in completely discordant positions on the older rocks, and in this highly altered tract are covered by dolomites of the Jurassic age. The order of the inferior strata in Sardinia is, therefore, the same as elsewhere, and is quite subversive of the idea that any true old coal plants existed during the Silurian or lower palæozoic era.

* It is right to state that whilst he has seen unquestionable upper Silurian fossils from Sardinia, which on the part of General della Marmora I submitted to him, M. de Verneuil is of opinion, that some of the fossils (of which, however, he only saw drawings) are of Devonian age. It is, indeed, probable that the succession will eventually be found to be similar to that of Spain.

† See my Memoir on the Alps, Apennines, and Carpathians. Quart. Journ. Geol. Soc. Lond., vol. v. p. 157., and particularly its translation into Italian by Professors P. Savi and Meneghini, entitled " Struttura Geologia delle Alpi, degli Apennini e dei Carpazi," followed by their " Considerazioni sulla geologia della Toscana" (Firenze, 1850), in which these distinguished Italian geologists announce the important discovery of old Carboniferous formations in the hills near Volterra.

Schists and conglomerates, though not so carbonaceous, but containing similar plants with those of Sardinia, have indeed been described by Professors Meneghini and Savi, in the older rocks of Tuscany.* This fact, which was unknown when I published upon the Alps and Apennines, is most important, and leads us to infer, that the traces of coal which are associated with certain land plants in parts of the Western Alps, are also of the old Carboniferous age. For, the conglomerates and schists with which such plants are associated in the Alps are precisely the same rocks as the 'verrucano' and schists of Sardinia and Northern Italy, which lie beneath all strata of Triassic, Liassic and Jurassic age.

It has, indeed, been already stated, that rocks containing fossils which are truly of Silurian, Devonian, and Carboniferous species, occur in the Austrian or Eastern Alps. With this fact before us, and looking to the prodigious amount of metamorphism, dislocation, and convolution to which the component parts of that chain have been subjected, particularly in their extension to the west, we can have little difficulty in imagining how the Silurian and Devonian strata have there passed into a crystalline state; while the sole remnants of the carboniferous rocks, identifiable through their organic remains, are the plant-bearing conglomerates and schists of the Valorsine and of the tracts around Mont Blanc.

* As long as the question seemed to me doubtful, it was my duty to state, after my own survey, that the apparent intercalation of the coal plants in the Tarentaise, among beds containing belemnites, was so striking, that M. Elie de Beaumont was fully justified in his original conclusion. When that eminent geologist reasoned on this apparent intermixture of old coal plants with Jurassic belemnites, the phenomena of the rapid interplication of strata of very dissimilar ages, and the entire reversal of order in certain tracts, had not been sufficiently delineated. I cannot now admit, that there is any evidence to disprove the inference that the old Carboniferous Flora, which began to prevail in the Devonian period, expired at the close of the Permian deposits. The fauna and flora of the palæozoic rocks are, therefore, entirely in harmony.

CHAPTER XVI.

SUCCESSION OF PRIMEVAL ROCKS IN AMERICA.

ORDER OF THE PALÆOZOIC STRATA IN SOUTH AMERICA, THE UNITED STATES, AND BRITISH NORTH AMERICA.

ALL the evidences which the researches of geologists have brought to light, indicate a wide diffusion of similar groups of animals over the globe during the primeval periods. A striking proof of this lies in the fact, that the palæozoic fossils which we have followed over the various countries of Europe, are found to have their exact equivalents in the continent of America.

Although the science of the United States has afforded the chief means by which this parallelism has been worked out, I cannot commence even this slight sketch of the older transatlantic rocks, without first alluding to that masterpiece of natural history, the work of Humboldt, wherein he gave to mankind a comprehensive view of the structure of the chain of the Andes,—the main axis of the western continent.

The· oldest of those slaty and quartzose formations, so eloquently described by the illustrious traveller, have subsequently been referred, through their organic remains, to the Silurian rocks. Following up the inquiries of his precursor, M. Alcide d'Orbigny has shown, in maps and sections, as well as by elaborate descriptions*,

---

* See Voyage dans l'Amérique Méridionale, tome 3me, Partie Géologique, Paris, 1842. In justice to my friend, Mr. Pentland, so well known to geographers by his measurements of the high peaks of the Peruvian Andes, around the lofty lake of Titicaca, let me say that he was the first person who made me acquainted with the occurrence of Silurian trilobites in the slaty rocks of this chain.

that these slaty rocks contain the fossil sea-weed Cruziana (or Bilo-bites), with Graptolites, Lingulæ, Orthidæ, and trilobites of the genera Asaphus and Phacops (Calymene). He has further pointed out, that these Silurian masses are succeeded by sandstones and silicious strata of Devonian age, and these by limestones and other rocks, charged with fossils of the Carboniferous era.

The Silurian slates and schists form enormous bands; and examples of them may be well seen on the declivities of the plateau of Bolivia, as well as on the flanks of the Cordillera, extending from Sorata to Illimanni. They are, in most parts of their range, in a metamorphic state; and in this form extend from Chili, on the south *, to the Rocky Mountains and the Sierra Nevada on the north. Even in Texas they have been recognized by F. Röemer.† Including the quartz rocks which are associated with them, they constitute the chief matrix of the gold and other metals, so extensively worked along this great chain. (See next Chapter.) It has been suggested, that the huge bands of stratified quartz rocks described by Humboldt may be the equivalents of the sandstones termed Devonian by D'Orbigny‡, and in which that author detected various characteristic fossil shells. But, whilst in the absence of sufficient proof, doubts may be entertained as to whether these sandstones and quartz rocks of the Andes may be of Upper Silurian rather than of Devonian age, there can be no hesitation in referring the next deposits in ascending order to the Carboniferous or true Upper Palæozoic group. At numberless places, limestones have been observed charged with well-known carboniferous types. Several of these, such as Productus Cora; Spirifer striatus; Athyris Roissyi, &c., are specifically identical with forms that characterize the strata of this era in Europe and other parts of the globe. And as these rocks are associated with, or followed by, accumulations of coal, the general relations of the series are clearly similar to those of other countries.

* In the admirable work on "South America," by Mr. Charles Darwin (1846), the only palæozoic fossils alluded to are those of the Falkland Islands (Silurian? and Devonian), though it is probable that some of the clay-slates, &c. in the chain of the Chilian Andes, to which he adverts, are also of similar age.

† Röemer. Texas, und die physischen, &c. 1849.

‡ See Cordier. "Rapport à l'Académie Royale des Sciences sur les résultats scientifiques du Voyage de M. Alcide D'Orbigny dans l'Amérique du sud, pendant les 8 années depuis 1826 jusquà 1833." 1842.

We now know, therefore, (and the recent explorations in California, Oregon, &c., have confirmed the view,) that sedimentary deposits of Silurian, Devonian, and Carboniferous age constitute the loftiest ranges and metalliferous plateaux of the American continent. These were, in ancient times, perforated by granites, porphyries, trachytes, and other eruptive matters; and in the modern era are in some localities the seat of active volcanic forces.

To bring, however, the older formations of America into an accurate parallel with those of Europe, we must quit the chain of the Andes, and the high grounds of Mexico, and turn to the eastern side of North America. In that region the older strata have been comparatively exempted from igneous disturbances, and as the arrangement of the strata is symmetrical, men of science have been able to demonstrate the order of succession to be the same as that of which numerous proofs have been recited in the preceding pages.

Many of the crystalline rocks of North America, composed of granites, porphyries, and other igneous masses, or of stratified gneiss, mica, and chlorite schists, and marbles, are unquestionably more ancient than the oldest rock to which the term Lower Silurian can be applied. For such rocks are seen in Upper Canada (just as in Scandinavia) to be the base on which the oldest of the fossiliferous formations rest—a subject to which we shall return in the sequel.

A glance at the geological map of North America by Lyell, as deduced from the various works of the United States' geologists, or at the map of Marcou which has just been issued, will show to what extent these ancient crystalline rocks are spread out along the eastern side of the continent. They there range from S. W. to N. E. along the flanks of the Apalachian and Alleghany chain; i. e. from South Carolina to Massachusetts, New Hampshire, and Maine : they also occupy large spaces, chiefly along the eastern sides, of the British colonies of Nova Scotia, Newfoundland*, and New Brunswick.

* See Jukes's General Report of the Geological Survey of Newfoundland, in 1839 and 1840, with map and sections.

Similar granitic rocks, and on an equally grand scale, trend from Lower Canada on the east, to Lakes Superior and Winnipeg on the west, forming a broad low watershed recently termed the Lawrentine Mountains.* It is on the western flank of the Apalachian chain, and to the south, north, and west of the Lawrentine Mountains, that the palæozoic formations are exhibited. They consist of Lower and Upper Silurian, Devonian, and Carboniferous rocks, and are repeated in broad undulations, forming basins of a grandeur in extent unknown, as yet, in any other part of the world excepting Russia. In this way they occupy large portions of the southern and western states of Alabama, Tennessee, Arkansas, Missouri, Illinois, Indiana, &c., as well as the newly settled territories of Iowa, Minnesota, and Wisconsin. Such deposits range, too, along the southern side of the great chain of lakes, and are spread over large tracts of New York, Pennsylvania, Maryland, and Virginia, where their stratigraphical features have been ably and elaborately worked out, and their organic remains described, by American geologists.

These native authorities have honoured my labours by assimilating their lower fossiliferous rocks to my Silurian System; whilst all my contemporaries† who have gone from Europe to explore the United

---

* See History of Canada, by Mr. Fox Garneau.

† See Bigsby, Trans. Geol. Soc. Lond. 2 ser. vol. i., and Quart. Journal Geol. Soc., vol. viii. p. 400.; Featherstonhaugh, Reports on the Countries between the Missouri and Red Rivers, and of Wisconsin, &c. Washington, 1835-6 [My old friend Mr. F. first made known to me the existence of the Silurian series in the United States]; Lyell, Travels in North America, with general map, 1841-42, London, 1843; Lyell's Geological Manual, 3d ed. p. 351.; Castelnau, Système Silurien de l'Amerique Septentrionale, Paris, 1843; De Verneuil, Bulletin de la Société Géologique de France, 2d series, vol. iv. Paris, 1847; Richardson, Narrative of an Arctic Searching Expedition, 2 vols. London, 1851; Logan, Geological Survey of Canada; Reports to Legislative Assembly, and maps; Logan with Salter on the Rocks of Lower Canada, Brit. Assoc. Rep. 1852, p. 59.; Quart. Journ. Geol. Soc. London, 2d ser. vol. ix.; Ferdinand Roemer, Texas und die physischen verhaltnisse des landes, Bonn, 1849, with geological map; Desor, Bull. Geol. Soc. Fr. 2d ser. vol. ix. p. 342.; and lastly, M. Jules Marcou, "Geological Map and Description of the United States and British North America," Boston, 1853. This last-mentioned author, a Swiss by birth, has condensed the results of the labours of the numerous American geologists, and has dedicated this work to his friend Agassiz, who has thrown so much new light on the natural history of his adopted country. After a joint personal survey, they have followed a

States, or the adjacent territories, have recognized Silurian rocks,
both Lower and Upper, followed by deposits of Devonian and Carboniferous age.   In the Canadas, where Bigsby and others were
his precursors, Logan, the director of the geological survey, has
mapped enormous tracts both of Lower and Upper Silurian rocks;
whilst Richardson, triumphing over all the obstacles of the incle-
ment north, has followed Silurian limestones along the western flank
of the granitic chain from Lake Winnipeg to the mouth of the
Mackenzie river.   He has also shown to how great an extent they
wrap round the huge granitic nucleus of North America, and form
the edges of the continent towards the Arctic Sea.   In many of the
polar islands, too, Silurian fossils have been detected by the skilful
navigators who have sailed in search of Franklin.

I shall cull, in the sequel, a few data from some of these
British authorities; but let me first allude to those tracts which,
through the labours of American geologists, have been rendered
classical in our science, and where the types of comparison have been
the most sedulously and accurately described.*   The state of New

classification which is in harmony with the published opinions of the above-mentioned writers
and of the native geologists.   The small book and map of Marcou embraces too the facts
derived from the extensive new surveys of Dale Owen, Logan, and others.   The work also
contains a list of all the writers on American geology, together with general transverse sections,
and plates of the fossils of each group, from the Lower Silurian or oldest, to the Tertiary or
youngest division.   It will prove, I doubt not, of great use to all travellers.

* My contemporaries in America, whose labours I so highly estimate, will readily understand that in a work which is limited to the general history of the palæozoic rocks, and specially to those of Europe, I have no space to render justice to the numerous able writings which
treat of those tracts of the United.States or the British Colonies, where such palæozoic rocks are
not developed.   Thus, the works of Hitchcock, whose description of the structure of his native
state, Massachusetts, and the accompanying map, are models of geological monography, — of
Dana, whose insight into the natural history of zoophytes, and whose philosophic reflections on
the outlines of the earth, have secured for him a wide reputation, — and the various contributions to the excellent volumes of my friend Professor Silliman, — must now be passed over;
although, if general geology were my object, they would be eagerly appealed to.   Nor will the
reader find in the text, any allusion to the labours of some authors who have been highly useful
in building up the now well-established series of the older formations; such as Maclure, Eaton,
Troost, Emmons, Perceval, Vanuxem, Conrad, Jackson, Foster, Dawson, Thompson, Whitney,

York, for example, presents a noble series of palæozoic deposits, laid open on the banks of the several rivers which there flow from west to east. Examining these river banks, the geologists of that state have been enabled to describe a detailed order, which is remarkable for its symmetry and unbroken condition from the base of the Lower Silurian to the coal measures included; the whole being arranged in slightly inclined and conformable strata.

It is beyond the scope of this work to endeavour to compare with European deposits all the local subdivisions (about eighteen in number, and of very unequal dimensions) of which the Silurian system, as described by Mr. James Hall, is composed in the state of New York. Independent of these numerous subdivisions, the lower portion of the series is admitted by the American authorities to be divisible, as in Europe, into two groups, each of which is characterized by peculiar fossils. Thus, from the "Potsdam Sandstone," or base of the whole fossiliferous series, up to the slates and arenaceous schists overlying the Trenton limestone, the group so composed represents the Lower Silurian. In this view, Lyell, de Verneuil, and Sharpe agree with the United States' geologists. In the lowest of these deposits, at Potsdam, a small Lingula (L. antiqua, Hall) was, for a long time, the only fossil known, except fucoids, and abundant traces of marine worms (Scolithus linearis). Subsequently, foot-prints of a large nondescript animal were discovered in rocks of a similar age in Upper Canada, on the north bank of Lake Ontario. In the first instance, those markings were supposed to have been made by the feet of a chelonian reptile. But a further examination of numerous casts taken from the footmarks, and brought to England by the zeal of their discoverer Logan, induced Professor Owen to refer them to crustaceans, — the class of

and others. The few works above referred to are necessarily those in which the authors have dilated on the subjects which are specially treated of in this volume, i. e. comparative views of palæozoic geology.

animals which is most commonly met with in rocks of this age. This view being now adopted, we have evidence of a gigantic protozoic crustacean, probably allied to the Limulus or king crab.

The great mineral distinction between the rocks which immediately lie upon this true Silurian base, as compared with those of like age in Britain, is the much greater prevalence of limestone. A calcareous sandstone, called the "calciferous sand-rock," is succeeded by the Chazy, Bird's eye, and Black River limestones, and these are at once followed by the great Trenton limestone with its associated schists (Utica slate and the Hudson River group). In the lowest of these calcareous masses — the Chazy limestone, the peculiar genus Maclurea (see the figure, p. 193.) is found, together with some corals, Bryozoa, and a few Trilobites (Illænus, Asaphus, &c.); and the limestones which succeed, and are more or less connected with it by their organic remains, contain many enormous Orthoceratites, the singular cephalopod genus Gonioceras of Hall, several univalve shells (Murchisonia, Scalites, &c.), besides some Orthidæ, and other brachiopods.

The deposit above these, or the Trenton limestone, has been fairly paralleled with the Llandeilo limestone; and, as in that British formation, the number of trilobites, gasteropods, and brachiopods increases vastly when compared with the inferior deposits. Besides the large Asaphus (Isotelus) gigas, the characteristic fossil of the stratum, Trinucleus concentricus occurs, with the genera Cheirurus, Lichas, Phacops, and Calymene. Among the shells, Orthis striatula (foss. 19. f. 3. p. 185.), Orthis biforatus or lynx (ib. f. 4.), O. porcata (ib. f. 5.) (occidentalis, Hall), and Bellerophon bilobatus (Pl. 10 f. 8.), are all Lower Silurian fossils, and characteristic of the same rocks in Britain, Scandinavia, and Russia. The Lower Silurian character of this limestone is maintained, too, by the presence of numerous species of Lituites, Orthoceratites, and large plaited Orthides, which are *like* those of Europe, without being identical.

The Trenton limestone is overlaid, throughout a considerable tract in North America, by the Utica slates, which are full of their characteristic trilobite Triarthrus Beckii; and these are covered by a greywacke schist (the Hudson River group) containing many graptolites and most of the fossils of the Trenton rock, with some new species.

M. de Verneuil, who has closely examined the collections and visited the localities of the American fossils, has compared them * with those of Europe. He considers the above-mentioned formations, which constitute the major part of the "Champlain division" of the United States' geologists, to be equivalents of the Lower Silurian group†; and that the graptolite schists of the Hudson River, above which the fossil types of the older era disappear, form the limit between Lower and Upper Silurian. He therefore connects the conglomerates and sandstones of Oneida and Medina with the Upper Silurian, of which they form the base. Conrad would include a part of this series — the Medina sandstone — in the lower division, to which it has a great resemblance in lithological character; while Hall and most other American authors are very distinctly of opinion that the "Clinton group" forms the base of the upper division in the United States. It is characterized, nevertheless, by the typical Pentamerus oblongus, which is never found in any of the succeeding strata. His figures, lately published, show that this persistent fossil attained dimensions quite as grand in the old seas of America as in our own country. (See Preface, and pp. 98, 99.)

The chief or central mass of Upper Silurian limestone in North

---

* Bull. Soc. Geol. Fr. 2d ser. vol. iv. p. 646.; a most instructive memoir.

† Out of 45 fossil mollusca, chiefly brachiopods, brought home by Sir C. Lyell from the Lower Silurian of North America, 14, or about 30 per cent., were identified by Mr. D. Sharpe as species common to Europe and America. Among these are the common British types, Leptæna sericea, Orthis biforata, O. parva (elegantula), Bellerophon bilobatus, &c. See Sharpe on the Palæozoic Rocks of North America, Quart. Journ. Geol. Soc. London, vol. iv. p. 151.

America is that of Niagara, which, unquestionably, is of the same age as that of Wenlock and Dudley in England, or of Gothland in the Baltic. This rock appears to contain a greater number, than the Lower Silurian strata, of fossils identical with those of Europe. Among these are such forms as Calymene Blumenbachii, Homalonotus delphinocephalus, Bumastus Barriensis, Rhynchonella cuneata, R. Wilsoni, Pentamerus galeatus, Orthis elegantula, O. hybrida, Orthoceras annulatum, Hypanthocrinus decorus Bellerophon dilatatus. (See chaps. 8 and 9.) Some of the same large corals, also, prevail, as in Europe, — those which may have been capable of forming reefs, such as Favosites Gothlandica, F. alveolaris, and the chain-coral. There are also many peculiar species.

The lower formations of the next overlying or Helderberg division, up to the higher Pentamerus limestone inclusive*, constitute (according to M. de Verneuil) the probable equivalents of the Ludlow rocks. But the line of demarcation between these deposits and those beneath them is also drawn with difficulty; as must, indeed, be the case in countries where the succession is symmetrical and undisturbed. Again, the much greater abundance of calcareous matter in this division than exists in our own Ludlow rocks, and the absence of that muddy and sandy matrix which characterizes those strata in Britain, have necessarily given to the whole group more the character of the Wenlock formation— a feature which, as we have already seen, is dominant in Gothland, Russia, Bohemia, &c.

Devonian Rocks. — The base of the Devonian rocks of the United States has recently been placed lower than it was in the earlier comparisons with European deposits. Seeing that a mass of sandstones and conglomerates had there been called " Old Red Sandstone," it was natural, in the first instance, for geologists to suppose, that the

---

* These consist in ascending order of the Onondaga Salt group, the Water-Lime group, Pentamerus limestone, Delthyris shaly limestone, Encrinal limestone, and Upper Pentamerus limestone.

Devonian series in North America did not descend far beneath this rock. But judging from the included fossils, such as goniatites, identical with those of the duchy of Nassau in Germany, with Murchisonia bilineata, Productus subaculeatus, Athyris concentrica, and other fossils, M. de Verneuil has included in this group the Chemung rocks, the Hamilton and Marcellus shales, and the Cliff limestone of Ohio and Indiana (Corniferous and Onondaga limestones). The last-mentioned contain fossil fishes analogous to those of Scotland, besides some characteristic Devonian shells. Again, the occurrence of an ichthyolite of the genus Asterolepis in the Schoharrie grit of New York, necessarily placed that rock also in the Devonian series; and, finally, Mr. J. Hall agreed with M. de Verneuil by even including in the same group, the Oriskany sandstone with its large Spirifers, analogous to those of the Rhine — fossils almost unknown in true Silurian rocks. The Devonian base in America is, therefore, similar to that of the Rhenish provinces (p. 367.).

As fossil fishes have everywhere proved the most exact chronometers of the age of rocks, the occurrence of Asterolepis, a genus common in the lower beds of this epoch in Scotland, and the intermixture of other ichthyolites, identical with those of Scotland, with shells of the Eifel and of Devonshire, are to me the most convincing proofs that these diversified deposits in North America (like those of Russia, the Rhenish provinces, and Devonshire) are simply equivalents, in time, of the very grand British deposits called Old Red Sandstone. For it must be recollected, that under this term, and particularly in Scotland, are included conglomerates, sandstones, grey schists, limestones, and flagstones, as well as sandstones of both red and whitish-yellow colours; in all, many thousand feet thick. (See p. 246., et seq.) A further reason for placing the Oriskany sandstone at the base of the Devonian rocks of New York consists in the clear indications of its having been deposited in excavations of the inferior stratum, as explained by Mr. J. Hall. This phenomenon marks a previous denudation, and the commencement of a new series. M. de Verneuil further reminds us, that the Pterinæa fasciculata, Spirifer macropterus, and Pleurodictyum problematicum of America, are fossils typical of the bottom Devonian beds of the Rhine.

In one portion of the Canadas, Mr. Logan has estimated the Devonian rocks to have a thickness of 7000 feet; and in the state of New York they occupy a space more considerable than the Silurian. But this Devonian series of micaceous sandstones and schists thins out, as it passes westwards into the states of Ohio, Indiana, and Kentucky, and disappears entirely on the Mississippi, where the carboniferous strata repose at once on those of Silurian age.

Carboniferous Rocks. — The Carboniferous rocks are, as is now well known, developed in the United States, in basins of vaster dimensions than in any part of Europe. Through their included fossils, both of the animal and vegetable kingdoms, the American strata are absolutely identified with the deposits of like age in Europe. Great masses of sandstone and schists, once considered to be Devonian, and so coloured in early geological maps of the United States, have been shown by M. de Verneuil to be, from their imbedded fossils, of Carboniferous age. These masses, which are of considerable thickness, form the true base of the carboniferous rocks of North America, and are probably the equivalents of much of the yellow sandstone of Ireland, as well as of certain rocks in Westphalia, which were formerly classed with the ' Grauwacke,' but are now known to be the inferior carboniferous beds of the Rhenish provinces. (See p. 376.)

Next follows the limestone (Mountain limestone of the early English geologists), including many varieties of Productus; viz. P. semireticulatus, P. Cora, P. punctatus, &c., with species of Bellerophon, Goniatites, Spirifer, and Terebratula, known not only in Britain, and in many parts of the continents of Europe and Asia, but also in latitudes extending from the polar regions far to the south of the equator.

The same may be said with regard to the Foraminifera, a group of which few traces occur in the lower palæozoic strata; for the very same species (Fusulina cylindrica), which so abounds in Southern Russia, is also found in the carboniferous rocks of N. America.

The lamelliferous corals (Zaphrentis, Lithostrotion, Lonsdaleia, &c.) are not absent; but they are chiefly of species distinct from those of Europe; and some of the forms of this group most characteristic of the old continent (Syringopora, Amplexus, &c.) appear to be absent from those of the new.

It is unnecessary here to dilate upon the vast overlying productive coal fields, which occupying distinct basins of stupendous dimensions, have been described by numerous native authors, and have been so well depicted to us by Lyell and Dawson. But I must advert to the great similarity — nay, identity — between the plants of these vast coal deposits and those of Europe. It is truly a feature highly worthy of the special notice of geologists, that at this early period the same species of gigantic plants were spread over an enormous area of the earth's surface, which, from the nature of its vegetation must, therefore, have been under the same conditions of atmosphere and climate, if not of physical outline.

This short notice of the development of the earlier fossiliferous rocks in the United States, cannot be concluded without a special allusion to some recently made important additions to our knowledge. The researches of Mr. James Hall, as recorded in his work on the palæontology of the state of New York*, have shown that, with increased observation, the difficulty is great (except for short distances and in limited tracts), in drawing an arbitrary line between the lower and upper divisions of the Silurian System. The thinning out or thickening of the subdivisions termed Medina sandstone, and the Clinton and Niagara groups, are accompanied by just the same kind of interlacing and overlapping of the fossil types,

* Part VI. vol. ii. 1852, p. 3.

which has been observed, to a greater extent, in the British Isles. Several species, says Mr. Hall, known before only in the lower division are now found at the base of the higher group. But, before we can speak of the number of species which range from the Lower into the Upper Silurian, all the older fossils of North America must have been examined and compared, as those of our own country have been. Even, however, in the one state of New York we learn, that Bellerophon bilobatus, Orthis lynx, Strophomena (Leptæna) alternata, and some other forms which are common to Europe and America, and are abundant shells in the Lower Silurian, are found in the higher part of the Medina sandstone, which in that country is intimately united with the upper division.

The same elaborate work of Mr. Hall must be consulted for the figures of many species of fucoids common to the Medina sandstone last noticed, and the associated Clinton group. Numerous tracks or trails of gasteropods and crustacea also occur, indicating that the sediment so marked was accumulated in shallow water, or under ebbing and flowing tides, and on sloping shores. To the consideration of these data we shall return in the last chapter, in explaining the conditions under which some of the oldest marine deposits were formed.

In the mean time, I may be permitted to doubt one only of the many valuable inferences of Mr. J. Hall, — viz. that any fossil fishes have yet been discovered in the lower members of the Upper Silurian rocks, or the Clinton and Niagara limestones of the United States. The supposed fragments of bones or fish defences * in the Clinton deposit are manifestly much too imperfect and too void of structure

* Pal., New York, Pl. 31. figs. 6 and 7. In speaking of the presence of fish remains in the Lower Silurian of Britain, Mr. J. Hall has, it would appear, been misled by what is now known to be an erroneous statement. (See p. 320.) It has been ascertained, that the supposed fish of the Llandeilo or Bala rocks is a zoophyte. No trace of a vertebrated animal, I repeat, has yet been found in any deposit beneath the upper Ludlow rocks. (See p. 205, ante.)

and form to be so considered. And, in reference to the more perfect specimens from the Niagara group (Onchus Dewii *), the author must forgive me if I suggest, that it presents good evidence of having belonged to a crustacean, and is very like those so-called fish-defences of the Ludlow rock, which Professor M'Coy has removed to the crustacean genus Leptocheles. (See p. 236.)†

Particular notice must here also be taken of the survey of the extensive territories of Wisconsin, Iowa, and Minnesota, which, ably conducted as it has been by Dr. Dale Owen, has extended true palæozoic classification on a broad scale. The area examined by this author and his associates Drs. Norwood and Shurman, is much larger than Great Britain; and by far the greater portion of its subsoil is referred by them to the Silurian, Devonian, and Carboniferous rocks.‡ This magnificent region is traversed by the Mississippi in its descent from the boundaries of British America and Lake Superior, and is included between that mighty river and its huge affluent the Missouri. Surrounding and supporting an enormous carboniferous tract, the older primeval rocks rise out successively in almost horizontal positions. The results of all others which have most interested me, as proceeding from this grand survey, are, first, the

---

* Pal., New York, Pl. 71.

† Having requested Mr. Salter to give me his opinion on this subject, I subjoin an extract from his letter to me: —

'The specimens figured by Mr. James Hall, and referred by him to the defences of fishes, seem at least equivocal. The fragment figured as such from the 'Clinton group,' as well as that doubtfully called 'the rib of a vertebrate animal?' are in all probability tubes of Serpulina — the latter certainly is so. The Onchus Dewii, Hall, from the 'Niagara' group, is undoubtedly one of the hollow crustacean spines, such as were figured in the Silurian System as Onchus Murchisoni, and which Professor M'Coy endeavoured to show was a part of the pincers of some Limuloid crustacean. The more perfect evidence, however, which M. de Barrande possesses, and which he has lately put forth, convinces me that they are the tail-spines of large Phyllopods, somewhat like Dithyrocaris. We are still, therefore, without evidence of any fish below the horizon of the Ludlow rocks.'

‡ This work, following so many other valuable publications, is a proof that the senate, as well as the state governments, have justly appreciated the value of the application of geological science as a prelude to the settlement of a new country.

discovery of two bands, containing trilobites, in the lowest Silurian
sandstone, and next that the organic forms of these beds are considered
to be referable to the primordial zone (Barrande), such as it exists in
Bohemia and Scandinavia, or in the 'Lingula flags' of Britain.
(See p. 343.)   For, we thus obtain proofs, that the Potsdam sandstone,
in its extended form, and as exhibited in these western territories of
the United States, does really contain the first recognizable group of
fossil animals. *

The Lower Silurian, so termed and laid down on the map by these
authors, consists in its lowest parts of pebble-beds, grits, and sandstone of red and greenish colours, which begin to alternate upwards
with magnesian limestone.   Then, in ascending order, the magnesian
limestones with green grains (resembling some of the Russian Lower
Silurian) predominate, and are in parts oolitic, in other parts quartzose and cherty.   Again, sandstones, usually white, recur, and are
surmounted by shelly beds, which, in Iowa and Wisconsin, represent
the Trenton limestone of New York, or the Llandeilo formation of
Britain.

The Upper Silurian rocks are poorly represented, in these un-
trodden tracts, by what Dr. Dale Owen terms the Upper Magnesian Limestone, consisting of beds with corals and Pentameri; all
traces of the equivalents of the Ludlow rocks being absent.   The
Devonian period is, however, clearly defined, particularly by its
broad-winged Spirifers; whilst the carboniferous limestone is so

* The Trilobites of this, the lowest fossiliferous stratum in America (the Potsdam sandstone), are termed by Dr. Owen,—Dikelocephalus, Lonchocephalus, &c
Judging from the drawings, these genera seem to be referable to the same group
as Paradoxides.   They are associated with Fucoids and Lingulæ.   M. Barrande
informs me that, after studying the work of Dr. Dale Owen, he is convinced, that
the author has established in the New World the proof of the existence of the
primordial zone of Bohemia and Sweden, and the same order of succession upwards
from it into the mass of the Silurian strata, as in Europe.   In this view M. de
Verneuil and Mr. Salter also coincide.

expanded, as to be distinctly referred to the Yoredale series of Phillips, in Yorkshire. The western explorations of Captain Stansbury to the Mormon territory and the Salt Lake of Utah, to which an early allusion was made (p. 18.), have further shown us how strata, charged with Devonian and Carboniferous fossils, extend to the edge of the Rocky Mountains, those great northern limbs of the Andes.

Silurian and other Palæozoic Rocks of British North America. — Old sedimentary rocks, similar to those of the United States, range into the British provinces of North America, where, after a long and able survey, Mr. Logan has shown that they follow precisely the same symmetrical and unbroken order, along the northern shores of the Lakes Superior and Huron, as in the contiguous western countries of the United States. They have, also, in their eastern division, been thrown into the same contorted and occasionally inverted order, as had been indicated by the Professors H. & W. Rogers, along this line, in more southern latitudes. In the opinion of those geologists, the causes of the rapid curvatures and partial inversions, by which younger strata have been so folded, as to pass under those beds which were formed before them, were grand and paroxysmal earthquakes, the greatest intensity of which has been exhibited where the folds are most rapid and numerous. As the intensity of the shock diminished, the strata are supposed to have rolled on in broad and regular undulations, thus resembling the waves of the sea at a distance from the area of strongest vibration. Although I can here simply advert to this ingenious explanation, it is satisfactory to me, as an historian of the older geological series, to find that, by whatever agency such grand contortions were produced, these inverted strata of North America, like those of Europe already spoken of (p. 377.), can be followed, until they fold out from their overturned positions and regain their natural order. This is explained in the uppermost of the two following sections.

IDEAL SECTION ACROSS THE APPALACHIAN CHAIN.

By Professor H. Rogers.

GENERAL SECTION ACROSS THE OLDER ROCKS OF LOWER CANADA NEAR THE MOUTH OF THE ST. LAWRENCE.

By Mr. Logan.

gm. Gneiss, &c. { *a.* Lower Silurian (consisting in Canada of Potsdam sandstone, Calciferous sand-rock, Trenton limestone, Utica slate, and Hudson River group).
t. Trap Dykes. } *b.* Upper Silurian (in Canada, Gaspé limestone, &c.). *c.* Devonian. *d.* Lower carboniferous rocks and sands. *d*.* Coal-fields.

The upper diagram is reduced from one of the numerous instructive sections by the Professors Rogers, and, although called ideal, is strictly based on facts observed in crossing the Appalachian chain, in the parallel of Pennsylvania. The oldest stratified rock on the S.E. of the section is gneiss, *gn*, being a portion of the crystalline and granitic rocks which range along the sea-board of the continent, and which, extending northwards into the states of Vermont, Maine, and Massachusetts, occupies such large portions of the British colonies of Nova Scotia and Newfoundland. The oldest member of the Silurian rocks, or the Potsdam sandstone, *a*, is seen to be coiled up in folds of the gneiss, together with the Trenton limestone and associated schists. The whole are in an inverted position, the younger dipping under the older rocks. In traversing the Appalachian chain, those Lower Silurian rocks are succeeded by the Upper Silurian, *b*, having the Niagara limestone as its great central mass, which, supporting the Devonian rocks, *c*, in a partially inverted basin, are followed by the last oblique axis. Further to the north-west, each anticlinal and synclinal arrangement gradually assumes its *vertical* and normal relations, the folds become grander, until the Devonian rocks, *c*, are seen supporting regular carboniferous basins of shale, limestone, and coal, *d, d**, which, further to the north-west than is here represented, open out into the vast coal regions before adverted to.

The lower diagram is drawn by Mr. Logan, to show how in the eastern portion of Canada the ancient crystalline rocks, *gm*, are similarly succeeded by the Lower Silurian rocks, *a*, consisting of Potsdam sandstone, Calciferous sand-rock, Trenton limestone, Utica slates, and the Hudson River group. The overlying rocks Upper Silurian, Devonian, and Lower Carboniferous (*b, c, d*) succeed transgressively. The strata have been much contorted, as in the eastern region of the United States. See also Logan, Quart. Geol. Journ, vol, viii. pp. 203. 206.

In Lower Canada a great dismemberment is seen to have taken place between the Lower Silurian rocks, *a*, and the Upper Silurian, *b*; the latter being conformably succeeded by the Devonian and Carboniferous strata. As in Britain and other countries, this unconformity is, however, a local phenomenon; for Mr. Logan himself assures us, that in another part of this region, just as in the west of the United States, all the palæozoic rocks, from the Potsdam sandstone or Silurian base to the coal inclusive, are strictly conformable. The only great break in that territory, *i.e.* along the Lakes Huron and Superior, is between the Potsdam or lowest fossil rock and the great *un*fossiliferous and inferior mass of bluish shales with sandstones, chert, and limestones, which are estimated at a thickness of about 12,000 feet. These, again, are seen, by following them in that tract in a descending order, to have been unconformably deposited on still older rocks, consisting of micaceous and chloritic schists, gneiss, and granitic rocks.*

The oldest of these masses have little connection with any rocks which have been spoken of, except in the first chapter of this work; but the next in ascending order, or the copper-slates and sandstones of the great lakes, have, to a great extent, the same structure as the Longmynd or 'bottom rocks' of England. They are, in truth, defined as such by Mr. Logan, being termed by him the equivalents of the Cambrian of De la Beche, Ramsay, and the British Govern-

---

* Mr. Logan, who has laid down a most elaborate and valuable geological map of the Canadas, has pointed out some of the chief incidents in the geology of these regions in his Reports of Progress to the Provincial Government. He also published a succinct view of the whole ascending order in Upper Canada. There, the Lowest Silurian rock does not repose at once on gneiss and granite, as in the above section across Lower Canada, but is separated from such rocks by the unfossiliferous or bottom slaty rocks above noted. In the same memoir Mr. Salter confirmed the views of Mr. Logan with regard to the age of the strata in Lower Canada, and identified many fossils as equivalents of the New York series. See Trans. Brit. Ass. Adv. of Science, 1852, p. 59., and also Quart. Geol. Journ. vol. viii. p. 199.

ment Surveyors. In England and North Wales these old slaty rocks
contain no limestone, but in North America they are interlaced with it.

Again, whilst in Great Britain and Germany these bottom rocks are
succeeded conformably by the Lower Silurian, they have been severed
therefrom in America, as in France and Ireland, by a grand dislocation.* In thus appealing to the conditions in different countries, we
learn the impracticability of classifying sedimentary deposits by the
lines of fracture and dismemberment alone. But in stating these facts,
I beg not to prejudge the question, so ingeniously developed by my
eminent friend, M. Elie de Beaumont, concerning the parallel and
discordant directions of mountain chains, as typical of certain epochs.
The discussion of this topic would require many chapters, and lead
me away from the special objects of this work.

In Nova Scotia the older palæozoic rocks have not been traced, but
the productive coal formation is there magnificently developed.†

In taking a general view of the physical structure of the northern-
most portion of America, Richardson the great Arctic traveller, considers that, on the whole, the granitic and crystalline rocks of the
central and eastern countries of the Hudson's Bay territories are surmounted by few other deposits on the west and north, except by masses
of Silurian age, and very young tertiary strata. In his graphic
description of the structure of Canada, Lyell expresses the same
opinion. 'I seemed,' says he, 'to have got back to Norway and
Sweden, where, as in Canada, gneiss and mica schist, and occasionally

* M. Jules Marcou speaks of various dislocations, one of the most violent of
which, in the States, has taken place between the Potsdam sandstone and the
Trenton limestone, a feature never seen in the regions examined by Mr. Logan.
The above mentioned Irish unconformity, between Cambrian and Lower Silurian,
has just been discovered.—March 1854.

† For full descriptions of these rocks, I must refer the reader to memoirs
in the Journal of the Geological Society, by Sir Charles Lyell, Mr. Dawson, and
Mr. Logan. I would refer specially to the interesting fact dwelt upon by Lyell
—the occurrence of a Batrachian reptile (Dendrerpeton Acadianum, Owen) in
the erect stump of a fossil tree; and to Mr. Dawson's last memoir.—Q. Journ.
Geol. Soc. vol. x. p. 1.

granite, prevail over wide areas, while the fossiliferous rocks belong, either to the most ancient or to the very newest — to the Silurian or to deposits so modern as to contain exclusively shells of recent species.'* From the western shore of Lake Winnipeg, a limestone with gigantic Orthoceratites (probably of the age of those described by Dr. Bigsby and Mr. Stokes†, from the Upper Silurian of Drummond Island), and the strange fossil Receptaculites, was traced in horizontal sheets stretching westwards over four or five degrees of longitude. Though this plateau of limestone is separated from the Rocky Mountains by a broad belt of the prairies of the Sasketchewan, the bed of that river is full of limestone blocks which indicate the persistence of the rock. After crossing Methy Portage, in lat. 56¾, Richardson again met with extensive limestone deposits. The fossils which were gathered from this tract (Productus, Orthis, Spirifer, &c.) seem, however, to indicate an ascending order into beds of Devonian and Carboniferous age‡, particularly in the wide spread of calcareous matter along the Elk and Slave rivers, and upon the banks of the Mackenzie. Where the last-mentioned river skirts the Rocky Mountains, the limestones, more disturbed than in the Winnipeg basin, occupy, says Richardson, inclined and elevated ridges, the chief of which he considers to be Silurian; — ridges which, on the Great Bear Lake and the Coppermine River, abut against granite. In a letter to myself, he adds, ' I believe the strata of sandstone and limestone on the north coast of America to be wholly Silurian, though fossils are scarce. Towards the mouth of the Coppermine River there are besides magnificent ranges of trap with ores of lead and copper, including much malachite.'

---

* Travels in N. America, vol. ii. p. 124.

† Bigsby and Stokes, Geol. Trans. 2nd ser. vol. i. pl. 25, &c. Similar Orthoceratites were found by Mr. Logan further to the east, in rocks of Upper Silurian age.

‡ A few specimens of the fossils of these rocks are in the British Museum; they are Upper Silurian, Devonian, and Carboniferous. Unluckily numerous fossils brought home by Sir John Richardson in 1826 have been mislaid.

To whatever extent it may be found possible to separate the Silurian rocks which range along the Rocky Mountains into a Lower and an Upper group, it would at least appear, from the specimens which have been brought home by our naval explorers employed in the Arctic expeditions, that the great mass of the most northern rocks belong to the Upper Silurian group. This is certainly the case, if we judge from the collections made during the voyages of Parry, Franklin, Ross, Back, Austin, and Ommanney, and the private expeditions of Lady Franklin, particularly those of Penny and Inglefield. The fossils brought home by these commanders, and the officers and gentlemen accompanying them, have been examined and described* by Mr. Salter; and from his scrutiny it results, that the crustacea and mollusca are very similar to, and some of them identical with, those of Dudley or Gothland. Among these occur Encrinurus lævis, Angelin; Leperditia Baltica, His.; Pentamerus conchidium, Dalm.; Chonetes (Leptæna) lata? V. Buch; with Upper Silurian forms of Orthoceras, Murchisonia, Strophomena, Orthis, Rhynconella, &c., besides Encrinites, and very numerous corals, including the chain-coral, the Favosites Gothlandica, and some other species characteristic of this division of the Silurian system.

The same inference has been drawn by the geologists who have surveyed the rocks, and the naturalists who have examined the fossils, of the northern edge of the great granitic region of the Lawrentine

* Appendix to Dr. Sutherland's Journal of Captain Penny's Voyage in 1850-51. London, 1852, vol. ii. with plates; see also Quart. Journ. Geol. Soc., vol. ix. p. 313, 1853. The extent to which the limestones of the Arctic regions are charged with Silurian fossils, was shown last year by my requesting Mr. Wyville T. C. Thomson (now Professor in the Queen's College, Cork) to break up and examine the common ballast brought home to Aberdeen in the Prince Albert, one of the private exploring ships of Lady Franklin. For, although these fragments of limestone were taken without the slightest reference to their organic contents, they were found by that naturalist to contain the following fossils: Rhynconella Phoca, Salter; Leperditia Baltica, His.; Murchisonia, 2 species; Atrypa reticularis; with fragments of other genera and species of shells and trilobites.

or Canadian Mountains. Their conclusion is, that whilst on the south side there is an inferior succession through crystalline and metamorphic strata into a copious and diversified Lower Silurian, as above explained, the northern or Hudson's Bay territory is chiefly occupied by Upper Silurian limestones. * This inference is based on the occurrence, at the base of the whole fossiliferous series in that district, of a profusion of corals, several of which are characteristic of the Niagara and Onondaga limestones (Wenlock or Dudley), together with the trilobite Encrinurus punctatus, the shells Atrypa reticularis and Pentamerus oblongus, and forms of Ormoceras, and several other mollusca, indicative of the lower portion of the Upper Silurian group.

Looking, indeed, to that vast Arctic region, so little accessible to any man of science, the geologist has good reason to be thankful for the knowledge already obtained, which has enabled him to classify the older sedimentary rocks of icy regions, never trodden by civilized man before the present century.

In terminating this outline of the succession of the older fossiliferous rocks of America, let me remind the reader of the vast extent to which this continent is composed of Silurian, Devonian, and Carboniferous rocks. In the Western hemisphere, as in Europe, the first clear signs of life are met with at the same low horizon in the crust of the earth; and the same great groups are clearly distinguishable. We also observe, that fishes, which were only called into existence at the end of the Silurian period, and were of such peculiar forms in the Devonian epoch, become conspicuous in the carboniferous deposits of America, and exhibit many new types, including the remarkable large sauroid fishes of Agassiz. Again, whilst one small, air-breathing reptile has alone been found in the uppermost band of the

---

* Mr. Logan has found that these limestones at the head of Lake Temiscamang include enormous blocks of the sandstone on which they rest; so that, in all probability, they are littoral deposits.

Devonian or Old Red rocks of Britain, the Carboniferous rocks of
America, as well as those of Europe, contain larger animals of this
class. The only essential difference between the elder rocks of the two
hemispheres is, that in America no geologist has yet met with any
indication of that termination of palæozoic life, which in Europe is
marked by the Permian deposits. It is probable, therefore, that the
early sediments of the West, having been raised into the atmosphere,
remained for a long time in that condition; whilst in our countries
the same strata continued beneath the waters which sustained the
last remnants of primeval beings.

CHAPTER XVII.

ON THE ORIGINAL FORMATION OF GOLD, AND ITS SUBSEQUENT
DISTRIBUTION IN DEBRIS OVER PARTS OF THE EARTH'S
SURFACE.

CONSIDERING the great quantity of gold which has recently been
found in distant regions, it may be expected that an author, who
during the last ten years has borne a part in the discussions relative to
this phenomenon*, should here devote a few pages to so engrossing a
topic. Avoiding altogether, for the present, the question, as to the
effect which a great abundance of the precious metal may exercise
over the destinies of mankind, the views now put forth will chiefly
relate to the geological and mineralogical conditions under which it
has occurred. As a clear understanding of this point may tend, in
some measure, to allay the fears of those who think that gold may be
discovered over regions vastly more enormous than the tracts to
which it is restricted, certain data and arguments that I have before advanced in detail, are here brought together.

Let us first reflect upon the general fact, that whilst all the
stratified formations are composed either of crystalline and palæozoic
rocks, or of secondary and tertiary deposits, gold has never been

* See Russia in Europe and the Ural Mountains, pp. 437., *et seq.* Trans. R. Geogr. Soc. President's Discourses, vol. xiv., 1844, 1845, in the first of which the Australian rocks were compared
with those of the Ural. Trans. Royal Geological Society, Cornwall, 1846, p. 324., *et seq.*, in
which Cornish tin miners were incited to emigrate and work for gold in Australia. Report of
the British Association for the Advancement of Science, 1849 (Trans. of Sections, p. 60.).
Proceedings of the Royal Institution, March, 1850. Quarterly Review, 1850, vol. lxxxvii., Article
" Siberia and California," p. 39. Quart. Journ. Geo. Soc. Lond., vol. viii. p. 134. And lastly,
" Further Papers on the Recent Discovery of Gold in Australia," presented to Parliament
Aug. 16. 1853, p. 43.; including my correspondence, in 1848, with Her Majesty's Secretary for
the Colonies, on Australian gold.

found in any appreciable quantity in either of the two last-mentioned classes of strata. The vast areas, therefore, which are covered by all the formations younger than those whose relations we have been considering, are excluded from the application of our reasoning; and every one who lives amid such rocks may at once be assured, that he can never profitably extract gold from them.

Having thus cleared the ground, we may first proceed to consider the nature and limits of the rich gold-bearing rocks, and then offer proofs, that the chief auriferous wealth, as 'derived from them, occurs in superficial detritus.

Appealing to the structure of the different mountains, which at former periods have afforded or still afford any notable amount of gold, we find in all a general agreement. Whether, referring to past history, we cast our eyes to the countries watered by the sources of the golden Tagus, to the Phrygia and Thrace of the Greeks and Romans, to the Bohemia of the Middle Ages, to tracts in Britain which were worked in old times, and are now either abandoned or very slightly productive, or to those chains in America and Australia which, previously unsearched, have, in our times, proved so rich, we invariably find the same constants in nature. In all these lands, gold has been imparted abundantly to the ancient rocks only, whose order and succession we have traced, or their associated eruptive rocks. The most usual original position of the metal is in quartzose veinstones that traverse altered, palæozoic slates, frequently near their junction with eruptive rocks. Sometimes, however, it is also shown to be diffused through the body of such rocks, whether of igneous or of aqueous origin. The stratified rocks of the highest antiquity, such as the oldest gneiss and quartz rocks (like those, for example, of Scandinavia and the Northern Highlands of Scotland), have very seldom borne gold; but the sedimentary accumulations which followed, or the Silurian, Devonian, and Carboniferous (particularly the first of these three), have been the deposits which, in the tracts where they have undergone a metamorphosis or change of structure by the influence of igneous agency, or other causes, have been the *chief* sources whence gold has been derived.

The British examples, comparatively small in produce as they have been, will be first briefly alluded to; because the reader can at once refer, in the coloured map of this work, to two districts of Wales wherein gold has been found, and in one of which it is now in the course of extraction.

Gold in Britain.—In the Lower Silurian rocks, about ten miles west of Llandovery, at a spot called Gogofau, near Pümp-saint, large veinstones of quartz in slaty masses were cut into by the Romans, who excavated lofty galleries, which are still open. That enterprising people evidently derived gold from the portions of the veinstones in which much sulphuret of iron is diffused; since many gold ornaments have been found at the adjacent Roman station of Cynfil-Cayo, with traces of former aqueducts, probably to convey water to wash the gold. Even the stones * and troughs used in grinding the hard matrix are yet visible.

In North Wales, where similar but older strata have been more crystallized, and infinitely more penetrated by igneous rocks, gold has not only been obtained in ancient times, but is still found to a certain extent. In Merionethshire some of the older slaty rocks were ten years ago announced to be auriferous by Mr. A. Dean.† The district then referred to, which lies to the north of Dolgelly, and to the north and west of the small river Mowddach, has lately been resurveyed by Professor Ramsay, who has described the precise geological relations and mineral character of several metalliferous lodes, which, though poor in lead and copper, are, more or less, impregnated with gold.‡ The lodes are subordinate to the

* See Silurian System, p. 367. At the time of the publication of that work (1839) I had not visited the Ural Mountains, and was little acquainted with the nature of gold-bearing rocks and the methods employed for the extraction of the metal, or I should at once have recognized as certain, what I only ventured to suggest, that the rock might formerly have been quarried for the gold it contained. Suspecting that some traces of gold might be detected in the pyritous refuse of the quartz rock, I submitted a little of it to the late eminent chemist, Dr. Turner; but he could not detect any trace of the precious metal. Recently, however, Prof. Warington Smyth, aided by Dr. Percy, has detected a small quantity of gold diffused in the quartz veinstone. See also Memoirs of the Geological Survey, vol. i., with an exposition of the various relations of gold, by Ramsay, E. Forbes, Jukes, Percy, Warington Smyth, Hunt, &c.

† Trans. Brit. Assoc. Adv. of Science, 1844, Proc. of Sect. p. 56.

‡ Quart. Journal, Geol. Soc. Lond. (about to be printed).

Lingula flags or lowest Silurian of the Government surveyors, and these slaty and arenaceous rocks being there traversed by trap dykes (including magnetic greenstone), and being also bounded by a large mass of eruptive rock, are much altered, and often in a talcose and chloritic state, with veins of quartz and much disseminated iron pyrites. The principal localities where the gold has been most observed are Cwm-eisen-isaf and Dol-y-frwynog. One of the veinstones at the latter place consists of white saccharoid quartz, in some of which small flakes of gold are distinctly visible to the naked eye.

Professor Ansted, who has examined the same gold veins *in situ*, had reported to me, that at Dol-y-frwynog, the gold is disseminated both in grains and in irregular bands or veins, parallel to the Lower Silurian schists, and contiguous to a poor lode of copper ore; the whole lying near to the junction of a greenstone with the slaty rocks. The auriferous bands, he says, are made up of numerous threads of quartz and sulphate of barytes, which, besides the grains and flakes of gold, contain crystals of galena and copper pyrites. The gold is partially present to such an extent (he adds), that in a small quantity removed by himself from one of the threads or thin veins (and richer specimens have since been found) the proportion, upon analysis, was that of 60 oz. to the ton !*

At a few places in Cornwall† and Devonshire, gold has been long known to exist in small quantities, both in the matrix of mineral lodes and occasionally in accumulations of rolled materials. One of the spots which now promises, as is said, to be most productive, is the Poltimore mine, near North Molton, Devon, where certain schists of the age of the uppermost Devonian or lowest Carboniferous strata are mineralized to some extent, and were formerly worked for copper. There, the matrix or gossan of the lode is suffused by particles of gold, but so minute as to be for the most part invisible to the naked eye. *If* this condition be found persistent throughout considerable masses of rock, the proprietors may doubtless derive considerable profit from such mines; particularly when the gold-

---

* As gold often occurs in small strings or threads which, though rich at the surface, are soon lost in the body of the rock, time alone can determine the value of these British mines — *i.e.* whether some of them are to prove remunerative, now that new processes for extracting the ore have been discovered.

† Mr. S. R. Pattison has just read before the Geological Society (February 1st, 1854) a notice on auriferous quartz rock near Davidstow, North Cornwall, the chief gold-bearing mass of which is the *gossan* of a dyke in a metamorphosed rock of Upper Devonian age, which, mantling round the granite of Roughtor, is also associated with dykes of trap.

bearing matrix is friable. On the other hand, the Cornish, coarse, ancient alluvia or gravel, from whence the tin ore has been extracted, as well as other portions of drift in that county, have afforded small portions of gold ; but although some of the largest fragments have been (though very rarely) of the size of a pigeon's egg, none of these superficial accumulations have been considered worthy of exploration.

In Scotland, whilst slender traces * only of gold have been perceived in the older crystalline rocks of the Northern Highlands, the metal was formerly found in the slates of the South of Scotland (Lead Hills), which, like those of North and South Wales, are of Lower Silurian age. There again it is in a region where the strata have been much penetrated by porphyries and other igneous rocks. These south Scottish gold mines, after affording a small sum, in the reign of James the Fifth, were, however, abandoned as soon as the cost of production exceeded the value of the ore extracted.†

In Ireland we read the same lesson. It is from the altered lower Silurian schists of Wicklow, which clasp round the eruptive granite of Croghan-Kinshela traversed by hornblendic greenstones, that the gold known in Ireland has been derived, and is still occasionally picked up in fragments which have been detached from the sides of that mountain, or are washed down by the rivulets which descend from the hill.‡

Now, if any portion of these old slaty British rocks, or their associated eruptive rocks, had been largely penetrated by gold, then most assuredly much more auriferous débris would have been recognized in the local adjacent gravel ;—just as it occurs in all really rich gold-bearing lands. *But, as no rich auriferous sand or gravel is known in any part of the*

---

* Near Loch Erne Head, a metalliferous veinstone on the property of the Marquis of Breadalbane has recently been found to be partially impregnated with gold. The gold occurs in a glossan, contiguous to the junction of trap with crystalline limestone and schist, and is associated with arsenical pyrites and lead ore.

† See Harkness on the "Lower Silurian Rocks of Scotland." Quart. Journ. Geo. Soc. vol. viii. p. 396.

‡ The Earl of Wicklow, whose property is in the vicinity of the mountain of Croghan Kinshela, has collected several 'pepitas' of this Irish gold, the largest being about two inches long. They are free from quartz or other rocky matrix, and have been picked out of the débris or coarse gravel on that slope of the hill where a rivulet descends through the property of the Earl of Carysfort. No veinstone *in situ* has ever been detected (Mills and Weaver, Trans. Dublin Society); and although poor persons have stealthily procured specimens during this century, the quantity has never been sufficient to lead to the belief that really productive diggings could be opened at the Royal Gold Mine of Croghan. Tinstone is said to have been found with the gold here as in Cornwall and other places. (See Fitton, Trans. Geol. Soc. Lond., first series, vol. i. p. 270. The phenomena are well developed by Prof. Warington Smyth, Records of the School of Mines, &c., vol. i. p. 3., with a map.)

British Isles, we may rest satisfied that in our own country, as in many others, the quantity of gold originally imparted to the rocks, was small and has, to a great extent, been exhausted.*

Even in Bohemia, which produced so much gold in the Middle Ages, and where the Silurian strata are, as we have seen, (chap. 14.) penetrated by many igneous rocks (and in parts much metamorphosed), there are now no gold works, though other ores (copper, &c.) are profitably extracted; and, just as in the rocky and mountainous tracts of Britain, very few places only can be cited which were auriferous.   The Thüringerwald, and some chains of Central Germany, also anciently afforded a little gold in rare and widely-separated localities†; but these regions, as well as the Peninsula and its golden Tagus, so auriferous in the classical era, have long since ceased to offer any notable quantity of the ore.

The Ural Mountains. — No country of the world furnishes a clearer example than Russia, of the dependence of gold on certain geological and mineral relations.   Her chief European territories are, as has been stated, occupied by slightly solidified, primeval deposits.   Under those conditions, and with a total absence of any crystalline rocks, whether of intrusive or of sedimentary and metamorphic characters,

---

* As these pages are going through the press, a work by Mr. John Calvert has appeared, entitled the "Gold Rocks of Great Britain and Ireland," wherein the author (who has recently returned from Australia) holds out incentives to mine largely for the precious metal at home, though he candidly admits that his main suggestions are at variance with the recorded opinions of Sir H. T. De la Beche, myself, and many geologists. The reader who may have attended to this subject will observe, that the chief argument I have employed in the writings adverted to in the note p. 431., to satisfy the public mind, that auriferous sites in the old countries of Europe would for the most part prove slightly profitable only, was that all such works had ceased in former times for want of remuneration. It is foreign to the purpose of this work to analyze theoretical views respecting the diffusion of gold and its associated minerals. Let me, however, say, that whilst I believe the old gold tracts of Europe have been on the whole exhausted of their wealth, there may still be found spots where *some profit* is attainable. I would further guard any inferences I have drawn from our previous state of knowledge, by saying, that my opinions were formed irrespective of the new discoveries in mechanical science. Crushing machines, and the improved application of mercury may, indeed, liberate a notable quantity of ore from a matrix of apparently slight value, and thus set at nought the experience of ages. Not pretending to enter into this mercantile part of the question, I cling to the belief expressed throughout the text, that, in the long run, gold will mainly be found in the rocks indicated, and will be worked to great advantage only, as during past ages, in the natural débris of those rocks.

† The sands of the river Rhine (which drains so vast a rocky region) are in one part slightly auriferous, but the cost of extraction of the gold was too great to repay the speculators.

not a particle of gold has been discovered in them, over an area larger than the rest of Europe. But where the same formations have been thrown up into inclined and broken positions in the Ural Chain, and have there been pierced by numerous eruptive rocks of porphyry, greenstone, syenite, and granite, in association with huge masses of serpentine, the very same deposits, so soft in European Russia, have been hardened, crystallized, veined, and rendered highly metalliferous; some even of the igneous rocks being also occasionally auriferous.

As the rocks in this chain, which separates Europe from Asia, are now known to be similar in character to those of numerous other auriferous ridges in Siberia and the Altai Mountains of Asiatic Russia, the present description may serve to explain the composition of vastly larger eastern tracts. The study of this Uralian Chain has, indeed, already thrown light, not only on the nature of metalliferous rocks of other countries, but has enabled us to define, within certain limits, the period when they were chiefly impregnated with gold, and also to suggest that, in this region at least, the metal appears to be of more recent origin than the ores of copper and iron with which it is there associated.

With a watershed for the most part not exceeding 2000 feet above the sea, their highest peaks rarely rising above 5,000 to 6000 feet, the Ural Mountains, extending from north to south through 18 degrees of latitude, are composed of rocks more or less crystalline; chiefly the metamorphosed representatives of the Silurian and Devonian, and occasionally of the carboniferous age. The Lower Silurian strata are, indeed, to be recognized in a crystalline state only. They are chiefly talcose schists, quartzites, and limestones; whilst the Upper Silurian, Devonian, and Carboniferous, though often also considerably metamorphosed (the limestones being frequently converted into marble occasionally dolomitic), offer here and there traces of their characteristic fossils. The flexures and fractures of the stratified rocks (Devonian and Carboniferous), as they approach the

western flank of the altered and metalliferous axis of the chain, have
been represented in the view of the gorge of the Tchussowaya, p. 330.

The following rough sketch will convey some idea of the wild,
central, and mineralized masses which, in the Northern Ural, peer
out, here and there, from amid the forests of the gigantic *Pinus Cembra*.
It was taken by myself from the summit of the Katchkanar*, a

VIEW FROM THE SUMMIT OF THE KATCHKANAR, N. URAL.

(From Russia in Europe, vol. i. p. 392.)

The snowy mountains seen in the distance to the north, are the much loftier peaks of
Konjakofski Kamen, &c.

* The following is the sketch of the approach to this mountain (visited, I believe, by no other
European travellers) through the forests, as given in the work "Russia and the Ural Mountains," vol. i. p. 392. "A large chaotic assemblage of loose angular blocks now lay around us,
from amid which rose the magnificent *Pinus Cembra*, towering above all its associates, the rocks
being overgrown with pæonies, roses, and geraniums. Such stony features alone would have
led us to suppose that we were at the foot of the object of our exploration, when in a few
minutes the broken and jagged outline of the Katchkanar burst upon our sight under a fine
bright sun and amid the merry song of birds. The dull, wet, and marshy woodlands, were now
exchanged for sunshine, and rocks. . . . . Accustomed as we have been to the wildest features
of the Highlands of Scotland and the Alps, we are unacquainted with any scene presenting a
finer foreground of abruptly broken rocks, and never certainly had we looked over so grand and
trackless a forest as that which lay around us, and from which some straggling distant peaks
(those on the north only being still capped with snow) reared their solitary heads."

rugged pile of stratified and jointed augitic greenstone, highly charged with magnetic iron. Platinum, as well as gold, has been washed down from the edges of this mountain and carried into the adjacent gorges on the east; though the sources from whence these metals have been drifted into the coarse alluvium, have not been detected.

While the traveller who simply crosses the Ural in the partial depression followed by the high road to Ekaterinburg, will scarcely be aware that he has passed over any dominant ridge, the geologist who explores the mountains along their line of bearing, soon perceives that, to the north, their crest is marked by lofty and usually snowy summits, as in the preceding drawing. To the south, the central mass near Zlataust, consists of altered sandstone and quartzites, forming sharp peaks, that separate Europe from Asia.*

Few chains offer more contrasting outlines than are seen upon the European and Asiatic flanks of the Ural. On the former the limestones and other stratified rocks are indeed contorted, fractured, and partially altered, as before represented (p. 330.), whilst in the centre, as on the eastern slopes, the masses consist everywhere either of highly altered and crystalline strata, or of the eruptive rocks which pierce them. There only, and particularly where the schists are traversed by veinstones of quartz, or cut by dykes of igneous rocks, has gold been originally imparted in any quantity to the slaty, talcose, and chloritic strata. Though some efforts were made by the earlier Russian miners to extract gold from the solid matrix by underground mining, such a process was not continued; it having been found infinitely more profitable to extract the ore from the broken accumulations of ancient drift, in which the indestructible gold has been carried down to the slopes of the hills or lodged in the higher valleys wherein small watercourses meander.

The only work at which subterranean mining in the solid rock is still

* See the frontispiece of " Russia and the Ural Mountains."

practiced, and at a very small profit, is at Berezovsk near Ekaterinburg. There the shaft traverses a mass of apparently metamorphosed and crystalline matrix, called 'beresite,' resembling a decomposed granite with veins of quartz, in which some gold is disseminated. The syenite of the Peschanka mines, near Bogoslofsk, is likewise impregnated with particles of gold, and the surface degradation of that rock affords profitable washings. Much farther to the east in Siberia, Colonel Hoffman long ago indicated a tract, where the schistose stratified rocks are equally permeated by the small diffused particles of the metal, imperceptible to the naked eye.

Considering, however, the practice which has hitherto prevailed, it is from the broken or drifted materials only, that the Uralian gold has been ground out by water mills and collected for use. This is notably the case on the east flank of the chain*, where the mixed and coarse detritus of all the hard rocks has, at certain spots, proved to be auriferous, and in some cases much more so than others ; there being very large tracts indeed where no trace of gold can be detected even in similar detritus.

On the east flank of the South Ural (S. of Miask), where the chain is still auriferous, conical igneous rocks have burst out, as represented in the opposite drawing, and constitute a picturesque scene. The low rich grassy grounds around this lake of Aushkul, in the country of the Baschkirs, are, to some little extent, auriferous ; the gold having been derived by debacles of former periods, which have denuded the surfaces of the slaty and quartzose chain (as seen in the distance), or the adjacent conical hills of greenstone, altered schists, porphyries, syenite, and serpentine which surround the lake. The slopes and depressions are partially occupied by the gold-bearing debris.†

---

* The reader must recollect that the superficies of all the localities of the Ural Mountains united, in which gold has been found, amount but to a very small part of the whole chain. Incalculably more is the proportion diminished between rocks which from their nature might be but are not auriferous, and those of similar structure which really contain gold, when we extend researches into Central and Eastern Siberia. Thus, gold-works are only known at one spot in the western side of the watershed, viz., at Chrestovosdvisgensk. It was there that most of the few diamonds found in the Ural chain were detected in coarse ancient drift. They were no longer observed when I visited that place, and thence traversed the wild, wooded chain by the Katchkanar to Nijny Turinsk. See "Russia in Europe and the Ural Mountains," pp. 391 —480. et seq.

† The rock in the foreground on which we stood is a compound of diallage and serpentine,

LAKE OF AUSHKUL, S. URAL.

(From a lithograph, p. 437.   Russia and the Ural Mountains, vol. i. p. 359.   The Holy Mount
of the Tatars is opposite, and the Ural range is seen in the distance.)

The extent to which limestones of the carboniferous age have been
altered on the eastern flank of the South Ural, is instructively seen
at Cossatchi-Datchi, a remote spot visited by my companions and
myself, and where also a little gold has been found.

HILLS OF COSSATCHI-DATCHI.

(From Russia and the Ural Mountains, vol. i. p. 439.)

The hills forming the background of this sketch, composed of eruptive

and is to some extent magnetic; the most striking of the conical mounts is the Holy Hill of
the Baschkirs. ("Russia and the Ural Mountains," p. 437.)

rocks and some old schistose strata, subtend a little basin in which a number of small conical hillocks of limestone, as seen in the foreground, are thrown about. Their form and mineralized condition are probably due to the action of gaseous vapours and change of the original substance; since not only the traces of bedding are obliterated, but the rock has been rendered fetid and saccharoid; breaking upon a slight blow of the hammer. In this case the metamorphic action, however it may have been produced, has been just of that degree of intensity, that, whilst in rendering the limestone pulverulent as sugar, it has left numerous organic remains so uninjured, that they are easily removed from the ambient matrix.*

At the Soimanofsk mines, south of Miask, great piles of ancient drift or gravel having been removed for the extraction of gold, the eroded edges of highly inclined crystalline limestones have been exposed, which, from being much nearer the centre of the chain than the above, are probably of Silurian or Devonian age. It is from the adjacent eruptive serpentinous masses and slaty rocks, *b*, that the gold shingle, *c* (usually most auriferous near the surface of the abraded rock *a*), has been derived.

DIGGINGS AT THE SOIMANOFSK MINES.

(From Russia and the Ural Mountains, vol. i. p. 487.)

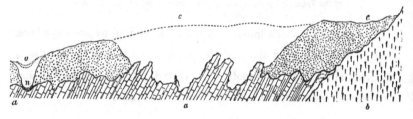

The tops of the highly-inclined beds *a*, are, in fact, rounded off, and the interstices between them worn into holes and cavities, as if by very powerful action of water. Now here, as at Berezovsk (of which hereafter),

---

* After enumerating upwards of thirty species of these fossil shells, my companions and self thus spoke of them:—"Those alone who have the same respect for a true characteristic fossil as ourselves, can imagine the feelings of delight with which we here found congregated in one natural Siberian storehouse so great a number of shells, some of which we could not distinguish from well-known forms of the mountain limestones of Yorkshire, Westmoreland, and Derbyshire, nor others from species which are abundant in the same formations in Belgium and France! . . . . Without this discovery we could not have ventured to affirm, that many other adjacent masses of crystalline limestone immersed among the granites and trappean rocks of these mountains belonged to similar or associated deposits." ("Russia and the Ural Mountains," vol. i. p. 44.)

mammoth remains have been found. They were lodged in the lowest part of the excavation, at the spot to which the small figure of a man is pointing, and at about fifty feet beneath the original surface of overlying coarse gravel *c* before it was removed by the workmen from the vacant space under the dotted line. The feeble influence of the streams (*n*) which now flow, in excavating even the loose shingle is seen at the spot marked *o*, the bed of the rivulet having been lowered by *human* labour from its natural level *o*, to that marked *n*, for the convenience of the diggers.

In some spots the alluvium in which the gold occurs is a heavy clay; in others it is made up of fragments of quartz veins, chloritic and talcose schist, and greenstone, which lie upon the sides of the hillocks of eruptive rocks.* It was from the infillings of one of the gravelly depressions between these elevations south of Miask, that the largest lump of solid gold was found, of which at that time (1824) there was any record.† The diggings by which the gravel or local drift was cleared away from around the vertical masses of rock whose surfaces have been so much eroded and chan-nelled out, are expressed in the following diagram.

No watercourse sufficiently powerful to transport a single block, much less to spread out broad accumulations of such coarse materials, now flows into this upland depression. Nor could the action, during millions of years, of such an agency as that of the puny rivulets, which now meander in parts of the gravel, account for the eroded and highly worn surfaces of the rocks, whether crystalline limestones, quartz rocks, greenstone, porphyry, or serpentine. We are thus necessarily compelled, by numerous evidences, to adopt the belief, that on the Asiatic side of the Ural, as in many parts of Europe, the translation of vast masses of drift was accompanied by powerful

* The auriferous shingle, gravel, or sand of the Ural Mountains is poor in per-centage in comparison with what has of late years been discovered in California and Australia. Though very large 'pepitas' or nuggets have occasionally been found, much of the auriferous ground considered worth working in Russia, where labour is cheap, and water power for crushing is every where at hand, would, if situated in Australia or California, be little heeded.

† This 'pepita' weighs ninety-six pounds Troy, and is still exhibited in the museum of the Imperial School of Mines at St. Petersburg.

and long continued aqueous abrasion of the summits and slopes of the adjacent auriferous hills.

GOLD DIGGINGS AT ZAREVO ALEXANDROFSK.

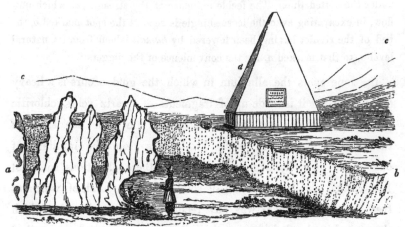

a. Ancient rocks consisting of talc schist, with veins of quartz (original matrices of gold), penetrated by concretionary felspar rocks, greenstone, &c. *b.* Coarse shingle and gravel about twelve feet thick, in which the great "pepita" was found. *c.* Hills from which the chief debris *b* has been swept. *d.* Pyramid erected to commemorate the visit of the Emperor Alexander.

Whatever may have been the period when the rock was first rendered auriferous, the date of this great superficial distribution of the gold is clearly indicated.   For it contains in many places the same remains of extinct fossil quadrupeds that are found in the coarse drift-gravel of Western Europe.   The Elephas primigenius or Mammoth, Bos aurochs, Rhinoceros tichorhinus, with gigantic stags, and many other species, including even large carnivores, were, unquestionably, before that period of destruction, the denizens of Europe and Siberia; and of these the Bos aurochs is the only one which has been preserved to our days.*

---

* The Bos aurochs was probably saved by having inhabited some isolated spot in Western Russia near the forest of Biela vieja in Poland, where the herd now lives, and by having been there locally exempted from the causes of that great destruction which befel their associates. This geological view is fully explained by me, accompanied by an excellent account of the Bos aurochs by Professor Owen.—("Russia in Europe and the Ural Mountains," vol. i. p. 503. *et seq.*)

This little diagram explains the relations of the coarse auriferous drift with Mammoth bones, as seen at the Bérézof mines near Ekaterinburg.

GOLD SHINGLE NEAR EKATERINBURG.

a. Auriferous rocks *in situ*. These rocks contain some gold in quartz veins in the mass of a metalliferous lode of soft granitoid composition called, as above said, Beresite, by the Russian miners. It is the only mine in this chain, or in Siberia, which is worked underground, and though very shallow, scarcely repays the expenditure. *b.* Auriferous debris with Mammoth bones. *c.* Alluvial clay covered by humus and bog earth.

Before we quit the consideration of the Ural Mountains the reader may be reminded that, throughout the length of 500 miles, the rocks are auriferous at wide intervals, and in limited patches only. Having clearly marked the geological period of the superficial gold drift, let me also here advert to a suggestion of my associates and self concerning the period at which the rocks were impregnated with gold.* It has been already stated, that no secondary formation contains veinstones charged with any notable quantity of gold, and that when the metal is found *in situ*, it is either in metamorphosed strata of Silurian, Devonian, and Carboniferous age or in associated eruptive rocks. Now, it would seem as if these rocks must, in the Ural, have been chiefly impregnated with gold, in a comparatively modern geological period. In the first place, the western flank of the Ural chain offers strong evidence, that this golden transfusion had not been effected in this region when the Permian deposits were completed. During that period, vast heaps of pebbles and sand, all derived from a pre-existing Ural chain, (the older stratified rocks of which had even then undergone much change,) were spread out over the lower country on the west.

* See the work on "Russia and the Ural Mountains," vol. i. p. 472. *et seq.*

Together with fragments of all the rocks, sedimentary or igneous, which are known in the chain, specimens of magnetic iron and of copper ore, which so abound in the range, are not uncommon in this Permian debris; but no where does it contain visible traces of gold or platinum. Had those metals then existed in the Ural Mountains, in the quantities which now prevail, many remnants of them must have been washed down together with the other rocks and minerals, and have formed part of the old Permian conglomerates. On the other hand, when the much more modern debacles, that destroyed the great animals, and heaped up the piles of gravel above described, proceeded from this chain, then the debris became largely auriferous. It is manifest, therefore, that the principal impregnation of the rocks with gold — *i. e.* when the chief lumps and strings of it were formed — took place during the intervening time.

What then was probably the geological period when these rich auriferous impregnations of the Uralian rocks took place? We cannot believe that it occurred shortly after the Permian era, nor even when any of the secondary rocks were forming; since no golden debris is found even in any of the older tertiary grits and sands which occur on the Siberian flank of the chain. *If, then, the mammoth drift be the oldest mass of detritus in which gold occurs abundantly,* not only in the Ural but in many parts of the world, we are led to believe that this noble metal, though for the most part formed in ancient crystalline rocks, or in the igneous rocks which penetrated them, was only abundantly imparted to them at a comparatively recent period; — *i. e.* a short time (in geological language) before the epoch when the very powerful and general denudations took place which destroyed the large extinct mammalia.*

. * In many instances gold is, I know, associated in the same veinstone with other ores — such as silver, or argentiferous galena, and with various ores of copper and iron — magnetic iron being, indeed, a very frequent accompaniment; whilst the association with tin-stone has before been alluded to. Such occurrences do not invalidate, but strengthen the view derived from the phenomena in the Ural Mountains. For, as copper and iron ores are frequently found in old

That the gold which occurs in quartz veins in the solid slate rocks resulted from an interior agency, in which heat and electricity were combined with water or vapour, seems to be a natural conclusion, if we judge from the appearance which the strings and expansions of the metal indicate as they ramify through the chinks of the hard rock, or are diffused in grains in its mass. We may also suppose, that the prevalent matrix of quartz, whether ejected from beneath, or poured in from above, was in a soft and gelatinous state when it filled the cavities, resembling the silicious " sinter " which now rises in a fluid spout from Hecla, and, falling, coagulates into a modern quartz rock around the volcanic orifice.

In viewing the widely attested fact of the dispersion of auriferous debris derived from the surface of certain rocks during some of their last great denudations, we are naturally led to favour the suggestion of Humboldt, that the formation of gold had some closer relation to or dependence upon the atmosphere than that of the baser metals lead, copper, and iron. An eminent metallurgist, Dr. Percy, who has detected minute quantities of gold in almost all lead ores, is, indeed, disposed to believe, that it may have been thrown down by deposition from an aqueous medium.

conglomerates or pebble beds of secondary age, and lumps of gold have never been detected in them, I see no means (explain the phenomena as we may) of evading the inference, that *no great quantity of gold ore* was formed (certainly not in the Ural Mountains) until the comparatively recent epoch indicated in the text. In the work "Russia and the Ural Mountains," vol. i. p. 473., the inference is thus stated: — "Whether, therefore, we judge from the total absence of auriferous matter in the ancient conglomerates on the west, and in the tertiary grits on the east, or from the absolute materials in the whole series of regenerated deposits, we conclude, that the chain became (chiefly) auriferous during the most recent disturbances by which it was affected, and that this took place when its highest peaks were thrown up, when the present watershed was established, and when the syenitic granites and other comparatively recent igneous rocks were erupted along its eastern edges."

The reader who wishes to have fuller information on the subject of Uralian and Siberian gold, must consult Humboldt's Asie Centrale; Reise nach dem Ural, &c., by Humboldt, Rose, and Ehrenberg, with the valuable mineral description of M. Gustaf Rose; various memoirs by Helmersen and Hoffman in the Annuaire des Mines de Russie; and Adolf Erman (Reise um die Erde), as well as an account of the general diffusion of gold and a valuable gold map of the world by that Author.

Whatever may be the correct hypothesis as to the original mode of formation, the fact is undeniable, that wherever the veinstones in the solid rock have not been ground down by former, powerful denudation, but remain as partial testimonials of the origin of gold, the portions which have as yet proved to be the richest, are those which are at or near the surface. Experience too, dearly bought in numberless instances, has taught the miner, throughout long ages, that in his efforts to follow the veinstones downwards, by deep shafts, into the body of the rock, he has either found the gold diminish in volume, or so difficult to obtain, that the cost of extraction has usually been greater than the value of the metal.

The points which have been alluded to, as drawn from personal observation in the Ural Mountains, are found to have a world-wide application in every tract which has been or is still auriferous. Thus, the giant chain of the Andes, which has for ages afforded much gold in its range through Chili, Peru, and Mexico, is essentially of the same composition. The Indians, who lived in tracts adjacent to those mountains, followed the simple process of picking the shining material from the gravel, sand, and shingle derived from the chain. So that when the Spaniards, the best miners of the sixteenth century, first colonized South America, they naturally inferred, that if ignorant natives could thus gather sufficient quantities of gold to roof the palaces of their sovereigns, they, as skilful Europeans, might extract incredible quantities from the bowels of mountains, the mere detritus of whose surfaces had contributed such a vast amount of gold. But, as frequently as deep mines enriched the speculators who sought for copper and silver, so surely gold mining in the solid rock proved abortive*, owing to the slender, *downward* dissemination of gold in a hard and intractable matrix.

---

* It has been too much the habit to underrate the capacity and skill of the old Spanish miners, though it is known from Humboldt, that during the government of the monarchy in South America, many of their works were well conducted; the operations having been paralyzed chiefly by the political revolutions which have occurred in those regions.

As the phenomena above described are common to many countries, the accompanying diagram is annexed, to convey, as far as possible at one view, a popular idea of the chief relations under which gold has been distributed over the surface, so as to be profitably collected by mankind.

IDEAL REPRESENTATION OF THE ORIGINAL FORMATION OF GOLD IN THE ROCKS, AND ITS SUBSEQUENT TRANSLATION INTO HEAPS OF GRAVEL.

(From a large diagram exhibited in 1850, at the Royal Institution.)

The rocks, *a*, are the ancient metamorphic masses, whether talcose, micaceous, chloritic, fel-spathic, or siliceous slates (altered primeval slaty rocks), which have been traversed by quartzose and other auriferous veinstones, *b*. All along their present dark summits *a** are seen lighter tinted lines, which represent the outline and condition of the auriferous ridges before their pinnacles were subjected to that agency of destruction and abrasion, by which great heaps of drift and gravel were transported from them, so as to form the hillocks, *c*, that cover the adjacent slopes and fill up the gorges and depressions, *d*. The first or highest of these drifts, &c., constitute the dry diggings of the miner; the lower heaps, *d*, in which streams meander, being the wet diggings. It being impossible to represent in one diagram all the conditions under which gold was originally formed in the rocks, I have merely selected the usual case of veinstones, the old or destroyed surfaces of which were the richest.

Gold of Australia.—The extraordinary quantity of gold which has been poured into Britain from her Australian colonies during the last two years, has been procured, like the chief masses in the other tracts described, from superficial accumulations of shingle,

If the former trials of Spaniards to procure gold with profit from deep mines in the solid rock, and which were proverbially failures, do not satisfy living speculators, let me refer them to similar results in our day, and trials by our own countrymen. Among these, I would specially allude to the well-known mine of Guadalupe y Calvo, near Durango, in Mexico, worked by British skill and capital, where, according to information I received from one of its ablest directors (my friend the late Colonel Colquhoun, R. A.), the works, which afforded a moderate profit near the surface, became less productive as the mine deepened, and finally failed alto-gether; the gold having thinned out, and its place being entirely taken by argentiferous galena. See Quarterly Review, Article, 'Siberia and California,' vol. lxxxvii. p. 410.

gravel, sand, and clay, derived from the wearing away of adjacent hard rocks, whether of aqueous or of igneous origin.

Having, in the year 1844, recently returned from the auriferous Ural Mountains, I had the advantage of examining the numerous specimens collected by my friend Count Strzelecki, along the eastern chain of Australia. Seeing the great similarity of the rocks of those two distant countries, I could have little difficulty in drawing a parallel between them; in doing which I was naturally struck by the circumstance that no gold 'had *yet* been found' in the Australian ridge, which I termed in anticipation the 'Cordillera.'*

* The announcement that 'no gold had yet been detected,' which was printed in my Presidential Discourse, Trans. Roy. Geog. Soc. 1844, is the clearest proof of my ignorance of a trace of the metal having been discovered by any one. In the last two years, however, facts have transpired, which were totally unknown to me when I ventured upon my comparison. Thus it appears, that Count Strzelecki himself discovered traces of gold in 1839; but on relating the fact to some friends and to the Governor of New South Wales Sir G. Gipps, secrecy was enjoined, and the Count never more reverted to the subject; not even in his own work, 1845. It now also appears, that the Rev. W. C. Clarke wrote to a friend in the colony (1841), mentioning that he had found gold ore; but this circumstance remained as much unknown to myself and all European men of science as the other. My views, whatever they may be worth, were therefore formed quite irrespectively of any such proceedings, as the following extract from a letter of my friend Count Strzelecki to myself, received when these pages are passing through the press, amply testifies:—'Nothing can give me greater pleasure and comfort at any time, than to bear my humble testimony to the inductive powers which you displayed on the occasion of your predictions in regard to the existence of gold in Australia; and consequently, I can affirm now, as I did and do whenever a necessity occurs, that I never mentioned my discovery or supposed discovery of Australian gold to you, prior to your papers on the subject, nor after their publication.'

A claim has recently been made in favour of Mr. G. Windsor Earl as one of the early indicators of Australian gold, which, with every respect for the accomplishments of that gentleman, seems to me to rest on no foundation. In June 1845, and therefore *after* both my Anniversary Discourses of May 1844—1845, in which the subject of the distribution of gold over the world was handled at some length, Mr. Windsor Earl communicated to the R. Geographical Society, a memoir on the general analogy between Australia and the Asiatic countries to the north of it ; (he having resided at Port Essington). Being led to believe, by the great geological authority Leopold von Buch (who was present on the occasion), that the insular tracts extending northwards from Torres Straits were chiefly volcanic, and not of the same mineral structure as any part of Eastern Australia with which we had then been made acquainted, I ventured (verbally) to suggest doubts as to the value of this portion only of Mr. Earl's memoir. But neither Leopold von Buch nor myself made any allusion to metalliferous productions, inasmuch as we only touched upon the broad outline of a general geological comparison; and no reference whatever having been made by Mr. Earl to *Australian gold*, no

Impressed with the conviction that gold would, sooner or later, be found in the great British colony, I learnt in 1846 with satisfaction that a specimen of the ore had been discovered. I thereupon encouraged the unemployed miners of Cornwall to emigrate and dig for gold, as they dug for tin in the gravel of their own district. These notices were, as far as I know, the first printed documents relating to Australian gold.

At that time, California, inhabited only by pastoral Indians and a few missionaries and Spanish herdsmen, was, it will be recollected, equally unknown to be auriferous. Its rich alluvial soil had not then been removed from the surface, and the accident at Sutter's Mill, in 1847, had not exposed the gold in the gravel and shingle beneath it. We can still better understand how this should have been the case with regard to vast tracts of Australia, where similar mineral constants' exist, but where, instead of a comparatively advanced people, like the Mexicans or Peruvians, a wretched race, incapable of appreciating the uses of the precious metals, had been for ages the sole inhabitants of a vast continent.

Unwilling to offer what must be a very imperfect epitome of the distribution of gold in Australia, I may, however, be permitted to say a few words on a subject to which I called the attention of my countrymen for several successive years previous to the opening out of the gold fields of that vast region.

The geological survey of the British colonies in Australia, to the advantages of which, in the opening out of *gold mines*, I directed the attention of Her Majesty's Government in the year 1848, or three years before their practical development, has been since ably carried into effect by very competent observers. To the general geological descriptions given by Mitchell, Strzelecki, Clarke and Jukes, voluminous details, published for

question respecting it was raised. I therefore repeat, that at that time no one, whether in Britain or the colonies, had *printed* any thing on the auriferous characters of the Australian rocks except myself, and I reiterate my conviction, that my memoirs of 1844, 1845, and 1846, are the earliest publications on the subject. See Trans. of the Roy. Geograph. Society, President's Discourses, 1844, 1845; and Trans. Roy. Geol. Soc. Cornwall, 1846.

the use of both Houses of Parliament, have been added within the last
three years, explanatory of the varied mineral condition under which gold
occurs in Australia.* Thus Mr. Stutchbury, the Government geologist,
gave us clear reports upon several tracts of New South Wales, after Mr.
Hargreaves had proved (1851) the value of the diggings. The Rev. W.
B. Clarke had, however, previously, or in 1847, publicly called the attention of the colonists of Sydney to the auriferous character of New South
Wales. This zealous geologist has since explored the largest range of
its gold-bearing lands over upwards of six degrees of latitude, or from the
Peel River on the north to the Australian Alps of Strzelecki on the south,
where the watershed or Cordillera, rising in Mount Kosciusko to 6,500
feet above the sea, trends south-eastwards into the province of Victoria.

From Mr. Clarke and other authors we learn, that the parallel which
I had drawn, in 1844, between the rocks of the Australian 'Cordillera'
and those of the Ural Mountains is well sustained. Just as in Siberia,
the greatest amount of gold is found in the heaps of debris, or old
alluvia, adjacent to certain intersections of the older slaty strata by
eruptive rocks, whether granites, porphyries, or greenstones. We may
also assume, that the chief auriferous slaty rocks of Australia are of the
same age as the central and older masses of the Ural, viz. Lower Silurian;
because, like them, they are overlaid by limestones, which contain Pentameri, Trilobites, and Corals, indicative of the Upper Silurian Group.
The last are followed by other strata, which are probably of Devonian age,
and then by true equivalents of the Carboniferous era.

Whilst the most prolific sources seem to have been the quartzose veinstones which traverse the older slates, we are further instructed, that in
Australia, as in the Ural Mountains, there are tracts wherein gold is
diffused in small and often imperceptible particles through the body of
certain granitic rocks†, specially those (according to Mr. Clarke) which are
hornblendic or syenitic; whilst the limestones in both countries are partially
auriferous. There is besides, not only this striking coincidence between the
Australian Mountains and the Ural, viz. that both have a main *meridian*

---

* See Papers relative to the recent Discovery of Gold in Australia, presented to both Houses
of Parliament, 1852—1853. For my own connection with this subject, see the Papers on the
same subject, presented August 16. 1853, p. 43.

† See Russia and the Ural Mountains, vol. i. p. 483., where it is stated, 'The fact is, then,
that though gold has frequently been, and is for the most part, formed in quartzose and other
veins which either have penetrated or been separated from the mass of the slate formation (and
of these the Ural affords countless examples), it has also been diffused in some tracts throughout the whole body of the rock, whether of igneous or of aqueous origin.'

direction ; but also the fact, that each chain is largely and profitably auriferous on one side of the watershed only ; in Russia on the eastern or
Siberian, and in Australia on the western slopes of the principal range.

Looking, then, at the Australian phenomena (including those of Victoria)
on a broad scale, there are no essential distinctions between them and the
mineral relations of gold in the Old World and America ; except that
certain slopes and hollows, notably those of Victoria*, have proved to be
much more lucrative than those of any known region, except parts of
California.

The great sources of wealth, as elsewhere, are depressions which have
been filled with debris from the mountains, whether coarse or fine.   The
sides of these depressions being once defined, and their bottoms being
easily ascertained, the period of their exhaustion may be estimated approxi-
mately.   It matters nothing to the statist whether this golden drift, or
accumulation of broken materials, was for the most part aggregated, as I
think, by causes now no longer in action, which powerfully abraded the
tops and sides of the hills (as explained in the previous pages, and particularly in the diagram, p. 449.), or whether the diurnal atmospheric
action alone for thousands of years has been the effective agent.   Both
causes have unquestionably contributed to spread out and make accessible
a material, the search after which in the solid rock, as I have already
stated, has hitherto usually been so profitless a speculation.

When the British public hear from our colonial authorities, that
16,000 square miles of a part of New South Wales are auriferous†, they

---

* See Memoir by Mr. Wathen, Quart. Journ. Geol. Soc. London, vol. ix. p. 74.   Mr. G. M.
Stephen, Vice-President of the Geological Society of Melbourne, recently arrived from Victoria,
has brought home instructive specimens of crystallized gold, and many small precious
stones from different parts of Australia, including topaz, chrysolite, garnet, zircon, tour-
maline, &c.   Besides some of these which also occur in the Ural Mountains (including a rare
diamond), the gold fields of Victoria have afforded a mineral unknown in Russia, i. e. tin, which
is found not only in the form of sand, as now worked by Messrs. C. Terry and Co., but also in
small lumps, and highly coloured crystals, described by Mr. G. M. Stephen in a Report published
at Melbourne.   The reader who wishes to catch a general view of these the most important
of the Australian gold diggings, and acquire an insight into the statistics and social condition
of the wonderful colony of Victoria, should peruse the recent work of Mr. Westgarth.

† Report of the Rev. W. Clarke to the Governor of New South Wales.   Parliamentary Papers,
presented Feb. 28. 1853.   Sir C. Fitzroy's Dispatch, p. 62.   See Wyld's Memoir and Maps ;
and Arrowsmith's new maps of the auriferous regions of Australia, as compiled from the
Government Surveys, on which (particularly that which is printed for the use of the Houses of
Parliament) the localities at or near which any appreciable amount of gold has been detected
are coloured yellow.   In stating that the gold from California and Australia has been chiefly

must not suppose that the geologist, who thus reports, intends to say, that profitable researches could be carried out over that vast area. The truly rich sites, he well knows, are very small as compared with the entire country. Again, in those districts where, by degradation for ages, fine particles of gold have been derived from the granitic or other rocks (the chief materials of the matrix perishing, but the gold remaining), we are not to infer, that any human endeavours to break up and grind down mountain masses of such original rock for the purpose of extracting the metal it contains, (seldom visible to the human eye) would be remunerative, except in some remarkable cases like those before alluded to (p. 437.) At all events, the extent to which such appliances can extend, will ere long, be ascertained through the spirit and intelligence of our countrymen : and in the mean time it is satisfactory to reflect, that a basis of all such researches and inquiries has been provided by the good geographical survey of a large portion of this new country, executed by Mitchell and his associates. On that important groundwork, geologists may now record all subsequent observations, respecting a region hitherto almost exclusively pastoral, but which has suddenly become a great and populous centre of commerce through the development of its very rich fields of gold.

Conclusion. — Notwithstanding the preceding sketch, it would ill become any geologist who throws his eye over the gold map of the world prepared by Adolf Erman, to attempt to estimate, at this day, the amount of gold which remains, like that of Australia, undetected in vast regions of the earth, as yet unknown even to geographers ; still less to speculate upon the relative proportions of it in such countries. At the same time, the broad features of the case in all known lands may be appealed to, to check extravagant fears and apprehensions re-

---

derived from drifted and broken materials, as in the Ural Mountains, I am quite aware that a small portion of ore has been obtained from veinstones which traverse the schists *in situ.* But, as few shafts have yet been sunk deep into those rocks (even in California), the speculators have yet to learn whether, when the richer surface strings of the metal have been worked out, the lower portions of the veins in those tracts will prove exceptions to the prevailing rule. As far as experience can teach, it is only where the mountains have been shattered, and great mounds of their debris thrown down in former ages upon their sides (as represented in the diagram, p. 449.), that gold can be extracted by human labour to *any considerable profit.* The chief exceptions, as in the Brazils, are where the deep mines are in a soft matrix, and are worked by slaves receiving very low wages. The *caveat,* however, which is inserted in the note * p. 437., must also be considered.

specting an excessive production of the ore. For, we can trace the boundaries, rude as they may be, of a metal ever destined to remain precious on account of those limits in position, breadth, and depth by which it is circumscribed in Nature's bank. Let it be borne in mind, that whilst gold has scarcely ever been found, and never in any quantity, in the secondary and tertiary rocks which occupy so large a portion of the surface, mines sunk down into the solid rocks where it does occur, have hitherto, with rare exceptions, proved remunerative;—and, when they are so, it is only in those cases where the rocks are soft, or the price of labour low. Further, it has been well ascertained, whatever may have been the agency by which this impregnation was effected, that the metal has been chiefly accumulated towards the surface of the rocks; and then by the abrasion and dispersion of their *superficial* parts, the richest golden materials have been spread out, in limited patches, and generally near the bottom of basin-shaped accumulations of detritus.

Now, as every heap of these broken auriferous materials in foreign lands, has as well defined a base as each gravel pit of our own country, it is quite certain, that hollows so occupied, whether in California or Australia, must be dug out and exhausted, in a greater or less period. In fact, all similar deposits in the old or new world have had their gold abstracted from heaps whose areas have been traced, and whose bottoms were reached. Not proceeding beyond the evidences registered in the stone-book of Nature, it may therefore be affirmed, that the period of such exhaustion in each country (for the deposits are much shallower in some tracts than in others) will, in great measure, depend on the amount of population and the activity of the workmen employed in each locality. Anglo-Saxon energy, for example, as applied in California and Australia, may in a few years accomplish results, which could only have been attained in centuries by a scanty and lazy indigenous population; and thus the

*present large flow of gold into Europe from such tracts, will, in my
opinion, begin to diminish within a comparatively short period.*

In defining the general character of the most productive auriferous
rocks, the geologist must, however, necessarily admit a considerable
number of exceptions to any prevailing rule. For, whilst the
chemist, as before said, has recently detected traces of gold in lead
and copper ores,—a discovery of considerable interest, doubtless, in
regard to the theory of the origin of the precious metal, — the researches of the miner teach us, that, in any auriferous region where
certain quartzose lodes are surcharged with ores of iron, particularly the oxides and sulphurets, there *some* amount of gold will
probably be found. Again, the diffusion or dissemination of small
particles of gold throughout the body of various rocks both of
igneous and aqueous origin is, as before said, a phenomenon dwelt
upon by certain authors. Humboldt, indeed, asserted long since, that
in Guiana ' gold, like tin, is sometimes disseminated in an almost
imperceptible manner in the mass itself of the granitic rocks without
the ramification or interlacing of any small veins.'* In Mexico
the gold mine of Guadalupe y Calvo, above alluded to †, was in porphyry. In Australia (districts of Braidwood, and others south of
Sydney), a peculiar variety of felspathic granite is described by Mr.
Clarke as being permeated by small particles of gold; whilst in
Siberia, Hoffman had some years before spoken of its distribution in
such minute quantities in clay-slate, that it was only by pounding
up large lumps of the rock that any perceptible quantity could be
extracted.‡

In all regions, therefore, where such rocks occur, we may find
gold either in the coarse debris or the fine alluvia resulting from
their decomposition. Felspar and quartz being their chief component parts, we can easily imagine how their former destruction

---

* Voyages, vol. ii. p. 238.   † Page 449.
‡ Reise nach dem Goldwäschen Ost Sibiriens, by Ernst Hofmann, St. Petersburg, 1847.

on a great scale would leave as a residue large heaps of that pipe-clay (the decomposed felspar), or those gritty pebbles (the abraded quartz), which with the accompanying ores of iron (particularly the black magnetic oxide) are so frequently the gold-bearing matrices in the drift of auriferous countries. But whilst it is an admitted fact, that gold has sometimes been so diffused in minute and imperceptible particles in certain rocks, we have yet to learn whether such diffusion extends far downwards into the body of any mountain. Even if it be so, the extraction of ore so diffused might, if the rocks were hard, prove too costly an operation. At all events, the indisputable fact is, that the *chief quantities of gold*, including all the considerable lumps and pepitas, having been originally imbedded in the upper parts of the veinstones, have been broken up and transported with the debris of the mountain-tops into slopes and adjacent valleys.

In conclusion, let me express my opinion, that the fear that gold may be greatly depreciated in value relatively to silver — a fear which may have seized upon the minds of some of my readers — is unwarranted by the data registered in the crust of the earth. Gold is, after all, by far the most restricted — in its native distribution — of the precious metals. Silver and argentiferous lead, on the contrary, expand so largely downwards into the bowels of the rocks, as to lead us to believe, that they must yield enormous profits to the skilful miner for ages to come; and the more so in proportion as better machinery and new inventions shall lessen the difficulty of subterranean mining.* It may indeed well be doubted whether the quantities both of gold and silver, procured from regions unknown to our progenitors, will prove more than sufficient to meet the exigencies of an enormously increased population and our augmenting commerce

---

* A recent report from Col. Lloyd, H. M. Chargé d'Affaires in Bolivia, communicated through H. R. H. Prince Albert to the Royal Geographical Society, shows to what an enormous extent silver may yet be extracted from Copiapò and other South American mines. This was, indeed, the view taken long ago by Humboldt.

and luxury.   But this is not a theme for a geologist; and I would
simply say, that Providence seems to have originally adjusted the
relative value of these two precious metals, and that their relations,
having remained the same for ages, will long survive all theories.
Modern science, instead of contradicting, only confirms the truth of
the aphorism of the patriarch Job, which thus shadowed forth the
downward persistence of the one and the superficial distribution of
the other: — ' Surely there is *a vein* for the silver. . . . . The earth
hath *dust of gold.*'

* The Book of Job, chap. 28.

P. S. — Since this chapter was in type, Mr. A. R. C. Selwyn, of the Geological Survey, whose
labours were so successful in North Wales (see p. 28.), having recently gone to Victoria in the
capacity of Government Geologist, has sent home an excellent sketch, accompanied by map and
sections, of the structure of the country around Mount Alexander, distinguishing the true
stratification of the sedimentary rocks (Lower Silurian?), whether micaceous felspathic sandstones, flagstones, or clay slates, from an intense and vertical slaty cleavage which runs through
them from north to south, and with which the course of the auriferous quartz veins and the
strike joints are coincident. He also shows how the granite has metamorphosed the contiguous
slaty rocks, and has injected ' elvan' dykes into them.   Mr. Selwyn coincides with Mr. Wathen
(p. 453.), and others, in stating, that the auriferous drift is local, and although lying at various
heights is usually richest at or near the bottom of these accumulations.

In reference to New Zealand, the Governor, Sir G. Grey, has communicated to myself a letter
addressed to him by Mr. C. Heaphy, giving an account of the gold diggings (not yet extensive)
at Coromandel, on the east coast of the Northern Island, where the chief rocks are quartzose
and granitic.

Lastly, Mr. J. S. Wilson, who after a residence in South Australia, passed three years as a gold
miner in the Sierra Nevada of California, has communicated a memoir, on the auriferous rocks
of that region, to the Geological Society of London.   The views resulting from his own observation and experience, confirm decisively the opinions stated throughout the text, in affirming
the fact of the downward impoverishment of gold-bearing quartz veins, and in demonstrating
that the richest produce is essentially derived from loose superficial debris, piled up on the
mountain sides and slopes, or in ravines, and at various and considerable altitudes above the
sea.   The quicksilver (cinnabar) which is extracted from the coast range, occurs in veinstones,
in clay slate (Silurian?), as in the Sierra Morena of Spain.   These memoirs will appear in the
Journal of the Geological Society of London.

In this chapter I have necessarily abstained from alluding to many tracts more or less
auriferous, but not *rich* in produce, which come under the same laws of distribution as those
described in the text.   Such, for example, are the phenomena of the southern provinces of
the United States (S. Carolina, &c.), and also of the British province of Canada, whence a
certain amount of gold has been derived from metamorphic rocks, which, according to Mr.
Logan, are probably of Lower Silurian age.

CHAPTER XVIII.

CONCLUSION.

RECAPITULATION — GENERAL VIEW OF THE SUCCESSION OF LIFE FROM A BEGINNING
AS BASED ON POSITIVE OBSERVATION — THE PROGRESS OF CREATION — THEORETICAL
SPECULATIONS DISTINGUISHED FROM ABSOLUTE GEOLOGICAL RESULTS.

REVERTING to the main object of this work, — the history of the
primeval rocks, — let us now see what inferences may be drawn from
data established by the researches of the geologist. Passing rapidly
over the earliest stages of the planet, which are necessarily involved
in obscurity, our sketch of ancient nature began with the first
attainable evidences of the formation of sediments composed of mud,
sand, and pebbles. It was shown, that the lowest accessible of these
deposits, though of enormous dimensions, and occasionally less altered
than strata formed after them, are almost entirely *azoic,* or void of
traces of inhabitants of the seas in which they were accumulated.
One solitary genus of zoophytes has been alone detected in such
bottom rocks; the heat of the surface during those earlier periods
having been, it is supposed, adverse to life.

Proofs were then adduced to demonstrate that in the next formations, scarcely differing at all in mineral character from those which
preceded them, observers in various regions had detected clear and
unmistakeable signs of a contemporaneous appearance of animal life,
as shown by the presence of a few genera of crustaceans, mollusks,
and zoophytes, occupying layers of similar date in the crust of
the earth. Proceeding upwards from this protozoic zone, wherein
organic remains are comparatively rare, we then ascended to other
sediments, in which, throughout nearly all latitudes, we recognize a

copious distribution of submarine creatures, resembling each other very nearly, though imbedded in rocks now separated by wide seas, and often raised up to the summits of high mountains. Examining all the strata exposed to view, that were formed during the first long natural epoch of similar life termed Silurian, we found that the successive deposits were charged with a great variety of forms, — of the trilobite a peculiar crustacean, — of the orthoceratite, the earliest chambered shell,—as well as with numerous exquisitely formed mollusks, crinoids, and zoophytes; the genus graptolite of the latter class being exclusively found in these Silurian rocks. In short, my contemporaries have assembled from those ancient and now desiccated marine sediments or repositories of primeval creatures, examples of every group of purely aquatic animals, save fishes. The multiplied researches of the last twenty years have failed to detect the trace of a fish, amid the multitudes of all other marine beings in the various sediments which constitute the chief mass of the Silurian rocks. Of these, though they are the lowest in the scale of the great division *vertebrata*, we are unable to perceive a vestige until we reach the highest zone of the Upper Silurian, and are about to enter upon the Devonian period. Even on that horizon, the minute fossil fishes, long ago noticed by myself, are exceedingly scarce, and none have since been found in strata of higher antiquity. In fact, the few fragments of cartilaginous ichthyolites of the highest band of Silurian rock, still remain the most ancient known beings of their class. (See Pl. 35.)

Looking, therefore, at the Silurian System as a whole, and judging from the collection of facts gathered from all quarters of the globe, we know, that its chief deposits (certainly all the lower and most extensive) were formed during a long period, in which, while the sea abounded with countless invertebrate animals, no marine vertebrata had been called into existence. The Silurian (except at its close) was, therefore, a series in which there appeared no example of that bony framework of completed vertebræ, from which, as approaching to the vertebrate archetype, the comparative anatomist* traces the rise of creative power up to the formation of man.

Whether, therefore, the term of 'progressive,' or of that of 'successive,' be applied to such acts of creation, my object is simply to show, upon clear and general evidence, that there was a long period in the history of the world, wherein no vertebrated animal lived. In this sense, the appearance of the first recognizable fossil fishes, is as decisive a proof of a new and distinct creation, as that of the placing of man upon the terrestrial surface, at the end of the long series of animals which characterize the younger geological periods.

Nor have we been able to disinter from the older strata of this long period of invertebrate life, any distinct fragments of land plants. But just in the same stratum wherein the few earliest small fishes have been detected, there also have we observed the first appearances of a diminutive, yet highly organized, tree vegetation. (See Appendix C.)

If it be granted that the position of the earliest recognizable vertebrata is good positive evidence on which to argue, still it might be contended that such forms may at a future period be found in lower strata.   In this work, however, I reason only on known data.   Nor is it on this testimony alone, strong, clear, and universal as it is, that my view is sustained; for as soon as we pass into the formation immediately overlying, and quit the zone wherein the first few small fishes are to be detected, we are furnished with collateral

---

* See Owen on the Homology of the Vertebrate Skeleton: Reports, Brit. Assoc. Adv. Science, 1846, p. 169. The general reader will find a powerful essay, embodying the opinions of the same high authority, on the proofs of a progression in creation, Quarterly Review, 1851, p. 412, *et seq.* The arguments there employed have been strengthened by subsequent discoveries alluded to in this volume. I would also specially refer the reader to Professor Sedgwick's 'Discourse on the Studies of the University of Cambridge,' for a masterly and eloquent illustration of several of the views which are here advocated.

proofs that this was the earliest great step in a progressive order of creation. In the following or Devonian period, we are surrounded by a profusion of larger fossil fishes, with vertebræ for the most part very imperfectly ossified, and with dermal skeletons of very singular forms; all differing vastly from anything of their class in existing nature. These fishes were thus clearly added to the other forms of marine life. Again, in this Devonian era, we are presented with well defined land plants, also of much larger dimensions than the very rare specimens in the uppermost Silurian; while, towards the close of the period, we meet with an air-breathing reptile. The little Telerpeton had groves of tree ferns or lycopodiaceous plants, and even of Coniferæ, amid the roots of which he could nestle. (See p. 254.)

Just as the introduction of cartilaginous fishes (*Onchus* e. g.) is barely traceable at the close of the long Silurian era, so becoming soon afterwards more abundant, they are associated in all younger formations with true osseous fishes, whose remains are found intermixed with the other exuviæ of the sea. Putting aside, therefore, theory, and judging solely from positive observation, we may fairly infer, first, that during very long epochs the seas were unoccupied by any kind of fishes; secondly, that the earliest discoverable creatures of this class had an internal framework almost incapable of fossilization, and so left in the strata their teeth and dermal skeletons only; and thirdly, that in the succeeding period, the oldest fishes having bony vertebræ make their scanty appearance, but become numerous in the overlying deposits. Are not these absolute data of the geologist clear signs of a progress in creation?

In like manner, there is a progress in the productions of the land; the great Carboniferous period being marked by the first copious and universally abundant terrestrial flora, the prelude to which had appeared in the foregoing Devonian epoch. This earliest luxuriant tree vegetation, the pabulum of our great coal-fields, is also specially remarkable for its spread over many latitudes and longitudes; and together with it occur the *same common species* of marine shells, all indicating a more or less equable climate from polar to inter-tropical regions; a phenomenon wholly at variance with the present distribution of animal or vegetable life over the surface of the planet.

Lastly, while the Permian era was distinguished by the disappearance of the greater number of the primeval types, and by essential modifications of those which remained, it still bore a strong resemblance, through its plants and animals, to the Carboniferous period ; whilst, in unison with all the great facts elicited by our survey of the older strata, it was marked by the appearance of an animal of a higher grade than any one in the foregone eras, — a large thecodont reptile — allied (according to Owen) to the living Monitor.

In speaking of the Silurian, Devonian, Carboniferous, and Permian rocks, let me however explain, that whilst each of the three latter groups occupy wide spaces in certain regions, no one of them is of equal value with the Silurian, in representing time or the succession of animal life in the crust of the globe. When the Silurian system was divided into lower and upper parts, our acquaintance with younger formations simply sufficed to show a complete distinction between its animal remains as a whole and those of the Carboniferous rocks, from which it is separated by the thick accumulations of the Old Red Sandstone. At that period, the shelly, slaty rocks of Devonshire were not known to be the equivalents of such Old Red Sandstone; still less had the relations and fossil contents of the strata now called Permian been ascertained. Judging from the fossils then collected, it was believed, that the Lower Silurian contained organic remains very distinct from those of the Upper Silurian; and yet the two groups were united in a system, because they were characterized throughout by a common *facies*. This so-called system was, in short, typified by a profusion of Trilobites and Graptolites, with Orthides and Pentameri of a type wholly unknown in the Carboniferous

rocks. And whilst fishes were seen to exist in the intermediate
masses of Old Red Sandstone, no traces of them could be detected
below the very uppermost zone of the Silurian rocks. Nineteen
years have elapsed, and, after the most vigilant researches in various
regions of both hemispheres, these great features remain the same
as when first indicated. The labours, however, of those who followed me, have infinitely more sustained the unity of that system*;
for its lower and upper divisions are now proved to be connected, not
only by such generic types and analogous forms, but further by the
community of a very considerable number of identical bodies.

In a broad classification of primeval life, one eminent naturalist†
views the Devonian, Carboniferous, and Permian rocks as simply
the Upper Palæozoic; the Silurian rocks constituting the Lower
Palæozoic. But, whether this ancient series be divided into double
or triple classes (some palæontologists preferring to hold the
Devonian as a separate and intermediate type), the result of the
researches of the numerous authors appealed to in this volume has
unquestionably justified the application of the term 'system,' to the
Silurian rocks.

At the close of the Permian era, an infinitely greater change took
place in life, than that which marked the ascent from the Silurian
system to the overlying groups. The earlier races then disappeared
(at least all the species), and were replaced by an entirely new
creation, the generic types of which were continued through those
long epochs which geologists term secondary or mesozoic (the medieval age of extinct beings). In these, again, the reader will learn,
by consulting the works of many writers, how one formation fol-

---

* See the works of John Phillips, E. de Verneuil, Edward Forbes, Joachim
Barrande, James De C. Sowerby, James Hall, J. W. Salter, D. Sharpe, T.
Davidson, John Morris. The most recent researches in Britain give the large
number of nearly 100 species of fossils, common to the Lower and Upper Silurian.
(See also the Preface, and Table A. in the Appendix.)

† Edward Forbes.

lowed another, each characterized by different creatures; many of them, however, exhibiting near their downward and upward limits certain fossils which link on one reign of life to another.

In surveying the whole series of formations, the practical geologist is fully impressed with the conviction, that there has, at all periods, subsisted a very intimate connection between the existence, or at all events the preservation of animals, and the media in which they have been fossilized. The chief seat of former life in each geological epoch, is often marked by a calcareous mass, mostly in a central part, towards which the animals increase from below, and whence they diminish upwards. Thus, the Llandeilo limestone of the Lower Silurian and the Wenlock of the Upper Silurian, are respectively centres of animalization of each of those groups. In like manner, the Eifel limestone is the truest index of the Devonian, the Mountain limestone of the Carboniferous, and the Zechstein or English Magnesian limestone of the Permian. Throughout the secondary rocks the same law prevails more or less; and wherever the typical limestone of a natural group is absent, there we perceive the deposits to be ill-characterized by organic remains. For example, the Trias, so rich in fossil contents when its great calcareous centre the Muschelkalk is present, as in Germany and France, is a miserably sterile formation in Britain, where, as in our New Red Sandstone, no such limestone exists.*

Whilst it is beside my present aim to enter upon descriptions of such secondary deposits, still less of those called tertiary, which intervened between them and the sediments of the present day, we may still cast an eye over the general order admitted by geologists, to

---

* The observation in the text does not apply to the oolitic series of England, throughout which lime is diffused at many stages; nor to many limestones which may have been formed by deposit from chemical solution. For a full explanation of the origin of such deposits and the associated phenomena, the reader is referred to the instructive work by Sir H. T. De la Beche, 'The Geological Observer,' 1853.

see how it harmonizes with what has been related of the succession
of animals belonging to the older rocks, as proving a succession from
lower to higher grades of life.

Proceeding into the secondary strata, which often constitute vast
mountain masses, we pass first through the Trias, then traverse
the Lias, and afterwards the long series of the Oolitic or Jurassic
formations, all of which are charged with many animal and vegetable
remains, and laden with a great profusion of curious and large Saurians,
very unlike the lizards which preceded them. Besides numerous land
plants, insects now become abundant,—a tribe which began to appear
in the Carboniferous era, or in the first great forests; and with
these are found bones of that large-winged reptile, the 'pterodactyle'
of geologists. But still, surrounded as we are in these secondary
strata by the spoils of the land, we have to journey through nearly
a half of them, before we obtain any other evidences of the existence of mammalia, than two or three teeth of a carnivorous animal
said to be found in a bone bed of the Trias* (but which may possibly be the bottom bed of the Lias), and the scarce fragments, in the
Stonesfield Oolite, of the Amphitherium, formerly named Didel-
phis Bucklandi by Broderip. This last-mentioned creature, which is
allied to living marsupials, though discovered many years ago, has
had no companions added to it by the hordes of collectors who have
sought for them, except the small but highly curious Phascolotherium
of Owen.†

* See Lyell's Elements of Geology, 4th edit. p. 13.

† See Owen, British Foss. Mammalia, p. 31. On the point, indeed, of the
absence of mammalia in the older rocks, I may be allowed to quote, if not the
words, at least the sentiments, of the same eminent contemporary. ' Had
mammalia existed in the same number and variety in the ancient forests,
that have contributed to the coal strata, as in the actual woods and swamps
of the warmer parts of the globe, — had armadillos and ant-eaters been then
created to feed on the insects, sloths on the leaves, monkeys on the fruits of
the coal plants, as they do now in the Brazilian forests where the mammals pre-

After this, we have plentiful evidences of successive shore deposits at many periods, with abundant examples of sediments formed in lakes and rivers (an order of things of which vestiges were first apparent towards the close of the primeval period), and with continual signs of adjacent lands affording numerous forms of plants and insects. The Wealden, of such vast thickness, is, indeed, exclusively tenanted by land and fresh-water remains; and yet, though its strata are full of plants and the gigantic fossil lizards of the period, not one bone of a mammal has been exhumed from them; while bones of birds make their first appearance only in the greensand and chalk.

We have to work through the whole cretaceous series and its prolific fauna, to take leave, in short, of the secondary rocks, and enter upon the tertiary epochs, before such remains are at all plentiful. Then, for the first time in this incalculably long series of formations, we have before us, on all sides, the bones of the higher order of mammalia; and these having been drifted from adjacent lands, are constantly associated with the exuviæ of marine creatures, which, though of classes known in the earlier formations, are entirely different in species.

Animals of every sort thenceforward abound in each succeeding formation; and exhibit an increasing quantity and variety of both sea and land mammalia as we approach the superficial accumulations. In the last are entombed the bones of gigantic quadrupeds,— quadrupeds which once inhabited our present continents, and which must have required for their sustenance a range over lands as extensive as those now occupied by man and his associates.

ponderate over reptiles, we might have expected the first evidences of an air-breathing vertebrate animal to have been a mammalian!' Quarterly Review, 1851, p. 423.

Let the reader dwell on these remarkable facts which the labours of geologists have elicited in the last fifty years. Let him view them progressively and in the order indicated by Nature herself. Let him execute a patient survey from the lower deposits upwards, and he will find everywhere a succession of vertebrated creatures, rising from lower to higher organizations,—a doctrine first promulgated by the illustrious Cuvier, but from much less perfect data than we now possess.* Guided by facts only, he will everywhere recognize signs of a similar primordial life registered on the same lower tablets of stone; and thence examining upwards, he will admit the proofs of the advancing steps above indicated.

These views of the successive creation of different races are, it is true, mainly based upon the progressive rise in the scale of the vertebrate sub-kingdom. Of that class we can best register a beginning, and thence trace the distinct progress which sustains our general inference. When we turn to other and inferior classes, crustaceans, mollusks and corals, naturalists assure us (and I willingly subscribe to their dicta) that many of the earlier leading groups were quite as highly organized as any of their representatives in subsequent ages or at the present day.

Nay, even whilst I close this volume, one of our leaders in the philosophy of Natural History has propounded a new theory, which he terms 'The Manifestation of the relation of Polarity in the Distribution of Generic Types in Time.'†

Regarding all the epochs after the Trias to be combinedly equivalent to those preceding (or the palæozoic), he maintains that the

---

* The palæozoic or primeval fossils were necessarily little known to that great comparative anatomist, who drew his conclusions from data within the scope of the knowledge of his day; *i.e.* from the higher orders of vertebrata which appear in the more recent epochs only.

† Address to the Geological Society of London, Feb. 17. 1854, by Professor E. Forbes, President.

development of types of life, as manifested by generic combinations during the long secondary and tertiary epochs, styled by him 'Neozoic,' is in opposition to, or in a relation of polarity with, the comparable phenomenon during the 'Palæozoic' period. This is indicated by the concentration of the maximum of what he terms 'generic ideas,' in some of the earlier stages of the older, and towards the termination of the more recent period; whilst, as we approach to the interlacing point, (i.e. the Trias) there is a poverty in the origination of generic types. This polarity, he contends, is evident in both directions within each great group of *invertebrate animals*, as well as in the assemblage of *plants;* but in the vertebrata, the main direction of development of generic ideas is, he admits, towards the newer or Neozoic pole.

This ingenious speculation does not interfere with my chief argument, as based on facts which indicate great periods wherein few or no animals existed, followed by a long, invertebrate epoch. At the same time, let me repeat, that the earliest signs of living things, announcing as they do a high complexity of organization, entirely exclude the hypothesis of a *transmutation* from lower to higher grades of being. The first fiat of Creation which went forth, doubtlessly ensured the perfect adaptation of animals to the surrounding media; and thus, whilst the geologist recognizes a beginning, he can see in the innumerable facets of the eye of the earliest crustacean*, the same evidences of Omniscience as in the completion of the vertebrate form.

Yet, however they admit the facts, some of my speculative contemporaries think that they can so explain them, as not to justify the inference of progressive creations. They suppose, that nearly all the strata of date antecedent to those in which the first signs of life have

---

* See Dr. Buckland's clear and beautiful illustration of this phenomenon in his Bridgwater Treatise.

been detected, are often in so crystalline a state, that if they originally
contained remains of animals, the traces of them must have been obliterated by changes since effected in the structure of the rock. Now,
if this supposition had been supported by the researches of late
years, we must doubtless have admitted that the origin of life in the
globe was buried in a hopeless obscurity. But the hypothesis has
been set aside, as before explained, by the fact of the existence
of deposits many thousands of feet thick, and scarcely at all
altered, which, made up of sand, mud, and pebbles, constitute the
very foundations of the fossil bearing strata. In these huge lower
sediments a zoophyte only has been detected; but immediately above
them, in various and distant countries, we perceive the oldest known
small group of animals. If the opposing argument had been derived
from the evidences collected in one region only, it might have been
suggested, that as the same formation which is barren in life
over one district, teems with signs of it in another, so the infra-
Silurian or bottom rocks (the Cambrian of the Geological Surveyors)
may still prove to be fossiliferous. But, even if a few types should
hereafter be discovered in those lower strata, the reasoning would in
no wise change its character, if such infra-Silurian fossils were
not found to pertain to higher forms of life, — a result which would
be in manifest contradiction to all the ascertained facts respecting
geological succession. Nor can we allow this hypothesis, founded on
an exceptional possibility, to countervail the universal data that have
been registered, and which demonstrate in many countries the unfossiliferous character of the lowest sedimentary formations.

Again, when the explorer of the older formations produces his
specimens of fossils from various parts of the world, to show that the
mass of the Silurian rocks contains all classes of marine life with the
exception of fishes, his antagonist might reply, that gelatinous fishes,
void of backbones (like the solitary little Amphioxus now living),
may have been the only creatures of their class which swarm in the

broad seas then prevailing: and if so, that no traces of them could exist; their boneless bodies perishing and leaving no sign of their former existence. As an old student of Nature's works, I cannot allow this bare possibility to be placed in opposition to the very numerous and well recorded facts, which announce the perfection of all the other classes of the ancient submarine kingdom. If thousands of invertebrate animals have left their coverings behind, is it rational to suppose, that *every form* of the great class of fishes should be wanting in that framework which, whether consisting of dermal plates or of bony vertebræ, characterizes them in the strata of all succeeding epochs? Nay more, we see that in this same long period in which no traces of fishes appear, there specially prevailed a superabundance of cephalopods; and as creatures of that structure are well known to be carnivorous, little doubt can be entertained, that they acted the part of fishes, and were the scavengers of the Silurian seas.

Another hypothesis which has been advanced in opposition to the mass of positive evidence is, that although such earlier rocks are void of ichthyolites, the sediments *may* have all been formed in limited zones around the earth; just as it is believed by some naturalists, that there are seas subject to certain currents and conditions of the bottom in which no fishes are now living. But here, again, the application of such a theory is, I would suggest, still more negatived by the facts adduced. Silurian rocks similar in structure, and containing the same organic remains, are not confined to any one segment of the earth's surface however broad, but are largely developed in nearly all known regions. The argument is, therefore, untenable in face of the knowledge we have acquired, that amidst a profusion of all the other forms of aquatic life, fishes only are absent from strata of this early age.

This prevalence of a widely spread, primeval ocean, and of a surface which had not yet been subjected to those innumerable variations of outline which have since changed and modified the different climates of the earth, when connected with a belief in the former greater radiation of heat from its interior, are the chief data required to satisfy us, that physical conditions then prevailed, with which the nature and extensive spread of the earlier groups of animals are in harmony.

Admitting that in the remote sedimentary periods, large areas of land (though probably of no very great altitude), as well as vast rivers, must have existed as sources of the enormous primeval deposits, we may still well believe that such lands were widely separated by seas; and that hence we ought necessarily to meet with a smaller number of shore animals and a greater number of oceanic forms. Taking advantage of this simple fact, some persons have suggested, that we may have as yet discovered only the deep-sea products of the Silurian period; and that when the true edges of its lands come to be detected, we may then find plants and many creatures now unknown to us. To this I confidently reply, that many proofs have already been adduced of lands which were contiguous to the marine Silurians, both Lower and Upper. Innumerable pebble beds, coral reefs, and trails of animals that crawled upon the mud, are the principal evidences required, and to these I will presently revert.

It has also been said, that the great number of floating shells or cephalopods, particularly the Orthoceratites, which abounded in the Silurian era, are in themselves indicators of deep seas, remote from land, into which, therefore, terrestrial spoils were little likely to be transported. Now, the Silurian chambered shells may, indeed, have required a certain depth of water, and yet many of them, like the Nautilus Pompilius of the present day, may have lived at no great distance from the land of that time. For, in the succeeding Carboniferous period, when the mountain limestone was formed, we find no lack of floating shells, and other mollusca equally characteristic of certain depths of water, and yet these are associated with

abundance of plants which were drifted from land; simply, as I
would say, because the earth then bore an arboreal or tree vegetation,
whereof we can discover no traces in the Silurian time, notwithstanding the certain existence of contiguous lands and of large streams
at that early period.   So is it, indeed, in every subsequent formation
of the geological series.   Thus, in the secondary deposits of the Lias
and Oolites, we meet with an equal, if not a greater, number of
floating shells than in the Silurian.   The Ammonites and Belemnites
of that period, requiring the same depth of water, have taken the
place of the Orthoceratites; and yet associated with them, we have
everywhere proofs of the proximity of the land, in the abundance of
fossil plants and wood derived from *terra firma,* doubtless then much
more extensive and diversified than in the earliest times, and also
clothed with a rich vegetation.

If the old continents and islands, which existed during the accumulation of the marine Silurian deposits had borne large trees, the
numerous researches of geologists in all quarters of the globe must
have brought to light some signs of them.   For, whilst we know
that there are rocks of considerable extent, which, from the fine nature
of their materials, may probably have been deposited in an ocean at
some distance from a shore (though we have as yet little or no
evidence as to the accumulation of sediment in deep seas, where no
currents prevail); there are, on the other hand, many Silurian districts
of the Old and New World, where the form and structure of the
deposits bespeak the action of waves and surge, and where the imbedded seaweeds, zoophytes, and other remains, compel us to adopt the
same view.   And if the primeval fauna does afford fewer spiral univalve shells, than are seen among the animals of the laminarian zones
of our modern seas, we may suggest that shore lines, as we understand them, must have been much less numerous in primeval epochs
than at the present day; now that the surface has been diversified
by lofty dividing ridges on the land and corresponding depressions

in the ocean. With this important reservation, we do, however, obtain as many of those signs of shores as we can expect to find in the earlier deposits.

Take, for example, the illustrations of this point furnished by the American geologists, from a very wide extent of their country, where the strata are nearly horizontal, and where, without any ambiguity, our kinsmen have traced life downwards in the successive crusts of the earth, to the same primordial zone as their contemporaries have done in Britain*, Scandinavia, and Bohemia. The Americans have evidences in their lowest Silurian beds of *numerous trails or tracks of animals,* whether crustaceans or gasteropods, which moved over a film of mud or sand formed by one tide before another covered the impressions, and left them as proofs to future ages of layers which were deposited on the shores and edges of former lands. Again, in other Silurian beds of the far West, there exists the same abundance of coral reefs as in Britain, and the still stronger evidence of pebbly shores, which, though they must have been beaten by waves, never contain the trace of a land plant. Why, therefore, wander from such plain facts into the region of theory? And why not admit what is, indeed, in accordance with all we have observed, that the very long Silurian era had nearly passed away before trees grew upon the land or fishes swam in the waters?

In the fundamental facts described in this volume, we cannot, therefore, but recognize arrangements, which, though perfect as respected all truly primeval creatures and plants, were essentially different from those of our own time. For, if the then existing continents or islands had borne trees, some fragments of them must have been transported into adjacent estuaries, and mixed in the mud and sand, like the vegetables of every subsequent epoch, by the

---

* Even after these pages were written, Mr. Salter has described a trail of the *Hymenocaris vermicauda,* in the Lingula Flags of North Wales (Lowest Silurian of the Geological Survey). See Quart. Jour. Geol. Soc. Lond. (May, 1854.)

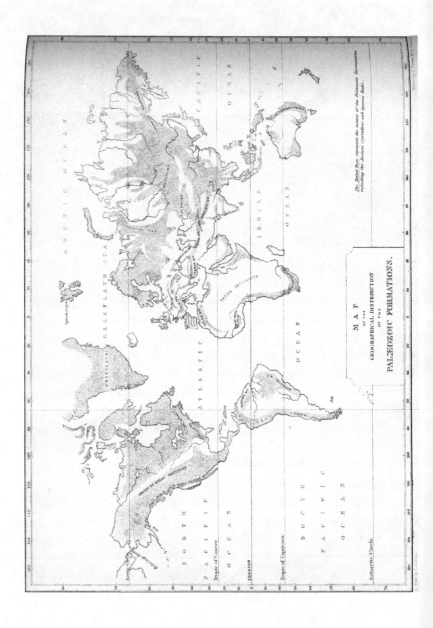

MAP
OF THE
GEOGRAPHICAL DISTRIBUTION
OF THE
PALÆOZOIC FORMATIONS.

The material originally positioned here is too large for reproduction in this reissue. A PDF can be downloaded from the web address given on page iv of this book, by clicking on 'Resources Available'.

agency of those great streams, of whose mechanical power we have such decisive proofs. The Silurian rocks extend over areas as large, if not larger, than any great system of the following periods; and yet in them alone, I repeat, is there an entire absence of an arborescent vegetation, derived from the then adjacent lands. (See Appendix C.)

And here it is well to remind the student of the wide, if not universal spread of the primeval strata. In the annexed small general map of the world are represented all the regions over which one or more of the primeval fossil groups are known to exist, as well as those crystalline rocks, which were formed before, or are associated with them.* In viewing the dark tint of this map we may suppose, that when such extensive palæozoic sea bottoms were raised into lands, the former continents, from which the sediments had been derived, were submerged. But be this as it may, it is a fact, that in all quarters of the globe, Silurian strata constantly lie in juxtaposition to the other overlying palæozoic formations; and hence it is impossible to apply to the lowest strata any reasoning which does not equally refer to those which repose upon them. For, as the Silurian rocks are constantly found in the same longitudes and latitudes as the Devonian and Carboniferous, why is it that in the one there are never found traces of vertebrata and land plants, and that in the same places remains of both these classes abound in the other? By no theoretical suggestion, therefore, can the fair inference be evaded, that things which did not exist during the Silurian period, were created in the very same tracts during the following ages.

The uniformitarian who would explain every natural event in the earliest periods by reference to the existing conditions of being, is thus stopped at the very threshold of the palace of former life,

---

* The reader must not be critical as to the exact demarcation assigned to the palæozoic and crystalline rocks by the dark tint of this map, which marks the relative proportions of the primeval to the secondary and tertiary rocks. The chief intention is to direct the eye to all the regions where the older rocks prevail.

which he cannot deprive of its true foundations. Nature herself, in short, tells him through her most ancient monuments, that though she has worked during all ages on the same general principles of destruction and renovation of the surface, there was formerly a distribution of land in reference to the sea, very different in outline from that which now prevails. That primeval state was followed by outbursts of great volumes of igneous matter from the interior, the extraordinary violence of which is made manifest by clear evidences. Fractures in the crust of the earth, accompanied by oscillations that suddenly displaced masses to thousands of feet above or beneath their previous levels, were necessarily productive of such translations of water, as to abrade and destroy solid materials, to an extent infinitely surpassing any change of which the historical era affords an example.

I could here cite the works of Leopold von Buch, Elie de Beaumont, Sedgwick, Studer*, and numerous other geologists, for countless proofs of this grander intensity of former causation, by which gigantic masses were inverted, and strata forming mountains have been so wrenched, broken, and twisted, as to pass under the very rocks out of whose materials they were constructed. The reader who may travel to the Alps and other mountain chains, will there see signs of such former catastrophes, each of which resulted from convulsions utterly immeasurable and inexplicable by any reference to those puny oscillations of the earth which can be appealed to during the times of history. (For my own views on this point see Appendix Q.) But restraining my pen on these collateral points, I must adhere to our immediate subject, the proofs of successive creation coincident with such former conditions apart from great physical revolutions.

If we reflect upon the succession presented to us in the primeval rocks, we have, I repeat, cumulative evidence, that the wide-spread

---

* See particularly Studer's new work, ' Geologie der Schweiz,' vol. ii. 1853.

and general diffusion of the same types of animal and vegetable life was due to a former temperature and outline of the surface essentially distinguished from those of our day. To whatever extent continents and islands may have existed during these long early periods, and however we may speculate on the extent of pristine shores, it seems certain, that the lands accessible to our survey and bearing any such vegetation, increased very considerably in the carboniferous times. In those days, the very same species of marine animals lived, from the latitude of Spitzbergen to the parallels of Peru and Australia.* Vast, low deltas then also extended themselves, which, on being desiccated, bore an absolutely uniform terrestrial vegetation nearly all over the globe. Many of the ancient tree ferns must have grown on tracts little above the water, and jungles larger than Britain must have been successively and repeatedly submerged and re-elevated to form new lands, covered by the vegetation that supplied the elements for the construction of great coalfields. These phenomena assure us, therefore, that a very great portion, if not the whole surface of the earth, enjoyed at that time an equable and warmer climate. (See Sketch, p. 268.)

Believing, as I do, with many geologists, that this former temperature of the earth was, in great measure, caused by the radiation of its internal heat, independently of solar action, other physical as well as zoological phenomena lead me further to suppose, that the land of those early days could not have been thrown up into very lofty mountains. For, if so, such great elevations must have been accompanied by corresponding deep chasms in the sea bottom, and these would necessarily have been impassable barriers to the numerous primeval submarine creatures which have been described as co-existent over the most remote region. Profound abysses of the

---

* See the observations of d'Orbigny, p. 409. Colonel Lloyd has just brought home from the Peruvian Andes several species of carboniferous mollusca identical with British forms. The same occur in Australia. See McCoy, Ann. Nat. Hist. 1847.

ocean are, it is well known, as complete barriers to the migration
of marine creatures, as lofty mountains are to inhabitants of the land.
The recent discovery, therefore, of the vast profundity of the ocean
midway between the Cape of Good Hope and Cape Horn, amounting to 7000 fathoms or nearly double the height of the loftiest mountains (even allowing for some amount of error in the sounding),
is, indeed, the strongest possible illustration of the impassable nature
of such subaqueous barriers. These have resulted from oscillations
of the surface, which occurring since primeval times, have necessarily
caused great variations in the marine provinces of life, placing them
in strong contrast with the uniformity of the ancient faunas. (See
Appendix F.)

Duly estimating the great value of the submarine knowledge
acquired and applied by a naturalist, whose researches, coupled with
those of Löven and other contemporaries, have thrown new lights
on many phenomena previously unexplained, let us guard against
the inference, that because such acquaintance with the natural
operations of our own era is applicable to the last geological data
or those of tertiary age, it should also apply to the *quasi* universal,
and therefore very different, physical conditions under which primeval creatures existed. During the tertiary period, the crust of
the earth had, as we know, approximated considerably to its present varied outline, and before it drew to a close, great changes
had taken place, by which regions formerly occupied by animals and
plants requiring a warm and equable climate were covered even by
ice and glaciers. Whilst, therefore, we thank Edward Forbes for
dredging the sea bottoms, and teaching us that deep sea mollusks
are now living near high lands in the Mediterranean, whence pebbles
may be so washed down as to lie in juxtaposition with the tenants of
the deep;—either this argument cannot bear upon the primeval era,
in which there is no evidence whatever of high lands and detached
mediterranean seas; or if applied, it is the strongest evidence to

show, that the supposed primeval cliffs and grounds were void of terrestrial vegetation.

On what data, it may be asked, is founded the beautiful and rational theory of Lyell, which explains the successive changes of the climate of the earth? Is it not mainly dependent on those diversified evolutions proceeding from beneath the surface, which have caused changes in the outline of former lands and seas, equivalent in extent, although different in position to, our present continents and oceans? And, if such a varied distribution of earth and water as the present had existed in the pristine periods we have been considering, how could the same groups of animals, manifestly requiring the same conditions, the same temperature, and the same food, have had an almost universal diffusion?

Although it is quite true, that specific distinctions are seen to have frequently prevailed in the fossils of the Lower Silurian rocks of countries situated at no great distance from each other, as explained in the chapters which treat of the distribution of the fauna of that age, the fact is by no means antagonistic to my reasoning. Should it, for example, be even said, that the variety in the distribution of Silurian species is as great as in the same areas of sea of the present day, I reply, that it is not to species but to the classes, groups, and genera, that the objector must appeal. If unable to explain why in the earliest primeval times, so many Trilobites, Orthides, Orthoceratites, and Corals of analogous forms were spread out over enormous distances, it is enough for us to feel assured, that the various associates of the Calymene Blumenbachii, or any other well-known Trilobite, must have required just the same temperature and surrounding media in whatever part of the world they lived. It is not because the land animals of Europe are dissimilar in *species* to those of Africa that the faunas of the two regions are so distinct, but that we have not among our European associates the same groups as those which live in hotter climes. I therefore conceive, that the fact of the diffusion of the same families and genera of Trilobites, Corals, and other fossils, however they may vary as to species, must have required an equably diffused temperature and similar conditions for their existence. Still more clearly is this inference sustained by the spread of the same old vegetation, and often of the same species of plants over half the globe.

Resting then on these universal facts as a firm basis, the geologist who explores his way upwards, sees, as before stated, that the formations which were next accumulated in the same latitudes and longitudes as the Silurian rocks, and sometimes in actual and conformable contact with them, do contain land plants mixed with marine remains. And yet the only unequivocal vegetables found in the elder strata are sea-weeds; whilst the later formed and contiguous rocks, though equally charged with exuviæ of the sea, are laden with many spoils of the land, both vegetable and animal.

Patient researches having thus demonstrated, that in the primeval eras all living things differed completely from those of our own times, so we see how the animals subsequently created, were adapted to new and altered physical conditions. Proceeding onwards from the early period in which we can trace no sign of land plants or vertebrata, and in which the solid materials, inclosing everywhere a similar fauna, were spread out with great uniformity, we soon begin to perceive the proofs of powerful revolutions, chiefly commencing after the coal formation*, by which the earth's surface was so corrugated,

---

* I by no means deny that much perturbation prevailed in the earlier stages of the planet, as explained in the two first chapters. On the contrary, I admit the powerful emission of much igneous matter, the transmutation of sedimentary into crystalline rocks, and the presence of great terrestrial masses anterior to the accumulation of any of the deposits which are now under consideration. But I deny that there are indications of the existence of any very lofty mountains, until the crust of the earth had undergone some of the subsequent mutations alluded to; particularly those which followed the accumulation of the carboniferous strata, and which were repeated at so many subsequent periods, notably and powerfully in the tertiary times.

that, after many perturbations, the groups of animals and plants were infinitely more restricted than before to given regions and climates. And, as the highly diversified conditions of the latest geological era and of the present day were wholly unknown in the primeval epochs, so it follows, that we should greatly err, if we endeavoured to force all ancient nature into a close comparison with existing operations. (See Alps, Appendix Q.)

The numerous, positive evidences of a former wide distribution of similar animals and plants, enable us fairly to bring before our mind's eye the physical geography of those great epochs when such large portions of our present continents were under the waters, and forests of tree-ferns occupied very extensive low lands, subjected during a long period to numerous oscillations. Not less clearly do we infer from other physical evidences, how eruptive forces subsequently acted; breaking out with great violence after the close of the carboniferous era, and throwing the strata into those grand undulations and contortions, accompanied by stupendous fractures, which have given to the coal basins their curvatures and limits.

Thenceforward was continued that long series of additional and repeated emissions of volcanic matter from within, of elevations of the sea bottom, and corresponding depressions of land, combined with the metamorphism of strata (these changes being often accompanied by corresponding new creations of animals suited to the existing conditions), during the secondary and tertiary periods. By these great physical operations, our planet was eventually brought to possess the climatal relations which have for so long a time prevailed. That these elevations and depressions finally produced a state of things altogether distinct from that of the earlier eras, is, in short, registered by a multitude of well-attested data.

Among the terrestrial changes to which science clearly points, there is no one which better deserves to be recorded in a few parting words than that great mutation of surface and its accompanying loss of warmth, by which extensive fields of ice were first formed upon the sea, and large glaciers upon the land. As very lofty mountains in moderate latitudes, and masses of land and water in Arctic or Antarctic regions, are now essentially the seats of glaciers and ice-rafts; so we know that these bodies alone have the power of trans-porting huge, erratic blocks from their native mountains to con-siderable distances by land, or for hundreds of miles over the sea in floating icebergs. Now, of the translation of such blocks we have no evidence whatever in any former geological period! On the contrary, whilst every boulder of the primary, secondary, or older tertiary rocks bears on its surface the signs of having been water-worn or rounded by aqueous or atmospheric agency, the great blocks of the later cold period (gigantic in comparison with all that preceded them) are often angular or nearly in that state in which they left the mountain side, before, in short, they were wafted over seas or lakes to be dropped, at remote distances from their parent rocks, upon sediments which by subsequent elevation have been made portions of our continents. Hence, independently of the indications of a more equably diffused and warmer temperature in older times than at the present day, these large erratics are in themselves decisive testimonials of that intense cold which, it is believed, was principally due to the great, elevated masses of land which specially characterize the modern period.

Receding backwards from this glacial phenomenon, which con-tinuing into our own times, has been so skilfully illustrated by eloquent writers*, it is specially my province to endeavour to impress upon the reader the importance of endeavouring to form an esti-mate of the physical geography of the earth during those remote periods when the Palæozoic deposits were accumulated. If very lofty mountains did not then exist, we have, indeed, in this single phe-

---

* Charpentier, Agassiz, and James Forbes. (See Appendix Q).

nomenon, what may have been one of the chief causes of that equable and warm climate so indispensably requisite to harmonize with the facts recorded in this volume. And if we add the inference adopted by many philosophers and geologists, that the earth, in cooling down from its original molten state, must, during long succeeding ages, have diminished in heat over its whole surface, we are enabled, by reference to physical changes alone, to satisfy ourselves, that we have in them the chief elements required to explain all climatal results.

Finally, if this retrospective survey of the changes of the earth be but the outline of a picture which must be filled up by an assiduous study of the works of nature, let me say that it has not been attempted without deliberate consideration and extensive researches amid the younger as well as the most ancient deposits. And, if any theoretical portions of the preceding pages be objected to, they can easily be separated from those historical facts which are established upon positive observation.

The leading object of this volume is simply to call attention to the oldest vestiges of life which have been discovered in the crust of the globe, and accurately to chronicle the order in which other early races followed them.

From the effects produced upon my own mind through the study of these imperishable records, I am, indeed, led to hope, that my readers will adhere to the views which, in common with many contemporaries, I entertain of the succession of life. For, he who looks to a beginning, and traces thenceforward a rise in the scale of being, until that period is reached when Man appeared upon the earth, must acknowledge in such works repeated manifestations of design, and unanswerable proofs of the superintendence of a CREATOR.

FINIS.

APPENDIX.

A.

Vertical Range of Silurian Fossils.

THE following table is designed to show the number of species ascertained to be common to the Lower and Upper Divisions of the Silurian System. It is compiled by Mr. Salter, chiefly from data in the Museum of Practical Geology, and those supplied by Prof. M Coy in his descriptions of the fossils in the Woodwardian Museum of Cambridge. A few other works have been referred to, but it has been thought best not to swell the list from less authentic sources. Where the species has been catalogued in the lists of the Geological Survey, the authority is marked G. S., for its occurrence either in the Lower or the Upper Silurian. W. M. signifies those species for which Professor M'Coy is responsible in the work above alluded to.

There being at present a difference of opinion as to the precise geological relations of the 'Upper Caradoc' (some authors regarding it as the base of the Upper, and others as the top of the Lower Silurian), this band has been omitted in the first estimate. If it be included in the Lower Silurian Division, as in the original Silurian System and in this work, the number of species common to the two divisions would be considerably increased above the number indicated in this Table.

NOTE.—A few of the species quoted by Professor M'Coy as occurring in the Lower Division are omitted; there being reason to doubt their *geological* position.

Lower Silurian (Llandeilo and Caradoc*).	Species which range from the Llandeilo into the Wenlock or Ludlow Formations.	Upper Silurian. (Wenlock and Ludlow).	Foreign Upper Silurian.
G. S.	Graptolithus priodon, Bronn. - -	G. S.	
G. S.	Retiolites Geinitzianus, Barr. - -	G. S.?	Barrande (Bohemia)

* Only the lower or typical Caradoc is here intended.

Lower Silurian (Llandeilo and Caradoc).	Species which range from the Llandeilo into the Wenlock or Ludlow Formations.	Upper Silurian (Wenlock and Ludlow).	Foreign Upper Silurian.
G. S.	Stromatopora striatella, D'Orb. -	G. S.	
G. S.	Heliolites interstincta, Wahl. - -	G. S.	
W. M.	—————— petalliformis, Lonsd. -	G. S.	
W. M.	—————— tubulata, Lonsd. - -	G. S.	
G. S.	—————— megastoma, M'Coy - -	Milne Edw. M'Coy	
G. S.	—————— inordinata, Lonsd. - -	Milne Edw.	
G. S.	Favosites alveolaris, Goldf. - -	G. S.	
G. S.	—————— Gothlandica, Linn. - -	G. S.	
G. S.	Stenopora (Favosites) fibrosa, Goldf., both the branched and amorphous varieties.	G. S.	
G. S.	Halysites catenulatus, Linn. - -	G. S.	
W. M.	Nebulipora papillata, M'Coy - -	W. M.	
W. M.	Omphyma turbinata, Linn. - -	G. S.	
G. S.	Sarcinula organum, Linn. - -	Milne Edw.	(Gottland)
Lonsdale (Llandovery)	Petraia bina, Lonsd. - - -	G. S.	
G. S.	Ptilodictya lanceolata, Goldf. - -	G. S.	
G. S.	Glauconome disticha, Goldf. - -	G. S.	
W. M.	Fenestella Milleri, Lonsd. - -	G. S.	
W. M.	—————— subantiqua, D'Orb. - -	G. S.	
G. S.	Cornulites serpularius, Schloth. -	G. S.	
G. S.	Tentaculites annulatus, Schloth. -	G. S.?	
W. M.	—————— ornatus, Sow. - -	G. S.	
W. M.	Serpulites dispar, Salter - -	G. S.	
W. M. ?	Beyrichia tuberculata, Klöden -	G. S.	
G. S.	Cyphaspis megalops, M'Coy · -	G. S.	
G. S.	Lichas bulbiceps, Phill. & Salter -	G. S.	
G. S.	—————— anglicus, Beyrich -	G. S.	
Eichwald (Russia)	—————— verrucosus, Eichw. - -	G. S.	
G. S.	Acidaspis Brightii, Murch. - -	G. S.	
G. S.	Cheirurus bimucronatus, Murch. -	G. S.	

Lower Silurian (Llandeilo and Caradoc).	Species which range from the Llandeilo into the Wenlock or Ludlow Formations.	Upper Silurian (Wenlock and Ludlow).	Foreign Upper Silurian.
W. M. Barrande	} Staurocephalus Murchisoni, Barr. -	G. S.	
G. S.	Encrinurus punctatus, Brünnich. -	G. S.	
G. S.	Phacops caudatus, Brongn. - -	G. S.	
G. S.	—— Stokesii, Milne Edw. -	G. S.	
G. S.	Calymene Blumenbachii, Brongn. -	G. S.	
G. S.?	Proetus latifrons, M'Coy - -	G. S.	
G. S.	—— Stokesii, Murch. - -	G. S.	
G. S.	Crania (Patella) implicata, Sow. -	G. S.	
W. M.	Spirifer crispus, Linn. - - -	G. S.	
Sowerby	—— plicatellus (radiatus), Linn.	G. S.	
G. S.	Atrypa reticularis, Linn. - -	G. S.	
G. S.	—— marginalis, Dalm. - -	G. S.	
G. S.	—— cuneata, Dalm. - - -	G. S.	
W. M.	Rhynchonella depressa, Sow. - -	G. S	
G. S.?	—— borealis, Schloth. -	G. S.	
W. M.	—— Lewisii, Davidson -	G. S.	
W. M.	—— Nucula, Sow. - -	G. S.	
W. M.	—— rotunda, Sow. - -	G. S.	
G. S.	Pentamerus undatus, Sow. (the same as P. linguifer, Sow.?) - - -	G. S.	
G. S.	Orthis biloba, Linn. - - -	G. S.	
G. S.	—— calligramma, Dalm. - -	G. S.	
G. S.	—— elegantula, Dalm. - -	G. S.	
W. M.	—— hybrida, Sow. - - -	G. S.	
G. S.	—— insularis, Pander - -	W. M.	
G. S.	—— reversa, Salter - - -	G. S.	
G. S.	—— testudinaria, Dalm. - -	- -	Dalman (Gottland)
G. S.	Strophomena rhomboidalis, Wahl. (depressa, Dalm. and of this work)	G. S.	
G. S.	—— pecten, Linn. - -	G. S.	
W. M.	—— simulans, M'Coy -	W. M.	
G. S.	—— corrugata, Portl. - -	G. S.	

Lower Silurian (Llandeilo and Caradoc).	Species which range from the Llandeilo into the Wenlock or Ludlow Formations.	Upper Silurian (Wenlock and Ludlow).	Foreign Upper Silurian.
G. S.	Strophomena antiquata, Sow. - -	G. S.	
G. S.	———— euglypha, Dalm. -	G. S.	
de Verneuil (Russia)	———— imbrex, Pander. -	G. S. Davidson.	
W. M.	Leptæna lævigata, Sow. - - -	G. S.	
W. M.	———— minima, Sow. - - -	G. S.	
G. S.	———— quinquecostata, M'Coy -	W. M.?	
G. S.	———— sericea, Sow. - - -	G. S.	
G. S.	———— transversalis, Dalm. - -	G. S.	
W. M.	Pterinea pleuroptera, Conrad - -	W. M.	
G. S.	———— retroflexa, Wahl. - -	G. S.	
W. M.?	———— tenuistriata, M'Coy - -	W. M.	
W. M.	Modiolopsis antiqua, Sow. - -	G. S.	
W. M.	Orthonotus semisulcatus, Sow. -	G. S.	
G. S.	Cardiola interrupta, Goldf. - -	G. S.	
W. M.	Arca Edmondiiformis, M'Coy -	W. M.	
W. M.	Cucullella (Clidophorus?) antiqua, Sow. - - - -	G. S.	
W. M.	Nucula Anglica, D'Orb. (ovalis, Sil. Syst.) - - - -	G. S.	
G. S.	Acroculia (Nerita) Haliotis, Sow. -	G. S.	
G. S.? (Harnage Grange, Shropshire.)	Theca Forbesii, Sharpe - - - (T. triangularis, Hall, is a distinct Lower Silurian species, found in Scotland.)	G. S.	
G. S.	Conularia Sowerbyi, Defrance (C. cancellata, Sandberger and M'Coy)	G. S.	
G. S.	Scalites (Trochus) lenticularis, Sow.	W. M.	
G. S.	Bellerophon dilatatus, Sow. - -	G. S.	
W. M.?	———— expansus, Sow. - -	G. S.	
G. S.	———— trilobatus, Sow. - -	G. S.	
G. S.	———— carinatus, Sow. - -	G. S.	
G. S.	Orthoceras angulatum, Hisinger -	G. S.	
W. M.	———— filosum, Sow. - -	G. S.	
G. S.	———— annulatum, Sow. - -	G. S.	
W. M.	———— laqueatum, Hall. - -	W. M.	

Lower Silurian (Llandeilo and Caradoc).	Species which range from the Llandeilo into the Wenlock or Ludlow Formations.	Upper Silurian (Wenlock and Ludlow).	Foreign Upper Silurian.
G. S.	Orthoceras primævum, Forbes -	G. S.	
G. S.	———— subundulatum, Portl. -	W. M.	
G. S.	———— tenuicinctum, Portl. -	G. S.	
W. M.	———— ventricosum, Sharpe -	G. S.	
G. S.	———— Ibex, Sow. - - -	G. S.	
G. S.	———— subannulatum - -	W. M.	
G. S.	Lituites cornu-arietis, Sow. - -	G. S.	
96 species			

Llandeilo Formation.	Additional Llandeilo Species which range into the 'Upper Caradoc' or intermediate formation, but not into the Wenlock or Ludlow rocks.	Upper Caradoc.	
G. S.	Petraia elongata, Phill. - - -	G. S.	
G. S.	Trinucleus concentricus - - -	G. S.	
G. S.	Spirifer (Atrypa) percrassus, M'Coy	W. M.	
G. S.	Atrypa hemispherica, Sow. - -	G. S.	
G. S.	Pentamerus globosus, Sow. - -	G. S.	
G. S.	———— oblongus, Sow. - -	G. S.	
G. S.	———— lens, Sow. - - -	G. S.	
G. S.	———— undatus, Sow. - -	G. S.	
G. S.	Orthis Actoniæ, Sow. - - -	Forbes	
G. S.	—— vespertilio, Sow. - - -	G. S.	
W. M.	Strophomena compressa, Sow. -	G. S.	
W. M.	Rhynchonella decemplicata, Sow. -	G. S.	
G. S.	Modiolopsis orbicularis, Sow. - -	G. S.	
W. M.	Clidophorus ovalis, M'Coy - -	W. M.	
Hall (America)	Nucula? post-striata, Emmons -	W. M.	
G. S.	Holopella cancellata, Sow. - -	G. S.	
G. S.	Pleurotomaria (like P. balteata,Phill.)	G. S.	A marked species.
W. M.	Bellerophon subdecussatus, M'Coy -	W. M.	
18 species			

If, instead of regarding it as a distinct intermediate formation, the Caradoc band be classed either with the Upper or the Lower Silurian, these species must, in that case, be added to those before given; thus making a total of 114 species common to the Lower and Upper groups. It is believed that many more species exist in the Upper Caradoc formation which occur also in the Llandeilo flags.

No notice has been here taken of the very peculiar assemblage of fossils found in the Silurian sandstones of Galway. These beds constitute, as explained in the text (p. 170.,) a considerable thickness of Silurian strata. If they be classed with the 'Upper Caradoc' or intermediate formation, 10 or 12 more *very characteristic* Lower Silurian fossils must be added to the above common list. Or, if they be regarded as equivalents of the Llandeilo or Bala rocks, to which, as developed in the South of Scotland, they bear a strong analogy, then the unity of the Silurian System is still more sustained, inasmuch as they are crowded with characteristic Upper Silurian species; no less than 63 common Upper Silurian forms, associated with the above-mentioned and other Llandeilo flag species, being found in this single district.

B.

Recent Surveys of the British Silurian Rocks.

There are some points of detailed research made while the work was in progress, to which I may here direct attention. Certain typical tracts of the British Silurian rocks have been re-surveyed by the Government geologists, and results have been obtained which seem to me to support the inferences recorded in the text. In re-visiting North Wales, and in searching for more complete fossil evidences in the 'Lingula Flags, or Lowest Silurian' of the Survey, Mr. Salter has made important additions to our knowledge by the discovery of fossils, which prove that this deposit, like the protozoic band described by M. Barrande in Bohemia, has a marked and peculiar facies. At the same time, it is to be remembered that two of the few genera of Trilobites in this lowest zone of life, (Agnostus and Olenus), occur also in the Llandeilo or Bala group.

A few words are also called for to remove any doubt respecting the zoological contents of the strata that constitute the intermediate or debateable ground, now termed Upper Caradoc, between the Lower and Upper Silurian. In truth, the whole of the formation which I grouped under the term of Caradoc, having undergone a special scrutiny by Professor Ramsay, and Messrs. Salter and Aveline (Quart. Journ. Geol. Soc. vol. x. p. 62.), has been divided by those authors into lower and upper parts, the former constituting by its fossils a portion of the Llandeilo or Bala formation. This fact is partially alluded to above (pp. 80. 86. 97. *et seq.*), and it has been rendered clearer since my pages were printed. The partition just effected, was, indeed, previously suggested, on fossil evidence, by Professors Sedgwick and M'Coy.* But in respect to Shropshire, the Government Surveyors have proved, that the older of these shelly sandstones, as exhibited along the flanks of Caer Caradoc, is, as I described it, laden with fossils which are identical with types of the Llandeilo formation, most of which were published in my original work. Even in the 'Silurian System,' p. 305. pl. 32. fig. 9. and elsewhere, the geologist will see that, although I was unacquainted with the division recently indicated of Lower and Upper Caradoc, the rocks of Meifod, described as Caradoc sandstone, are truly the same in fossil contents as the greater mass of the rocks on the *flanks of Caer Caradoc itself*, which are now classed by the Government Surveyors as Llandeilo. My general term for all such rocks, whether sandy, flag-like, schistose, or slaty, was and is simply 'Lower Silurian.' The same fossils, indeed, not only pervade the districts of Shropshire and Montgomeryshire, or the Lower Silurian of my old map, but spread over North Wales, down to the Snowdon slates, and even below that horizon; and thus the reasoning I formerly employed in urging the broad application of the term 'Lower Silurian' is sustained to an extent much beyond my early anticipations. For, until the physical structure and fossil contents of the whole region had been precisely elaborated, no one could have imagined, that the soft shelly sandstones on the sides of Caer Caradoc were of the same age as the hard, slaty rocks of Bala and Snowdon!

The Upper Caradoc, on the contrary, as seen under the shale of Wenlock Edge, where it is now shown to be transgressive to the older strata, and conformable to the Wenlock shale, or in the tract north of Bishop's Castle, where it flanks unconformably the southern ends of the Longmynd and Shelve districts, or as ranging along the west flank of the Abberley and Malvern Hills to May Hill, or, again, as re-appearing in Radnorshire and adjacent parts of Wales, is unquestionably of an intermediate character. I classed this rock with the Lower Silurian, chiefly from its mineral characters and *infra* position to the Wenlock shale and limestone, and as indicating the highest sandstone of the lower group; and also because two or three of its fossils which seemed to me most characteristic

* Quart. Journ. Geol. Soc. Lond. vol. ix. p. 215.

are not found in the Upper Silurian division. Hence I have adhered to my old tabular arrangement. But I must now clearly state, that whilst in Shropshire and Radnorshire this upper sandstone retains such forms of the Lower Caradoc and Llandeilo strata, as Pentamerus oblongus and Atrypa hemispherica, it is charged in many places, particularly at May Hill, as indicated by Professors Sedgwick and M'Coy, with numerous species of the overlying Wenlock formation, which here make their appearance, for the first time, in the ascending order. Whatever, therefore, may be its name, whether Upper Caradoc, May Hill Sandstone, or Lower Wenlock grit, it is truly a transition band, which unites the Lower and Upper Silurian, and is to be regarded as identical with the stratum I have described as occupying the same position in Russia, and with that in America which is known by the name of the Clinton Group.

It must here also be noted, that the transgression recently detected in Shropshire, between the lower and upper Caradoc, is a local phenomenon, which does not extend to large portions even of the annexed map. The Government Surveyors have, indeed, published sections across the Silurian rocks of Wales, which, as in America, Bohemia, and many foreign countries, exhibit conformable passages between all the strata, from those of azoic age, through the Llandeilo and Caradoc, to the true Upper Silurian of Wenlock and Ludlow inclusive.

C.

Traces of Vegetable Matter in the Silurian and older Rocks.

Although, as stated in the preceding pages, no remains of plants have been discovered below the Ludlow rocks, which are *recognizably of terrestrial origin*, I have long known of the presence both of bitumen and anthracite in even the oldest greywacke, or the Longmynd rocks (see Sil. Syst. p. 261. *et seq.*) Theoretically I have inferred that these substances were derived from masses of sea-weeds. (See Russia and Ural Mountains, p. 14., Nicol, Quart. Journ. Geol. Soc. Lond. vol. iv. p. 204., and Harkness, *ib.* vol. viii. p. 393.)

In justice, however, to my friend Professor Nicol, it is right to state, that he detected under the microscope, a tubular, fibrous structure in the ashes of anthracite derived from the Lower Silurian greywacke of Peebleshire, and also observed, among those rocks in Liddesdale, fragments of what he supposed might be reeds, like the imperfect remains of such vegetables in the coal sandstones. (Quart. Journ. Geol. Soc. Lond., vol. vi.) In any endeavour to form a correct view of the origin of life, it is doubtless important to try to settle this point by a rigid scrutiny. But, even if these very rare vegetable remains should prove to belong to the land and not to the sea, my general argument would not be invalidated by the existence of a very scanty terrestrial Flora (not amounting in size to arborescence) which was coincident with the first traces of animal life.

Mr. John Kelly informs me that anthracite has been discovered in the old greywacke of Cavan, Ireland, the courses of which dip with the strata.

D.

Unconformity in Ireland of the Longmynd or Bottom Rocks ('Cambrian' of the Survey) to the Lower Silurian.

An excellent and clear account of this phenomenon, with sections, has been published by Mr. J. Beete Jukes, the director of the Survey in Ireland, and Mr. Andrew Wyley, entitled, "On the Structure of the N. Eastern part of the County of Wicklow." (Proc. Geol. Soc. Dublin, December, 1853.) It would appear that the black slates ranging from the counties Wicklow and Wexford, to Dublin, and forming the inferior portion of the Lower Silurian rocks of the S. E. of Ireland, rest quite unconformably on the upturned edges of the so-called 'Cambrian' rocks. The Lingula flags seem to be absent. The physical relations of these rocks in Ireland are therefore distinct from those of Wales, Siluria, Bohemia, Scandinavia, and parts of N. America, where such inferior masses are perfectly conformable to the Lowest Silurian with fossils. On the other hand, their position in Ireland accords with the physical disruption noticed in France by Elie du Beaumont and Dufrenoy, where the 'primordial zone' of Lingula flags is also wanting.

E.

Barrande's Silurian System in Bohemia.

In addition to what is stated (p. 340. *et seq.*) I may here take notice of a few memoranda made by myself last summer in the cabinet of M. Barrande at Prague, though they imperfectly allude to phenomena which he has described or will describe in full detail. Among the Bohemian graptolites the C. priodon (*Ludensis*, Sil. Syst.) is beautifully preserved,—its smaller end is incurvated. Of corals there occur quite as many Silurian forms as are known in England, perhaps seventy or eighty. The Cystideæ of the Lower Silurian (stage D. of Barrande) are occasionally of great size, and some of them, resembling the Echino-encrinites of Scandinavia and Russia, ascend into the stage E. or Upper Silurian.

Among the Cephalopoda the Orthoceratites (Foss. 59. p. 326.), with a large lateral siphuncle, are as rare as in Britain — shells which are very common in Russia and Sweden. In reference to one group of Orthoceratites, M. Barrande shows how a field geologist may erroneously recognize forms as identical, from a perfect similarity of outline, but which are clearly *distinct* when the ornamental features of the shelly covering are magnified. The profusion of chambered shells in the basin of Prague has enabled him to trace many gradations between the generic types Gomphoceras, Phragmoceras, Trochoceras, Cyrtoceras, &c.; and in the centre of two specimens of these he has detected animal matter in the state of Adipocere! These bodies are, therefore, he justly considers, the oldest *mummies* ever exhumed; since they occur in Lower Silurian rocks with Trinuclei!

The most ancient Nautilus yet found is probably the N. tyrannus Barr., a huge form of a genus which, as mentioned in the text, the author has traced through all its stages of growth. There are some Goniatites of forms approaching Nautilus; but no species is a known Devonian type; not even in the very highest strata near Prague.

It may also be remarked, that the singular genus of bivalves of the Lower Silurian rocks of Busaco, described as Ribeiria by Mr. D. Sharpe, occurs also in the middle of the Lower Silurian of Bohemia. Besides the genera of Trilobites enumerated at p. 342. of this work, as occurring in the primordial or lowest zone, that band also contains the minute Hydrocephalus, so named by Barrande from its monstrous head, and the Arionellus, a large flat form, as yet unknown elsewhere. In the mass of the Lower Silurian, or stage D. of the author, occur the Beyrichia (Agnostus) lata of Sweden and America, and the Acidaspis Buchi, described also by M. Bertrand Geslin from rocks of this age near Nantes; whilst the Cytherinæ (of which there are about twenty species) range from the Lower Silurian to the highest of the Upper Silurian limestones.

Space does not permit of an examination of the analogies and identities prevailing among the Brachiopods and Gasteropods of the basin of Prague as compared with those of Silurian age in other countries. Suffice it to say, that when fully elaborated, the whole of the fossils will probably amount, under the discrimination of M. Barrande, to about 1500 species!

F.

Depth of the Primeval Seas, and colouring of the Shells of the Mollusca which lived in them.

In discussing the subject of the probable depth of the primeval seas, it might have been inferred, that assuming the quantity of water upon the surface of the earth to have been constantly the same, it necessarily follows that during those long periods when the crust had undergone comparatively few variations of outline, there must have been a great uniformity in the depth of the waters. This inference has been incidentally strengthened by Professor E. Forbes, who having shown in a short memoir recently read before the Royal Society, that no mollusca which live beneath the depth of 50 fathoms exhibit those distinct *pattern-colours* which mark the shells of the animals inhabiting higher zones, brought this observation to bear on the fact, that some palæozoic types (specially those of the carboniferous age) still retained their colours, and hence that such animals could not have lived at any great profundity or far removed from the influence of light. MM. d'Archiac and de Verneuil did, indeed, advert many years ago (1839, Trans. Geol. Soc. Lond. vol. vi. p. 360.) to shells of Devonian age which were partially coloured, and Mr. Salter has called my attention to Silurian fossils in the Government Museum which also exhibit colour.

G.

Extension of Silurian Rocks in Spain.

Since the text was printed, I learn that M. Casiano de Prado has discovered Graptolites in several parts of the Guadarrama chain, where they occur in black schists associated with large masses of silicious grits (quarzites). This enterprising geologist has also traced such schists to near Salamanca, and has further satisfied himself, that there is an absolute continuity of Lower Silurian rocks from the Sierra Morena into Gallicia ; the range passing through Estramadura and by the Sierras of Gata and Francia. Most of the mountains which, extending from S. to N. form the boundary between Spain and Portugal may thus be considered to be of the same age.

The Sierra Morena and the environs of Almaden have again been specially explored by M. Casiano de Prado, who has prepared a geological map of that tract, which will shortly appear in the Bulletin of the Geological Society of France, and has discovered many Silurian and Devonian fossils which will be described by MM. Barrande and de Verneuil.

The Silurian rocks (lower) are very rich in Trilobites, and have already afforded about 20 species of those crustaceans, among which occur the greater number of the forms described by Mr. Sharpe from Portugal (p. 404.) ; such as Calymene Tristani, Phacops socialis, P. Dujardini, Placoparia Tournemini, Illænus Lusitanicus, &c.

An important suggestion too has been made by M. de Prado, and partially confirmed by M. Barrande, viz. that the equivalent of the primordial zone, with its peculiar fauna, exists in Spain. One imperfect fossil only has however been yet found.

MM. de Verneuil and de Loriere have also detected Silurian rocks in two ridges directed from N.N.W. to S.S.E., which range from near Moncayo towards Montalban in Aragon.

H.

Foraminifera in the Silurian Rocks.

Mr. H. C. Sorby has the merit of having been the first to discover Foraminifera under the microscope in the Silurian Rocks.

The forms in the Aymestry or Ludlow Limestone (the black limestone of Sedgley) are very clearly recognisable ; those of the Wenlock Limestone (from Easthope) being less so. These Foraminifera are related to the genus Endothyra, found by Professor Phillips in the Mountain Limestone of Yorkshire and Westmoreland.

I regret that I was unacquainted with an able memoir of Mr. Sorby on Slaty Cleavage when the earlier chapters were printed. (See Edin. New Phil. Journ., 1853.)

I.

Supposed Anomaly at Neffiez in France explained.

Allusion having been made (p. 393.) to a physical intermixture of Silurian with Carboniferous fossils at this one little spot in the South of France, I have the satisfaction to announce, that M. Fournet himself, who first described the phenomenon, is now convinced that it is simply due to a violent contortion and inversion of the strata, as suggested by de Verneuil, and supported by myself, (p. 394.)

K.

Australian Gold.

The reader who may wish to be acquainted with the amount of the produce, as well as with the history and nature of the Australian Gold Mines, will find a clear and able digest of the whole subject by M. Delesse, of Paris, printed in the Annales des Mines, 1853, Tom. iii. Serie 5, p. 185. The same memoir was printed in the "Moniteur."

L.

Produce of Silver.

In confirmation of the view taken (p. 457.) of the future augmentation of the quantity of silver, and in allusion to Humboldt's anticipation of that increase mentioned in the note of that page, let me say, that the same illustrious friend again expressed his conviction to that effect to me last autumn, in Berlin, and reminded me that Spain (the seat of the richest silver mines in the days of Hannibal) had of late years proved to be argentiferous to a prodigious extent.

M.

Silurian Rocks in Silesia

Since this work was printed, Professor Göppert, of Breslau, has informed me, that Graptolites have been found in rocks underlying the well-known Devonian formations of Silesia (see p. 338.), thus leading us to suppose that adequate researches will elucidate a similar succession in other parts of Germany.

K K

N.

Palæozoic Rocks in Asia Minor.

M. Pierre de Tchihatcheff has just communicated to the French Institute a sketch of the succession of Silurian, Devonian, and Carboniferous rocks in Asia Minor. (Comptes Rendus, 3 Avril, 1854.) The newly discovered large coal-field near Eregli (Heraclea) on the Black Sea, is in the Carboniferous Limestone.

O.

Discoveries along the Flanks of the Malvern Hills.

My friend the Rev. W. S. Symonds, F. G. S., (who has obligingly prepared the Index of this work,) has recently informed me that, in sinking shafts on the new railroad from Worcester to Hereford, the Woolhope beds have been observed in nearly vertical position opposite Winning Farm, on the west flank of the ridge. The strata at this place are charged with Bumastus Barriensis, and other characteristic fossils. Again, in a deep excavation for the foundation of the new house of Messrs. Burrows' at Great Malvern, a bright red sandstone was found beneath the detritus of the hill, and quarried to the depth of twenty feet, dipping to the east at an angle of 70°. The age of this rock, whether Triassic or Permian, has not yet been determined.

P.

Other Reptilian Remains in the Permian Rocks of Russia.

My friend Major Wangenheim von Qualen, to whom my associates and myself were much indebted for information respecting the structure of the Permian Rocks of Orenburg, has since our visit detected bones of another reptile which occurs, not like the remains of the Rhopalodon (Fischer) in the sandstone and conglomerate, but in the limestone of the formation. The preservation of the head of this animal has enabled M. Eichwald to assign to it the name of Zygosaurus Lucius, and to compare it with the crocodile. On referring M. Eichwald's description (Bull. Soc. Imp. Nat Moscow, 1852, No. 4.) to Professor Owen, that celebrated comparative anatomist thus writes to me:— " The characters which M. Eichwald points out in his fossil, as resembling those in the crocodile, also occur in the fossil skulls of the Labyrinthodont reptiles; and the observation upon the number and comparatively small size of the teeth, in the Permian reptile, would lead to a suspicion that it may really belong to the Labyrinthodont family. All doubt would be removed by an inspection of the occiput of M. Eichwald's fossil; if that part presented a single condyle for articulation with the neck-vertebræ, it would determine the accuracy of his views of its affinities: but if the occiput showed a pair of condyles it would prove the fossil to be a Sauroid Batrachian."

Q.

On the former Changes of the Alps.

(This abstract of a discourse addressed to the Royal Institution, March 7th, 1851, is given, to illustrate briefly the chief mutations of a mountain chain in comparatively recent geological periods.)

The complicated structure of the Alps so baffled the penetration of de Saussure, that after a life of toil, the great and original historian of those mountains declared, 'there was nothing constant in them except their variety.' In citing this opinion, Sir Roderick Murchison explained how that obscurity had since been gradually cleared away by the application of modern geology, as based upon the succession of organic remains. He then proceeded to specify the accumulations of which the Alps were composed, and to mark the changes or revolutions they had undergone, between the truly primeval days when the earliest recognizable animals were created, and the first glacial period in the history of the planet.

His object being to convey, in a popular manner, clear ideas of the physical condition of these mountains at different periods, he prepared for the occasion three long scene-paintings, which represented a portion of the chain at three distinct epochs. The first of these views of ancient nature exhibited the Alps as a long, low archipelago of islands, formed, in great part, out of the Silurian and older sediments which had been raised above the sea; the lands bearing the tropical vegetation of the carboniferous era. (See woodcut, p. 268.)

He then stated that there were in the Alps no relics of the formations to which, as marking the close of the primeval or palæozoic age, he had assigned the name of Permian, and rapidly reviewed the facts gathered together by many geologists from all quarters of the globe, maintaining that they unequivocally sustained the belief, that there had been a succes-

sion of creations gradually rising from lower to higher types of life, as
we ascend from inferior to superior formations. He carefully, however,
noted the clear distinction between such a creed, as founded on the true
records of creation, and the theory of transmutation of species; a doctrine
from which he entirely dissented.

In the second painting (an immense lapse of time being supposed to
have occurred), the Alps were represented as a mountainous ridge in
which all the submarine formations from the secondary to the older tertiary or Eocene had been elevated upon the flank of the primeval rocks.
Each rock system was distinguished by a colour peculiar to it, and the
nature of the animals contained in each of these deposits was succinctly
touched upon. Between the youngest of the primary formations and the
oldest of the secondary rocks, it was stated, that there is not one species
in common in any part of Europe; the expression being that "an entirely
new creation had succeeded to universal decay and death."

In speaking of the Alpine equivalents of the British Lias and Oolites,
Sir R. paid a sincere tribute to Dr. Buckland, who led the way, thirty
years ago, in recognizing this parallel. Leopold von Buch was also particularly alluded to as having established such and other comparisons, and
as having shown the extent to which large portions of these mountains
have been metamorphosed from an earthy into a crystalline state. In
treating of the cretaceous system it was shown that the Lower Green Sand
of England, so well illustrated by Dr. Fitton, was represented in the Alps
by large masses of limestone, since called Neocomian by foreign geologists.

Emphasis was laid upon the remarkable phenomena, that every where in
the south of Europe (as in the Alps) the Nummulite rocks, with the
'flysch' of the Swiss, and the 'macigno' of the Italians, have been raised
up into mountains together with the chalk (Hippurite beds and Inoceramus
rock) on which they rest. Hence it was, that before Sir Roderick Murchison made his last survey of the Alps, the greater number of geologists
classed the Nummulite rocks with the cretaceous system, and considered
them both to be of secondary age. But judging from the mass of fossils,
which differ entirely from those of the chalk, and also from their superposition, he had referred these Nummulite rocks to the true lower tertiary
or Eocene of Lyell. Beds of this age, though once merely dark-coloured
mud, have been converted into the hard slates of Glarus with their fossil
fishes (among which eels and herrings first made their appearance). Some
strata of this date at Monte Bolca contain well known fishes, and others

again have been rendered so crystalline amid the peaks of the Alps as to resemble primary rocks; so intense has been the metamorphosis!

Dwelling on the atmospheric conditions which prevailed after the elevation of the older tertiary, Sir R. inferred that a Mediterranean and genial climate prevailed during all the long period while the beds of sand (Molasse) and of pebbles (Nagelflue) were accumulating under the waters both of lakes and of the sea, derived from the neighbouring slopes of all the pre-existing rocks. The marine portions of the Molasse and Nagelflue contain the remains of some species of shells now living in the Mediterranean; whilst in alternating and overlying strata, charged exclusively with land and fresh water animals, including numerous insects, not one species among many hundreds is identical with any form now living. This point, on which he first insisted on his return from the Alps in 1848, Sir R. considered to be of paramount importance in proving, that terrestrial life was much less endowed with the capacity to resist physical changes of the surface than submarine life; for here we have a land fauna which is Pliocene in the order of the strata, and yet is scarcely Eocene when we regard its organic contents.

A certain number of the more remarkable animals that lived during this younger tertiary age were then adverted to, such as the Rhinoceros and other large quadrupeds, the fossil Viverrine fox *, collected by himself (the original of which was on the table), the huge Salamander (Andrias Scheuchzeri) and a Chelydra which Professor Bell has described as analogous to the snapping turtle of the southern states of North America. These creatures, with quantities of plants, including small palms, were all indicative of a warm and genial climate. On such sure grounds the second diagram represented the Alps as covered with a subtropical vegetation; several of the above-mentioned animals being sketched in the foreground.

The author had satisfied himself †, in common with Studer, Escher, and all the geologists who have well explored the Alps, that everywhere along their northern flank terrific dislocations have occurred, amounting in many places to a total inversion of mountains, whereby tertiary deposits accumulated beneath the waters during the period he had just been describing, *were thrown under the strata of former date.* He then briefly pointed out what he had demonstrated in detail elsewhere: viz. that the sands and

---

* Now in the British Museum.
† Quart. Journ. Geol. Soc. London, vol. v. p. 157.

pebble-beds of that age had been suddenly heaved up from beneath the waters along the outer or northern flank of the chain, so as to form mountainous masses, the inverted and truncated ends of which had been forced under the edges of the very rocks out of whose detritus they had been formed.

Before this great revolution had taken place no large erratic blocks were known, but after it they became common, and were the necessary products of that intensely cold climate to which the Alps were then subjected—a change of which their surface bears many distinct evidences.

During the same period the low countries of northern Europe were, it is well known, covered by an Arctic sea. The Jura and the Alps having also then, as it is believed, been subtended by water, icebergs and rafts must have been detached from the higher range, carrying away blocks of stone northwards to be dropped at intervals; just as it has been demonstrated that the Scandinavian blocks, which floated southwards, were dropped in Prussia, Poland, and the low lands of Russia, when all those regions were under the influence of an Arctic sea. In short, Bavaria and the lower parts of the Cantons Vaud, Neufchatel, and Berne, must then have been covered by waters which, whether salt or fresh, bathed the foot of the Alps.

The inference that the change from a former genial climate to the first great period of cold was sudden, was further sustained by the fact, that the inclined strata in which the Mediterranean animals are buried, are at once covered *transgressively* and *unconformably* by other beds of gravel, shingle, and mud, in which the remains of plants and animals are those of a cold climate.

The third scene, therefore, exhibited the beds of sand and pebbles of the genial tertiary period thrown up into mountains on the flanks of the chain, the peaks of which were probably then covered for the first time with snow; and from the ravines, whether opening upon the sea-shore or into deep fiords or bays, into which glaciers and their moraines advanced, giving birth to icebergs on which rocks were floated away.

In concluding, Sir Roderick thus expressed himself: — "Having thus now conducted you rapidly through the most prominent changes which the Alps have undergone, from the first period when they had emerged, probably as an archipelago of low islands in a hot climate, to that epoch when the animals and plants living upon them indicated a Mediterranean temperature, and then to that Arctic period, the conditions of which I have just been discussing, I have no longer to call for your assent to any inferences of the geologist, which all of you are not perfectly competent to understand.

" To convert the Alps of the earliest glacial period into the Alps of the present day, you have only to figure them to yourselves, as raised 2000 or 3000 feet above the altitude which they are supposed to have in the diagram last exhibited. All their main features remaining the same, you would then have before you, the present Alps and their valleys bathed by lakes and rivers instead of bays; and in place of the waters sketched in beyond them in the painting, with ice-bergs floating upon them, you will then have dry mounds of gravel, sand, and blocks, which were accumulated beneath these waters: such, in a word, as now constitute the low hills and valleys and all the richest land of Switzerland and Bavaria, where man has replaced the rhinoceros and turtles of one period, and the ice-bergs of another. You who have not visited this noble chain, and who wish to judge of its gorges, peaks, and precipices, may consult the views sketched by our associate Brockedon, and thus have Nature, in her present mood, brought in a telling manner before you. But those of you, who really wish to grapple with the geological wonders of former days, may look at the flanks of the Rigi from the lake of Lucerne, where even from the deck of the rapidly passing steamer, you may see how that great pile of pudding-stone, every pebble of which has been derived from more ancient rocks in the chain, has been lifted up from beneath the waters in the manner here pictorially represented; whilst, if you continue the same traverse up the lake to Altorf, you will pass by extraordinary folds, inversions, and breaks of the secondary limestones, and of the older Tertiary or Nummulitic rocks. Such a doubling or crumpling up of these strata, you may then perchance agree with me in thinking, was in a great measure the result of lateral pressure between two great masses; viz. the axis of the crystalline rocks of the chain towards the south, and the newly upraised deposits, of which the Rigi is a small part only on the north. The latter having been intruded upon the terrestrial surface, necessarily compressed the pre-existing formations into a smaller compass. If more adventurous, you should climb to the peaks, rising to 8000 or 9000 feet above the sea, and flanking the central summits, you may there satisfy yourself, that deposits which were once mere mud, formed during the same time as our slightly consolidated London clay, have been in many parts converted into schists and slates, as crystalline as some of the so-called primary rocks of our islands!

" In speaking of the last changes of the Alps as stupendous, I know it may be said that, in reference to the diameter of the planet, the highest of these mountains and the deepest of these valleys are scarcely. perceptible corrugations of the rind of the earth. But, when we contrast such asperities with many other external features of this rind, they *are* truly stupendous. Let the observer, for example, travel over vast surfaces such as those of Russia, where he cannot detect a single disruption—not one great fracture, and no outbursts whatever of igneous and volcanic rocks, but, on the contrary, a monotonous and horizontal sequence of former aqueous deposits, which, simply dried up, have never been disturbed by any violent revolutions from beneath,—and then compare such features with the adjacent Ural mountains, or still better with the loftier Alps, and he will be truly impressed with the grandeur of such changes.

" And here my auditors will recollect, that even beneath and around this metropolis they can be assured, by finding extinct fossil mammalia, that such also have been the changes, though on a less striking scale, in our own country. The large extinct British quadrupeds necessarily required a great range for their sustenance. They had doubtlessly roamed from distant tracts to our lands before the straits of Dover were formed and before the region of Britain was broken up into isles. One of our great insulating dislocations was, I conceive, coincident with that striking phenomenon in the Alps on which I have tried to rivet your attention, when the first glacial and icy period affected so large a portion of this hemisphere, and when large portions of our northern lands formed the bottoms of an Arctic sea. But such tracts were bidden to rise again from beneath the waters and constitute the present continents and islands before man was placed on the surface. Our race, in short, was not created until the greater revolutions of which I have treated had passed away.

" These grand dislocations belong, therefore, distinctly to former epochs of nature, and their magnitude is enormous when compared with any thing which passes under our eyes, or has been recorded in human history. At the same time geologists have shown upon clear evidences, that during the long periods which were repeatedly interrupted by great revolutions, there was a constant exhibition of diurnal agencies precisely similar to those which now prevail in the world. In those elder times, rain must have fallen—volcanic forces must have been active in scattering ashes far and wide, and in spreading them out together with sheets of lava beneath the waters—gradual movements of oscillation, and moderate elevations and depressions must also have occurred— and long continued abrasion of the sides of mountains must have produced copious accumulations of ' débris ' to encroach upon lakes, the overflow or bursting of which may have sterilized whole tracts.

" All such ordinary changes of the surface of the globe, with occasional slight breaks in the long career, were doubtlessly common to all epochs. But, whilst no such operations can be compared as to effect with those phenomena of disruption and overturning of mountain masses which have been specially dwelt upon this evening, so also, according to my view, it is impossible, that any amount of these small agencies, though continued for millions of years, could have produced such results.

" In thus attempting to shadow out in the space of little more than an hour the chief formations and transmutations of a chain like the Alps, I have probably laboured to effect what many persons may deem impossible ; but I have thought that some at least of these evening discourses should awaken the mind to the larger features of a science, the details of which must be followed out by a careful survey of nature. I would beg, therefore, those persons who have not studied geology practically, to dwell on the facts brought forward, and to believe that they are indisputably and clearly proven. They tell us unmistakeably how different animals and vegetables are entombed in these vast sepulchres of ancient nature, and they reveal to us, that the successive inhabitants of each creation lived during very long periods of time. They announce to us, in emphatic language, how ordinary operations of accumulation were continued tranquilly during very lengthened epochs, and *how such tranquility was broken in upon by great convulsions.*

" Being thus led to ponder upon the long history of successive races and also upon some of the most wonderful physical revolutions this chain has undergone, we cannot avoid arriving at the belief, that, in addition to many other great operations, the disruption which upheaved the middle and younger Tertiary formations from beneath the waters, and threw them up into mountain masses, accompanied by the production of the first great arctic period known in the history of the planet, was a change of immeasurable intensity. That change, in short, by which a period of snow, ice, glaciers, floating ice-bergs, and the transport of huge erratics far from the sources of their origin, suddenly followed a genial and Mediterranean clime! "

INDEX.

Colmers end, Woolhope limestone at, 110.
Colquhoun, Colonel, on Mexican gold, 449.
Combe Martin (N. Devon), 260.
Comus wood, Ludlow, 126.
Conditions, primeval, of the earth's surface, 473.
Conglomerates, in the Silurian rocks of Wales, 70.; of Malvern hills, 94.; in the Silurian rocks of Ayrshire, 160.
Coniston limestone, equivalent of Llandeilo limestone, 146, 147.; grits, equivalent of Caradoc sandstone, 147.
Connemara, Silurian rocks, 168—171.; Bins of altered Lower Silurian? 171.
Constantinople, Silurian and Devonian rocks near, 335.
Continents, aboriginal, mostly submerged, 475.
Conularia, 196. 231.
Conybeare, Rev. W. D., geology of England and Wales, 6.
Copper ore, veins of penetrating from Longmynd rocks into Pentamerus limestone, 27.; veins of at the Longmynd, 38.; at the Stiper stones, 39.; in Lower Silurian, 53.
Coprolites in Upper Ludlow rock, 142. 238.
Corals, of Wenlock shale, 113.; of Wenlock limestone, 119.; Lower Silurian, 178.; palæozoic, unknown in present seas, 210.; Upper Silurian, 212. 215.; Devonian, 256. 260. 371.; carboniferous, 282.; Permian, 308.
Coral reefs, evidences of contiguity of land, 472.
Corbières, palæozoic formations, 394.
Cordier, Baron L., report of, 409.
Cornbrook coal basin, 273.
Cordillera, Silurian rocks, 14. 409.; term why applied to the range of eastern Australia, 450.
Cork, carboniferous rock, 276.
Corndon, Salop, Lower Silurian schists, 28.; Hill, sketch of, 57.; Upper Caradoc near, 82.
Cornstone, of Old Red Sandstone, 244.; fragments of fishes in, 245.; of Murray Firth, 251.; of the Permian deposits, 301.
Cornulites of Lower Silurian, 200.
Cornwall, Lower Silurian and Devonian of, 145. 260.; inversion of strata, 145. 379.; gold in, 434.; tin miners of, recommended by the author to dig for gold in Australia, 451.
Corton, Upper Caradoc, 90.; Lower Wenlock (or Woolhope limestone), 103.
Cotta, B., on the geology of Saxony, 351.
Cossatchi Datchi, hills of, 441.
Coul-beg, mountain of west coast of Highlands, 249.
Coul-more, mountain of west coast of Highlands, 250.
Crania. in Lower Silurian, 187.
Creation, progress of, 460.; of different successive races, 468, 469.
Credner, Herr, map of Thuringia, 299. 360.
Criffel, granite hills of, 162.
Crinoids of Lower Silurian, 180.; of Upper Silurian, 219.; rare in Permian strata, 308.
Cromarty, fossil fishes of, 251.
Crustacea, of Lower Silurian, 200.; of Upper Silurian, 234.; of Devonian rocks, 264.

Crustacea, gigantic traits of in Lower Silurian 413.
Crossopodia, annelid in Lower Silurian, 199.
Cruziana (Bilobites), in England, 42.; in Lower Silurian of Spain, 396.
Cuenca, fossils near, 399.
Culm, or fractured stone coal, Pembroke and Devon, 258, 259.; slash of, in Pembroke, 275.
Cumberland, graptolite schists, 47.; Silurian rocks, 145.; Devonian rocks, 257.
Cumming, Lady Gordon, on fishes of Old Red Sandstone, 251.
Cuvier, Baron, on the succession of vertebrata 468.; Old Red fishes described by, 253.
Cwm-eisen-isaf (N. Wales), gold at, 434.
Cwm Cwyn, section across the hill of, 79.
Cwm Dwr (Brecknockshire), junction of Silurian and Old Red Sandstone, 141.
Cyclas, in Old Red Sandstone, 254.
Cyclopteris, plant of Old Red Sandstone, 255.
Cyrtoceras, genus of cephalopod, 197.
Cytheropsis, a bivalve Crustacean, 236.
Cystideæ, of Lower Silurian, 181.; of Upper Silurian, 217.
Dago, island of, 327.
Dale, Owen D.—(See Owen) works, 15. 422.
Dartmoor, eruptive granite, section to, 256.
Darwin, C., Silurian fossils and works, 15. 409.
Davidson, T. on Devonian fossils in China, 14.; on Silurian brachiopods, 176. 222. 389.
Davis, E., Lingula in oldest Silurian beds discovered by, 44.; on fossils of Lower Wenlock, 105.
Dawson, on coal-fields and a reptile therein, 426.
Dean, Mr. A., gold found by, 433.
Dean, Forest of, carboniferous rocks, 270.; coal basin of, 244.
Dechen, H. von, works of, 338. 378.
Delanoue, geological map by, 390.
De la Beche, Sir H. T.—(See *Dedication, Preface, Appendix.*) On the Silurian rocks of Wales, 8.; on fossiliferous Cambrian being the equivalent of Lower Silurian, 8.; and associates on unconformity of stratification, 50.; on a fault at Musclewick Bay, 65.; sanctions the Silurian classification, 138. 168; on iron oxides. 241.; on Devonian rocks, 258.; sanctions the term Permian, 292.
Desor, on American azoic rocks, 20. 411.
Denbighshire, slaty Wenlock shale, 111.; altered mudstones, 101.
Dendrerpeton Acadianum, 287.
Denudation, valley of, 118.
Derbyshire, coal field, 270.; development of carboniferous limestone, 271.; thickness of coal strata, 273.
Devon, North, section across, 256.
Devonian rocks, formerly merged in 'Grauwacke', 6.; term why applied, 11.; rocks of Britain, 241.; fossils in Ireland, 255.; rocks described, 256. 263.; term proposed by Sedgwick and Murchison, 257.; of Devon and Cornwall, 257.; of South Devon, 259.; of Rhenish provinces, 261. 365.; shells of Lower,

INDEX. 519

Ragleath Hill, sketch of, 81.

Ramsay, Prof., Lower Silurian rocks of North Wales identified with those of Salop, *Preface.* 8. 19.; section on map prepared by, 28. 54.; memoirs in Geological Journal, 79.; Permian rocks of England, 301.; additional observations, *Appendix* B.

Rastrites peregrinus, *woodcut,* 46.

Ratlinghope, 26.

Reise nach dem Ural, 317.

Reptile of Old Red Sandstone, 254.; of coal strata, 287.; of the Permian rocks of Russia, 297.; Permian strata, 313.

Rhayader, Lower Silurian sandstones, 70.

Rhenish Provinces, Devonian rocks, order of older rocks in, 365. 382.

Rhine, greywacke of compared with Devonian strata, 256.; crustaceans in Devonian of, 264.; rocks of, 364.; ascending series on, 369.; Upper Devonian of, 371.; Carboniferous of, 375.; sands slightly auriferous, 436.

Rhynconella in Lower and Upper Silurian, 189. 222.

Ribeiro, Carlos, Silurian rocks in Portugal, 404.

Richardson, Sir J., narrative of Arctic expedition, 411.; rocks of Hudson's Bay, 426.

Richter, R., of Saalfeld, 351.; plants, &c. discovered by, 358.; on Silurian and Devonian of the Thüringerwald, 357. *et seq.*

Rothenberg, Devonian of, 358.

Riesen Gebirge (S. of Breslau), Palæozoic rocks of, 338.; Permian rocks of, 339.

Riley, Dr., on Permian reptiles, 301.

Ringerigge (Norway), section across, 320.

Ripple marks, in Lower Silurian, 38.

Ripon (Yorkshire), Permian sandstones, 303.

Rocks, oldest crystalline effects of great heat, 2.; nuclei of the oldest sedimentary deposits, 21.; general order of the primeval, *woodcut,* 22..; oldest sedimentary in England, 28.; igneous, near mineral veins, 53. 55.; Lower Silurian and eruptive, 61. 77.; palæozoic of Alps and Apennines, 407.; carboniferous, 418.

Rocky Mountains, Silurian rocks of, 18. 409. 427.

Röemer, Adolf, on the Hartz, 362, 363.

Röemer, Ferdinand, on Rhenish rocks, 368. 370. 371. *et seq.;* on American Silurian rocks, 409.

Rogers, Prof. Henry D., on formation of coal, 288.; section by, 424.

Rogers, Profrs. H. and W., United States, 379. 423.

Rose, Gustav, on Russia, 316. 447.; mineral characters of rocks in Prussia, 338.

Ross, county of Scotland, 250.; railroad cutting to, 237.

Rotch, provincial term for mudstone, 101.

Rothe-todte-liegende (Lower Permian deposit), 12.; Flora of, 297, 298.; succession of, 300.

Roundstone Bay (Ireland), section across, 169.

Rowley Hills, basalt of, 116.

Roxburghshire, section across, 152.

Ruppersdorf, Permian rocks and fishes of, 315. 339.

Russell, Lord John, address at Leeds, 250.

Russia, types of British Lower Silurian, 9.; Silurian rocks, their sequence in an unaltered state, 15.; oldest deposits only partially hardened, 17.; forming level plains in, 19.; Cystideæ of, 19. 181.; Silurian strata unsolidified, 34.; organic remains of Old Red Sandstone identical with those of Scotland and Devon, 264.; southern steppes of, 280.; Permian rocks, 292. 300.; primeval rocks of, 322.; Lower Silurian, 323.; Upper Silurian, 325.; Devonian, 327. 332.; Carboniferous, 333, 334.

"Russia in Europe, and the Ural Mountains," 5. 9. 11. 17. 138. 293. 295.; work translated into Russian, 317.

Saalfeld, Germany, Lower Silurian of, 353 ; Devonian, 339. 357.

Saarbrück (Rhenish Bavaria), coal reptile of, 287.

Sabero (Spain), coal apparently in Devonian rocks, 400.

Sablé, section to, 385.

St. Abb's Head, view of, 151.

St. Bride's Bay (Pembroke), Lower Silurian, 65.

St. David's, Longmynd rocks, 62.

St. Omer, beds of coal near, 390.

St. Petersburg, rocks same age as those of Snowdon, 18.; Lower Silurian, 322.

Salt in Permian deposits, 295.

Salter, J. W., contributions to this work, *Preface* and *passim;* on fossil-bearing rocks of N. Wales, 8.; Cruziana and Hymenocaris described by, 39.; on Silurian encrinites, 113.; organic remains of the Lower Silurian, 176. *et seq.;* Upper Silurian, 207, *et seq.;* on North American fossils, 421.; table of fossils common to Lower and Upper Silurian, *Appendix* A.

Sanchez, Eusebio, on Spanish geology, 397.;

Sandberger, F., works on fossil remains, 264. 345. 365.; (the brothers) on slaty rocks, 369. 371.

Sandstone, Lower Silurian, 51. 71.

Sardinia, Silurian and other palæozoic rocks, 13. 405.

Sarthe, anthracite, 392.

Savi, P., memoir, 406.

Saurians Carboniferous, 313.; Permian, 313.; abundant in the Oolitic group, 466; *Appendix* P.

Saxony, Permian tree ferns of, 309.; Silurian, Devonian, Carboniferous, and Permian rocks, 12. 349, *et seq.;* contains no Upper Silurian, 361.

Scandinavia, oldest traces of life in, 9. 15. 20. 43. 317. 321.

Schists altered into slates, 20.; contorted at Anglesea, 24.; black, of Malvern, 92.; in the Highlands of Lower Silurian age?, 163. 248.

Schleitz, geological changes near, 354.

Schrenk, A. G., on Upper Silurian of Russia, 326.

Schultz, M., on rocks in Gallicia, 395.

Scotland, south of, Silurian rocks and fossils, 47. 149, *et seq.* ; section of Silurian rocks, 152. ;

THE END.

LONDON:
A. and G. A. SPOTTISWOODE,
New-street-Square.

PLATE I.

(The figures in this and the following plates are all transferred from the original engravings in the "Silurian System." In every case where the old names have been changed in accordance with the modern nomenclature, the name used in the original work is inserted in brackets, for the purpose of reference.)

LOWER SILURIAN CRUSTACEA.

Fig. 1. TENTACULITES ANNULATUS, Schlotheim. *a*, portion magnified. Acton Scott, Shropshire.

2. ————————————, the same. Interior casts in sandstone. (T. scalaris.)

3. ASAPHUS TYRANNUS, Murch. Tail of a young specimen. Llandeilo.

4. The largest specimen known. Same locality. The original of this figure is in the British Museum. It is a mould in sandstone of the exterior ornamented surface.

PLATE II.

LOWER SILURIAN CRUSTACEA.

Fig. 1. ASAPHUS TYRANNUS, Murch. Tail, and two thorax segments, of a moderate-sized individual. Llandeilo, Caermarthenshire.

2. ASAPHUS POWISII, Murch. Body and tail of full-sized specimen. Welshpool.

3, 4. HOMALONOTUS (Asaphus) VULCANI, Murch. Cornden Hill, Shropshire.

LOWER SILURIAN.

CRUSTACEA.

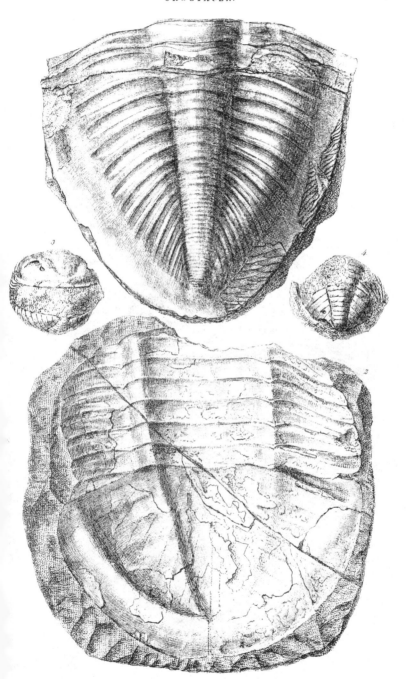

J.D.C. Sowerby

Ford & West, Imp.

CRUSTACEA.

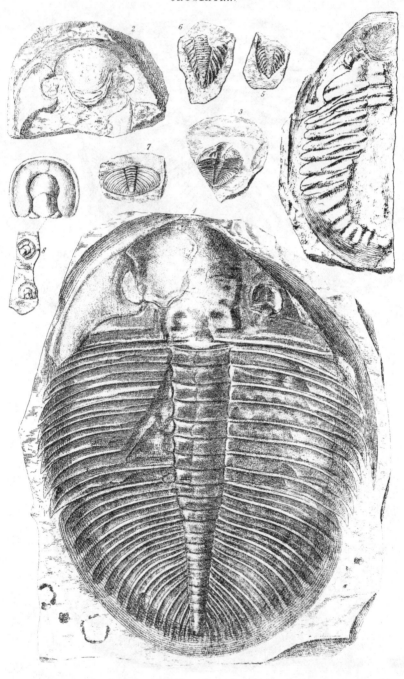

PLATE III.

LOWER SILURIAN CRUSTACEA.

Fig. 1. OGYGIA (Asaphus) BUCHII, Brongniart. Builth, Radnorshire.

2. Hypostome, or upper lip of ditto, fixed to the underside of the head. Llandeilo.

3. A very young specimen. Builth. (Trinucleus Asaphoides.)

4. OGYGIA (Asaphus) CORNDENSIS, Murch. From near the flanks of the Cornden Hill, Shropshire. The figure (4) is omitted, it applies to the right hand specimen, at the top of the plate.

5, 6. ENCRINURUS (Calymene?) PUNCTATUS, Brünnich. Upper Caradoc sandstone, Bogmine, Shropshire.

7. CALYMENE (Asaphus) DUPLICATA, Murch. Wilmington, Salop.

8. AGNOSTUS McCOYII, Salter. (A. pisiformis.) Builth, Radnorshire. Natural size, and the head magnified.

PLATE IV.

LOWER SILURIAN CRUSTACEA.

Fig. 1. STYGINA (Ogygia) MURCHISONÆ, Murch. Mount Pleasant, Caermarthen.

2, 4, 5. TRINUCLEUS CONCENTRICUS, Eaton. (T. Caractaci.) Welshpool.

3. Portion of fringe, highly magnified.

6. TRIN. LLOYDII, Murch. Llangadock, Caermarthenshire.

7. TRIN. FIMBRIATUS, Murch. Builth.

8. TRIN. RADIATUS, Murch. Welshpool.

9, 10. AMPYX (Trinucleus) NUDUS, Murch. Builth, Radnorshire.

11, 12. PHACOPS CONOPHTHALMUS, Bœck. (Head of Asaphus Powisii, Sil. Syst.) Horderley, and Acton Scott, Shropshire.

13, 14. ILLÆNUS PEROVALIS, Murch. N. E. end of the Cornden Hill, Shropshire.

CRUSTACEA.

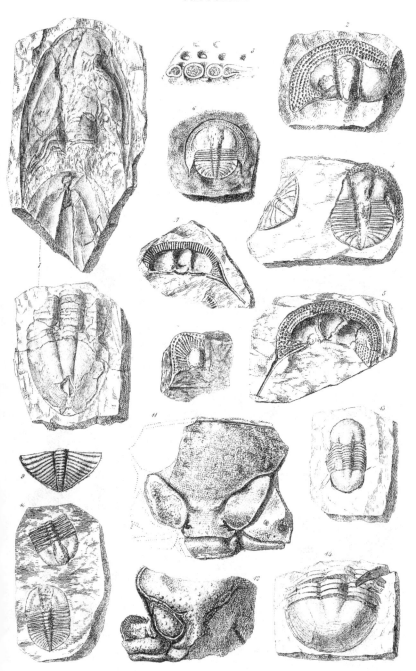

BRACHIOPODA.

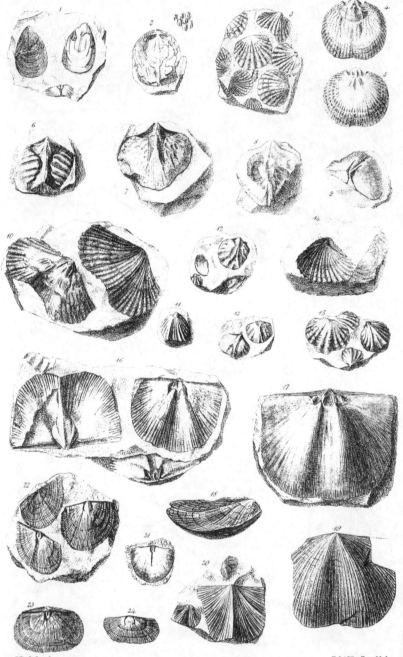

JD C Sowerby

Forä & West, Transf. lith

PLATE V.

LOWER SILURIAN BRACHIOPODA.

Fig. 1. LINGULA ATTENUATA, Sow. Llandeilo.

2. ORBICULA PUNCTATA, Sow. Cheney Longville, Shropshire.

3. ATRYPA HEMISPHÆRICA, Sow. Upper Caradoc of May Hill; Abberley, &c.

4, 5. ATR. RETICULARIS (affinis), Linn., var. orbicularis, Sow. Upper Caradoc, Presteign.

6, 7, 8. ATRYPA? CRASSA, Sow. Llandovery.

9. SPIRIFER PLICATELLUS (radiatus), Linn.—variety. Castell Craig Gwyddon, Llandovery.

10. RHYNCHONELLA (Terebratula) TRIPARTITA, Sow. Goleugoed, Llandovery.

11. RHYN. (Terebratula) PUSILLA, Sow. Cefn Rhyddan, Llandovery.

12, 13. RHYN. (Terebrat.) FURCATA, Sow. Upper Caradoc, Bogmine, Shropshire.

14. RHYN. (Terebrat.) NEGLECTA, Sow. Llandovery.

15. RHYN. (Terebrat.) DECEMPLICATA, Sow. Upper Caradoc, Longmynd; Malverns, &c.

16, 17, 18, 19, 20. ORTHIS VESPERTILIO, Sow. Shropshire; Bala. Including (O. bilobata.)

21. ORTHIS TRIANGULARIS, Sow. Near Chirbury, Salop; in Llandeilo flags.

22, 23, 24. O. LATA, Sow. (Including O. protensa.) Llandovery.

PLATE VI.

LOWER SILURIAN BRACHIOPODA.

Fig.   1, 2.   ORTHIS TESTUDINARIA, Dalman. Horderley ; Meifod.

      3, 4.   ORTHIS (Terebratula) UNGUIS, Sow. Horderley and Cheney Longville, Shropshire.

        5.   O. ELEGANTULA (canalis), Dalman. Meifod ; Bala.

        6.   O. (Spirifer) ALATA, Sow. Mount Pleasant, Carmarthen.

    7, 8, 9.   ORTHIS CALLIGRAMMA, Dalm. (Spirifer plicatus, Orthis callactis, O. virgata.) Llandovery ; Horderley ; Bala. Fig. 8. is from Upper Caradoc beds, Shropshire.

     10.   O. RADIANS, Sow. Near Builth.

     11.   O. ACTONIÆ, Sow. Acton Scott, Shropshire ; Bala.

     12.   O. FLABELLULUM, Sow. Hopesay, Shropshire ; Bala.

  13, 14.   LEPTÆNA SERICEA, Sow. Bala ; Whittingslow and Horderley, Shropshire ; Meifod ; Llandovery.

     15.   L. TRANSVERSALIS (duplicata), Dalm. Robeston Wathen, Pembrokeshire.

BRACHIOPODA.

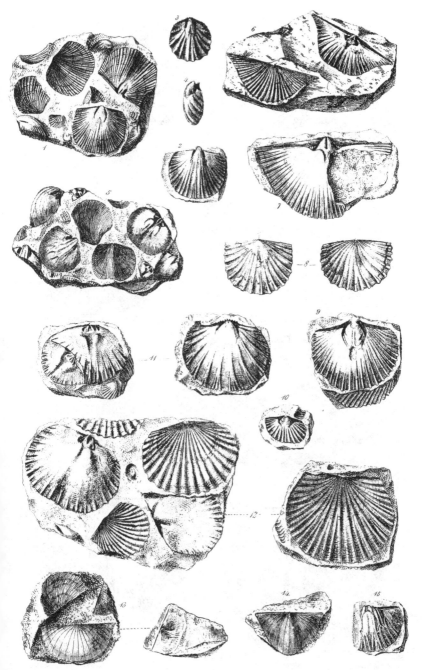

Ford & West Imp.[?] lith.

BRACHIOPODA

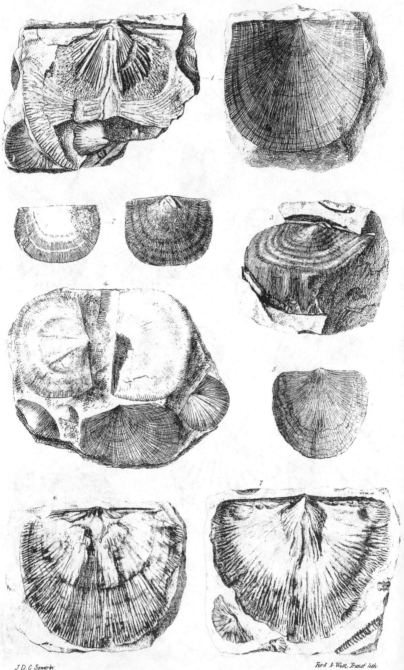

J. D. C. Sowerby

Ford & West, Trans.lith.

PLATE VII.

LOWER SILURIAN BRACHIOPODA.

Fig. 1. STROPHOMENA EXPANSA, Sow. (Orthis expansa, and O. Pecten.) Meifod.

2. STROPH. COMPRESSA, Sow. Upper Caradoc, Hope Quarry Shropshire.

3. S. (Leptæna) TENUISTRIATA, Sow. Guilsfield, Welshpool.

4. ORTHIS ALTERNATA, Sow. Cheney Longville, Shropshire; Bala.

5. STROPHOMENA (Leptæna) COMPLANATA, Sow. Acton Burnell, Shropshire.

6, 7. S. (Orthis) GRANDIS. Soudley and Acton Scott, Shropshire; Bala.

PLATE VIII.

LOWER SILURIAN BRACHIOPODA.

Fig. 1, 2, 3. PENTAMERUS OBLONGUS, Sow. Uppermost Caradoc, Shropshire; Llandeilo flags, Llandovery, &c., in S. Wales.

    4. Broader variety of the same (P. lævis). The Hollies, Shropshire; in Upper Caradoc.

5, 6, 7. PENTAMERUS (Atrypa) UNDATUS, Sow. Llandovery; and Pembrokeshire.

    8. P. (Atrypa) GLOBOSUS, Sow. Gorllwyn fach, Llandovery.

9, 10. P. (Atrypa) LENS, Sow. Llandovery; in Llandeilo flags. Lickey Quartz, Bromsgrove (Upper Caradoc).

    11. Portion of ditto (Spirifer? lævis). Upper Caradoc, May Hill.

LAMELLIBRANCHIATA & GASTEROPODA

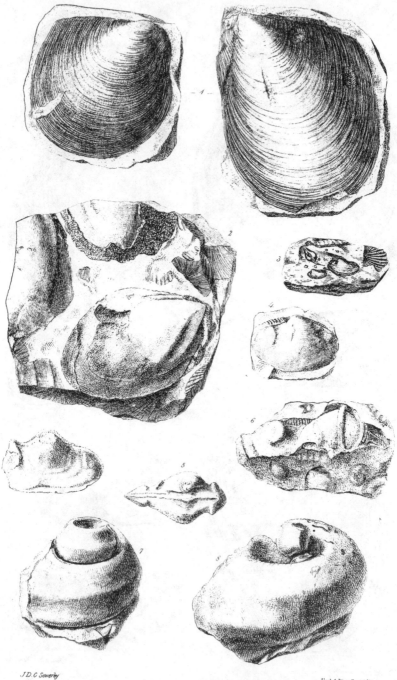

PLATE IX.

LOWER SILURIAN BIVALVE AND UNIVALVE SHELLS.

Fig. 1. MODIOLOPSIS or MODIOLA (Avicula) ORBICULARIS, Sow. Horderley, Shropshire.

2. MODIOLOPSIS? (Avicula) OBLIQUA, Sow. Soudley, Shropshire.

3. NUCULA LÆVIS, Sow. Pensarn, near Caermarthen.

4. NUCULA? (Arca) EASTNORI, Sow. Upper Carad c, Malverns.

5, 6. NUCULA? SUBÆQUALIS (Arca Eastnori), McCoy. Same locality.

7. PLEUROTOMARIA ANGULATA, Sow. Mandinam, Llandovery.

8. TURBO? PRYCEÆ, Sow. Llandovery. Same locality.

PLATE X.

LOWER SILURIAN UNIVALVES.

Fig. 1. HOLOPEA (Littorina) STRIATELLA, Sow. Horderley.

2. SCALITES (Trochus) LENTICULARIS, Sow. Old Storridge Hill, Worcester; Upper Caradoc.

3. MACROCHEILUS (Buccinum?) FUSIFORME, Sow. Upper Caradoc, Nash Scar, Presteign.

4 HOLOPELLA (Turritella) CANCELLATA, Sow. Llandovery.

5. BELLEROPHON PERTURBATUS, Sow. (Euomphalus tenui-striatus.) Middleton, Cornden Hills.

6. EUOMPHALUS CORNDENSIS, Sow. Leigh Hall, Cornden. (a, magnified.)

7. BELLEROPHON ACUTUS, Sow. Horderley.

8, 9. ———— BILOBATUS, Sow. Horderley.

10. ———— (Euomphalus) PERTURBATUS, Sow. Pensarn, Caermarthen, in Llandeilo flags. (See also fig. 5.)

GASTEROPODA & HETEROPODA.

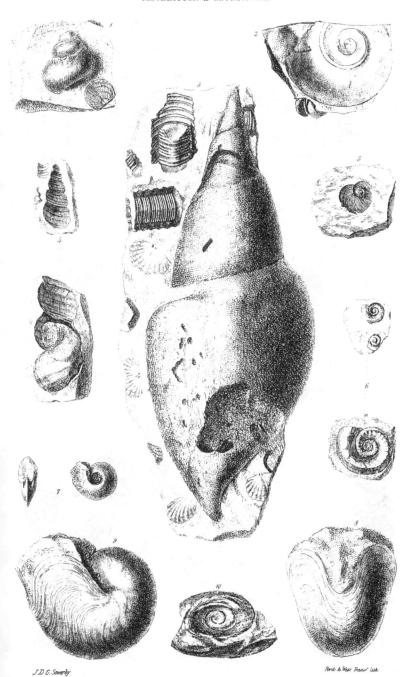

CEPHALOPODA.

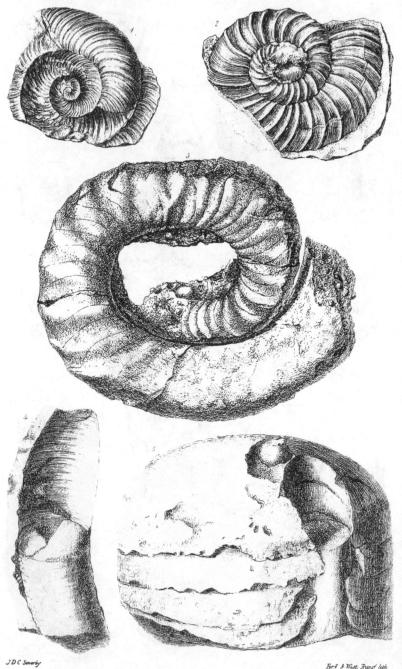

J.D.C. Sowerby.

Ford & West, Imp^t lith.

PLATE XI.

LOWER SILURIAN CEPHALOPODA.

Fig. 1. LITUITES CORNU-ARIETIS, Sow. Var. α, placed here for comparison, is from the very base of the Upper Silurian at Corton, Presteign ; but it is also found in the Upper Caradoc of Bogmine, Shropshire, and at Bala in Llandeilo flags.

     2. ———————————————————— Var. β. Cefn-y-Garreg, Llandovery; in Llandeilo flags.

     3. LITUITES (Nautilus) UNDOSUS, Sow. Blaen-y-cwm, Llandovery.

     4. CYRTOCERAS (Orthoceras) APPROXIMATUM, Sow. Eastnor Park, in Upper Caradoc.

     5. DIPLOCERAS (Orthoc.) BISIPHONATUM, Sow. Gorllwyn fach, Llandovery.

PLATE XII.

GRAPTOLITES AND "INCERTÆ SEDIS," LOWER SILURIAN.

Fig. 1. DIDYMOGRAPSUS (Graptolithus) MURCHISONÆ, Bœck. Llandrindod Hills, Radnorshire.

2. DIPLOGRAPSUS (Grapt.) FOLIACEUS, Murch. Meadow Town, near Shelve, Shropshire. Accompanied by the small ORBICULA PORTLOCKI, Geinitz.

3. GLYPTOCRINUS —— sp. Upper Caradoc, Nash Scar, Presteign. The disjointed stems of this genus are very common, both in the sandy portions of the Llandeilo flags, and in the Caradoc sandstone. The central canal is always very large.

UPPER SILURIAN.

4. GRAPTOLITHUS PRIODON (Ludensis), Bronn. 4a, magnified. Llanfair, Welshpool.

5. SPONGARIUM INTERRUPTUM Milne Edw. (a Calciphyte). Upper Ludlow Rock, Birchen Common, Aymestry.

6. ISCHADITES KÖNIGII, Murch. Probably a Cystidean. Lower Ludlow Rock, Ludlow.

LOWER SILURIAN.

GRAPTOLITES & INCERLÆSEDIS.

Pl.12.

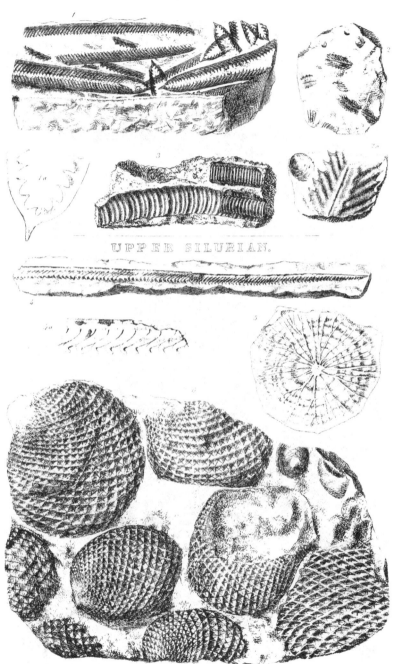

UPPER SILURIAN.

JDC Sowerby

Ford & West, Imp.t lith

CRINOIDEA.

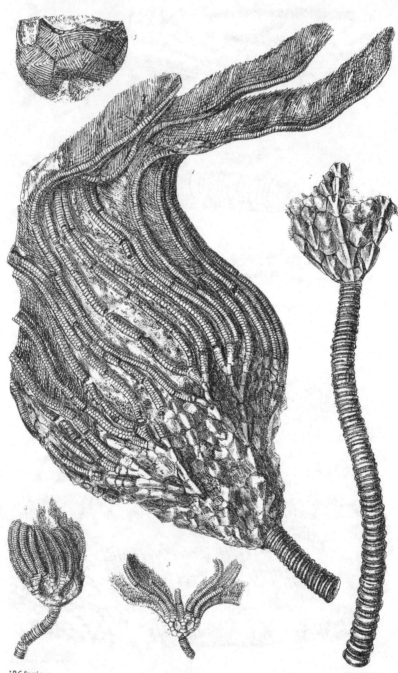

PLATE XIII.

UPPER SILURIAN CRINOIDEA.

PLATE XIV.

UPPER SILURIAN CRINOIDEA.

Fig. 1. MARSUPIOCRINUS CŒLATUS, Phill. Dudley.

2. EUCALYPTOCRINUS (Hypanthocrinus) DECORUS, Phill. Dudley. 2a, outlines of the pelvic plates.

3. ICTHYOCRINUS ? (Actinocrinus) GONIODACTYLUS, Phill. Dudley.

4. CYATHOCRINUS TESSERACONTADACTYLUS (Actinocrinites simplex), Hisinger ? Dudley.

5, 6. CYATH. TUBERCULATUS, Miller. Dudley.

7. ICTHYOCRINUS ? (Actinocrinus) ARTHRITICUS, Phill. Dudley. 7*, a few joints with tentacles, magnified.

8. ICTH. (Cyathocrinus) PYRIFORMIS, Miller. Dudley.

9. ACTINOCRINUS RETIARIUS, Phill. Dudley.

CRINOIDEA

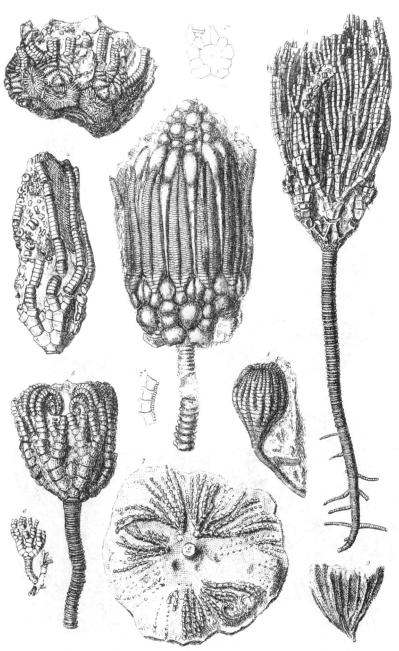

CRINOIDEA.

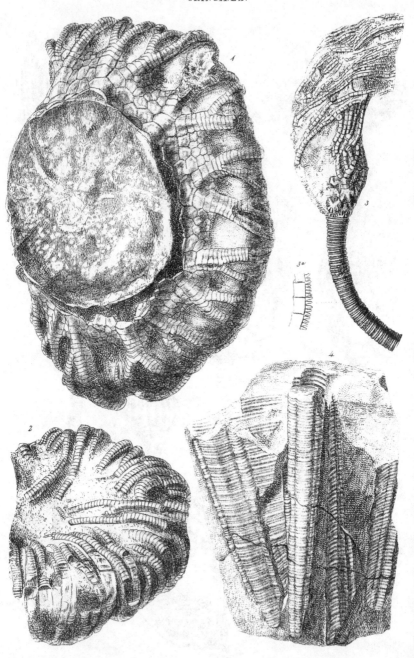

PLATE XV.

UPPER SILURIAN CRINOIDEA.

Fig. 1, 2. GLYPTOCRINUS (Actinocrinus) EXPANSUS, Phill. Dudley.

  3. ICTHYOCRINUS? (Actinocrinus) CAPILLARIS, Phill. Fig. 3*a*,
      a magnified portion of the finger. Wenlock and Dudley.

  4. Irregular inverted pyramid, formed in the mudstone of the
      Upper Ludlow Rock, by the rotary motion of Encrinite
      stems (one is seen at the extreme left hand of the fossil) in
      the soft micaceous matrix: see p. 137. (Cophinus dubius, Sil.
      Syst.) Upper Ludlow Rock, Ludlow.

PLATE XVI.

UPPER SILURIAN ANNELIDA.

Fig.   1. SERPULITES LONGISSIMUS, Murch. Ludlow.

      2. SPIRORBIS LEWISII, Sow. Natural size, on the inner surface of
a Lituites, and 2a magnified. Wenlock Limest. Ledbury.
Common in Upper Ludlow Rock.

3 to 9. CORNULITES SERPULARIUS, Schloth.

3 to 7. Are of the natural size, and show the tubes attached in twos or
threes to shells, &c. Fig. 8. is a longitudinal section of a full
grown tube, and shows the cellular structure at the successive
varices of growth.

3a, and 9. Are magnified portions of young tubes, shewing the outer
striated coat covering the cellular layers. These layers are
better seen in fig. 10. as well as the pits left by them on the
edges of the internal cast; also the impressed longitudinal
lines running down the inner surface. Malverns.

     11. TENTACULITES ORNATUS, Sow. Natural size, magnified, and a
section. Dudley.

     12. T. TENUIS, Sow. Natural size, and magnified. Usk, Monmouth
shire, in Upper Ludlow Rock.

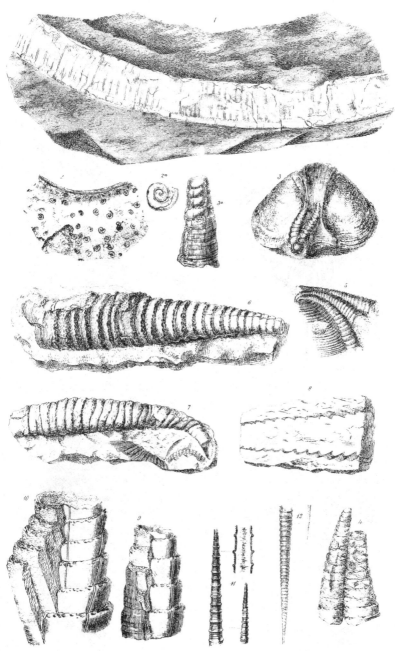

CRUSTACEA.

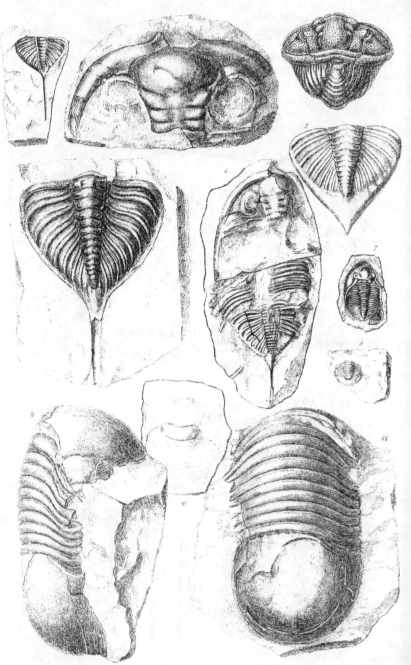

PLATE XVII.

UPPER SILURIAN TRILOBITES.

Fig. 1. CALYMENE BLUMENBACHII, Brongn. W. L.   Dudley.

2. PHACOPS (Asaphus) CAUDATUS, Brongn. W. L. Dudley.

†3 to 6. P. LONGICAUDATUS, Murch. W. Sh. Wistanstow and Burrington,
Shropshire.

7. PROETUS (Asaphus) STOKESII, Murch. W. L. Dudley.

8. PROETUS—sp. Wenlock Limestone, Ledbury, Malverns.

9 to 11. ILLŒNUS (Section—Bumastus) BARRIENSIS, Murch. 10. Side View
of the Head and Eye. Woolhope Limest., Barr, Staffordshire;
Malverns, &c.

* In this and the succeeding pages, U. L. stands for Upper Ludlow;
A. L., Aymestry Limestone; L. L., Lower Ludlow; W. L., Wenlock
Limestone; W. Sh., Wenlock Shale.

† Figs. 3 to 5 are omitted — they are at the left hand upper corner.

PLATE XVIII.

UPPER SILURIAN CRUSTACEA.

Fig. 1. PHACOPS (Asaphus) CAUDATUS, Brongn. W. L. Dudley.

2, 5. PHAC. (Calymene) DOWNINGIÆ, Murch. W. L. Dudley.

3, 4. Tails of young Specimens, (Asaphus Cawdori, and A. subcaudatus, Sil. Syst.) From Ludlow Rocks, Pembrokeshire.

6. PHACOPS STOKESII, Milne Edw. (Calymene macrophthalma.) W. L. Dudley and Wallsall.

7, 8. ACIDASPIS BRIGHTII, Murch. (8. Paradoxides quadrimucronatus.) W. L. Dudley.

9. ENCRINURUS (Calymene) VARIOLARIS, Brongn. W. L. Dudley.

10. CALYMENE BLUMENBACHII, Brongn. W. L. Dudley.

11. C. TUBERCULOSA, Salter. (C. Blumenbachii var., Sil. Syst.) W. Sh. Burrington, Shropshire.

CRUSTACEA.

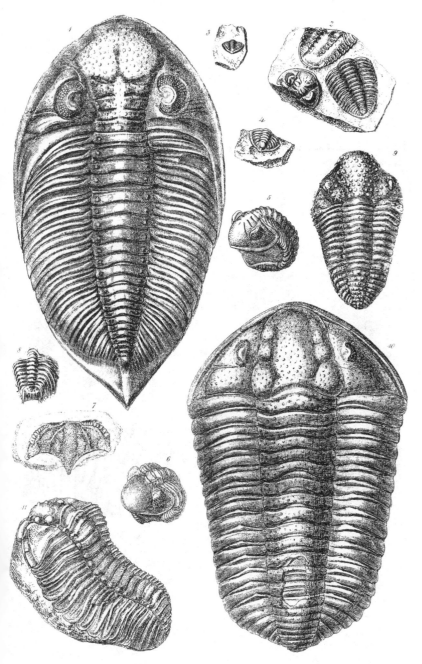

CRUSTACEA.

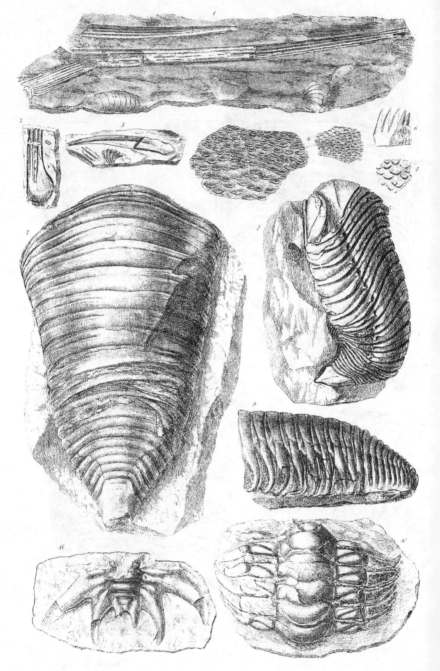

PLATE XIX.

UPPER SILURIAN TRILOBITES, AND OTHER CRUSTACEA.

Figs. 1, 2. LEPTOCHELES (Onchus) MURCHISONI, McCoy.—spines of the Tail. Uppermost Ludlow Rock, near Ludlow.

3. LEPTOCHELES—sp. fragment of tail spine. Same locality.

4, 5. PTERYGOTUS PROBLEMATICUS, Agass. Portions of the Carapace, nat. size, and magnified. Ludford, in the 'Bone Bed.'

6. Portion of the Base of one of the jaw-feet. Same locality.

7 to 9. HOMALONOTUS KNIGHTII, (including H. Ludensis, S. Syst.) Upper Ludlow Rock, Ludlow.

10, 11. CHEIRURUS (Paradoxides) BIMUCRONATUS, Murch. 10, the Body Segments; and 11, the tail': the former is placed in an inverted position, the anterior end downwards. (It was so engraved in the original work.) Wenlock Limestone, Malverns.

PLATE XX.

UPPER SILURIAN BRACHIOPODA.

Figs. 1, 2. ORBICULA RUGATA, Sow. U. L., Ludlow.

    3. —————— STRIATA, Sow. U. L., Delbury, Shropshire.

    4. CRANIA (Patella) IMPLICATA, Sow. W. L., Abberley.

    5. LINGULA LEWISII, Sow. A. L., Mary Knoll, Ludlow.

    6. L. LATA, Sow. L. L., Elton.

    7. L.? STRIATA, Sow. L. L., near Aymestry.

    8. CHONETES (Leptæna) LATA, Von Buch. U. L., Ludlow.

    9. ORTHIS ELEGANTULA, var. ORBICULARIS. (O. orbicularis, Sow.)
        Exterior and internal cast. U. L., Ludlow.

   10. ORTHIS CALLIGRAMMA, Dalm. Var. RUSTICA. (O. rustica, S.
        Syst.) W. L. Valley of Woolhope.

   11. O. LUNATA, Sow. Exterior and internal casts of both valves.
        U. L., Ludlow.

   12. O. ELEGANTULA (canalis), Dalm. W. L., Dormington, Woolhope.

   13. O. HYBRIDA, Sow. W. L., Wallsall.

   14. O. BILOBA, (sinuata), Linn. W. L., Wallsall.

   15. LEPTÆNA LÆVIGATA, Sow. W. Sh., Burrington.

   16. L. MINIMA, Sow. Natural size and magnified. W. Sh., Burrington.

   17. L. TRANSVERSALIS, Dalm. W. Sh., Buildwas.

   18. STROPHOMENA (Leptæna) ANTIQUATA, Sow. W. Sh., Woolhope.

   19. S. (Leptæna) EUGLYPHA, Dalm. W. L., Dudley.

   20. S. (Leptæna) DEPRESSA, Sow. W. L., Dudley.

Pl. 20.

BRACHIOPODA.

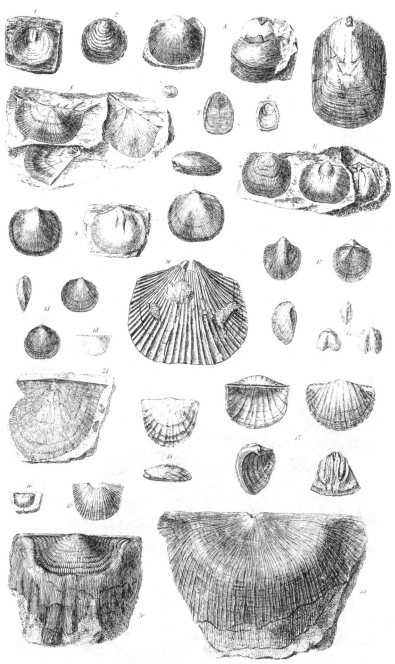

BRACHIOPODA.

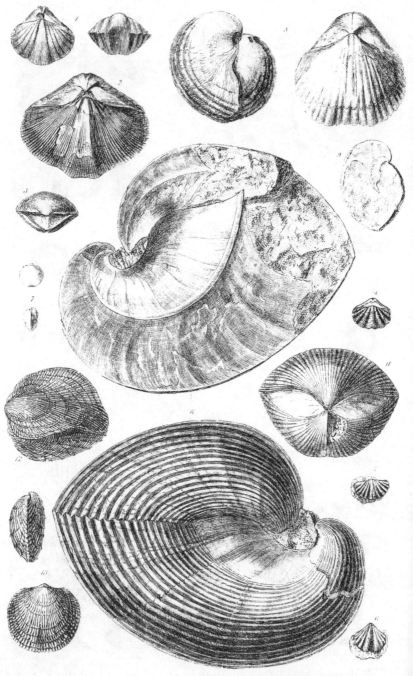

PLATE XXI.

UPPER SILURIAN BRACHIOPODA.

Fig. 1. SPIRIFER PLICATELLUS, Linn. var. INTERLINEATUS, (Sp. interlineatus). A. L., Aymestry.

2. SPIRIFER PLICATELLUS (radiatus), Linn. W. L. Malverns.

3. S. TRAPEZOIDALIS, Dalm. U. L., Usk.

4. S. CRISPUS, Linn. W. L., Dudley.

5, 6. S. ELEVATUS, Dalm. (octoplicatus). W. L., Abberley.

7. S.? PISUM, Sow. W. L., Wallsall.

8, 9. PENTAMERUS (Atrypa) GALEATUS, Dalm. W. L., Wenlock Edge; Dudley; Pembrokeshire.

10. P. KNIGHTII, Sow. A. L. Downton on the Rock; Sedgley.

11. ——— var. AYLESFORDII. Same localities.

12, 13. ATRYPA RETICULARIS (affinis), Linn. A. L., W. L., Shropshire; May Hill; Dudley, &c. Woolhope L. Woolhope.

PLATE XXII.

UPPER SILURIAN BRACHIOPODA.

Fig. 1, 2. RHYNCHONELLA NUCULA, Sow. (Terebratula lacunosa & T. nucula.) U. L. Radnorshire; Shropshire.

  3. R. (Terebrat) PENTAGONA, Sow. U. L., Delbury, Shropshire.

  4. R. BOREALIS, Schloth. (Terebrat. lacunosa.) W. L., Wenlock Edge.

  5. ———— var. DIODONTA (Ter. bidentata.) W. L., Dudley.

  6. R. (Ter.) CRISPATA, Sow. Woolhope limestone, Nash Scar.

  7. R. (Ter.) CREBRICOSTA, Sow. W. S., Tynewydd, Llandovery.

  8. ATRYPA (Ter.) CUNEATA, Dalm. W. L., Dudley.

  9. RHYNCHONELLA (Ter.) SPHŒRICA, Sow. W. S., Wallsall.

  10. R. DEFLEXA, Sow. (Terebr. deflexa, & T. interplicata). W. L., Wenlock Edge; W. S., Wallsall.

  11. R. (Ter.) STRICKLANDI, Sow. W. L., Longhope. W. S., Usk.

  12. R. (Ter.) NAVICULA, Sow. A. L., Ludlow; Usk. W. S.,Builth.

  13. R. (Ter.) WILSONI, Sow. A. L., Aymestry.

  14. TEREBRATULA † LÆVIUSCULA, Sow. W. S., Tynewydd, Llandovery.

  15. RHYNCHONELLA DIDYMA, Sow. (Ter. didyma & T. canalis) U. L., A. L., Usk, Abberley.

  16. R. (Atrypa) OBOVATA, Sow. L. L., Malverns.

  17. R. (Atrypa) DEPRESSA, Sow. W. S., Stumps Wood, Malverns.

  18. R. (Atrypa) ROTUNDA, Sow. W. L., Wenlock Edge.

  19. ATRYPA (Ter.) MARGINALIS, Dalm. (including T. imbricata, W. L.). Wenlock, &c.

  20. ATHYRIS TUMIDA, Dalm. (tenuistriata). W. L., Malverns.

  21. PENTAMERUS (Atrypa) LINGUIFER, Sow. W. Sh., Stumps Wood, Malverns; Wallsall, &c.

  22. RHYNCHONELLA† (Atrypa) COMPRESSA, Sow. Woolhope Limestone, Woodside & Nash, Presteign.

BRACHIOPODA.

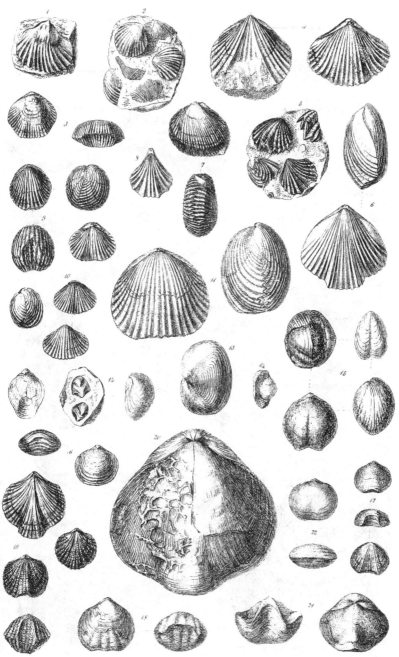

LAMELLIBRANCHIATA.

PLATE XXIII.

UPPER SILURIAN LAMELLIBRANCHIATA.

Fig. 1. MODIOLOPSIS (Pullastra) COMPLANATA, Sow. U. L., Near Bridgnorth.

    2. GONIOPHORA (Cypricardia) CYMBÆFORMIS, Sow. U. L., Ludlow.

    3. ORTHONOTA (Cypricardia) IMPRESSA, Sow. U. L., Delbury.

    4. O. (Cypric.) UNDATA, Sow. U. L., near Aymestry.

    5. O. (Mya) ROTUNDATA, Sow. A. L., Caynham Camp, Ludlow.

    6. (Two figures,) O. (Cypricardia) AMYGDALINA, Sow. U. L., Ludlow, &c.

    7. Variety of Ditto, (Cypricardia retusa, Sow.) U. L., Delbury.

    8. O. (Psammobia) RIGIDA, Sow. L. L., Garden House, Aymestry.

    9. O. (Cypricardia?) SOLENOIDES, Sow. L. L., near Ludlow.

    10. NUCULA ANGLICA (ovalis), D'Orb. U. L., Trewerne Hills, Radnorshire.

    11. CARDIOLA FIBROSA, Sow. L. L., Mary Knoll Dingle, Ludlow.

    12. C. INTERRUPTA, Broderip. L. L., Aymestry; Radnorshire, &c.

    13. C. (Cardium) STRIATA, Sow. A. L., Aymestry.

    14. MODIOLA ANTIQUA, Sow. W. Sh., Glass House Hill, E. of May Hill, Gloucestershire.

    15. PTERINEA SOWERBYI, McCoy. (Avicula reticulata), A. L., Croft, Aymestry.

    16. P. LINEATULA, D'Orb. (Avic. lineata), U. L., Ludlow.

    17. P. (Avic.) RETROFLEXA, Wahl. U. L., Malverns.

PLATE XXIV.

UPPER SILURIAN GASTEROPODA.

Fig. 1. TURBO CORALLII, Sow. U. L., Radnorshire.

2. MURCHISONIA (Pleurotoma) ARTICULATA, Sow. U. L., Ludlow.

3. LOXONEMA SINUOSA, Sow. L. L., Aymestry.

4. TURBO OCTAVIUS (carinatus), D'Orb. U. L., Trewerne Hills, Radnorshire.

5. MURCHISONIA (Pleurotomaria) LLOYDII, Sow. L. L., Shelderton; Aymestry.

6. PLEUROTOMARIA UNDATA, Sow. L. L., near Ludlow.

7. MURCHISONIA (Pleurotoma) CORALLII, Sow. U. L., Bradnor Hill, Kington.

8. ACROCULIA PROTOTYPA, Phill. (Nerita spirata). Woolhope Limestone, Presteign.

9. ACROC. (Nerita) HALIOTIS, Sow. W. L., Ledbury

10. TURBO CIRRHOSUS, Sow. W. S., Wenlock.

11. EUOMPHALUS CARINATUS, Sow. A. L., Aymestry.

12. E. DISCORS, Sow. W. L., Wenlock Edge.

13. E. RUGOSUS, Sow. Same locality.

Pl. 24

GASTEROPODA.

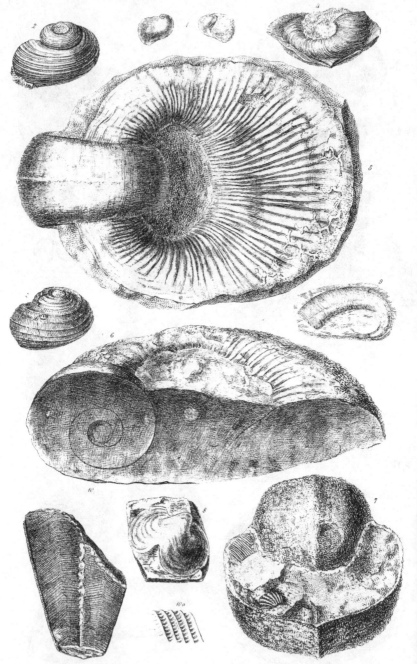

PLATE XXV.

UPPER SILURIAN GASTEROPODA, PTEROPODA, AND HETEROPODA.

1. NATICA PARVA, Sow. U. L., Fownhope, Herefordshire.

2. EUOMPHALUS SCULPTUS, Sow. W. L., Ledbury.

3. E. FUNATUS, Sow. W. L., Wenlock Edge.

4. E. ALATUS, Sow. W. S.

5, 6. BELLEROPHON DILATATUS, Sow. W. L., Burrington, Ludlow.

7. B. WENLOCKENSIS, Sow. W. L., Croft, Malvern.

6. B. EXPANSUS, Sow. U. L., Ludlow.

9. CYRTOLITES? (Cyrtoceras) LÆVIS. L. L., Abberley.

10. CONULARIA SOWERBYI (quadrisulcata), Sow. The figure (10) is placed too near fig. 6. W. L., Wenlock Edge.

PLATE XXVI.

UPPER SILURIAN CEPHALOPODA.

1. ORTHOCERAS ANNULATUM, Sow. W. L., Hay Head, Wallsall.

2. —————— var. FIMBRIATUM. (O. fimbriatum, Sow.) W.S. Aston, May Hill ; Malverns, &c.

3. O. ATTENUATUM, Sow. W. S., Onny River, Shropshire.

4. O. DISTANS, Sow. L. L., near Aymestry.

5. O. NUMMULARIUS, Sow. W. L., Whitfield Quarry, Tortworth.

CEPHALOPODA.

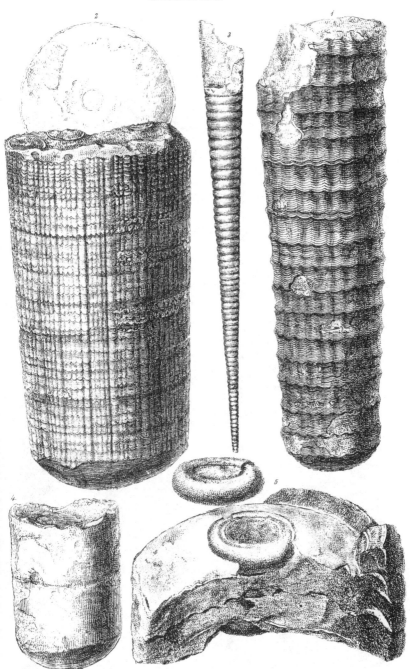

CEPHALOPODA.

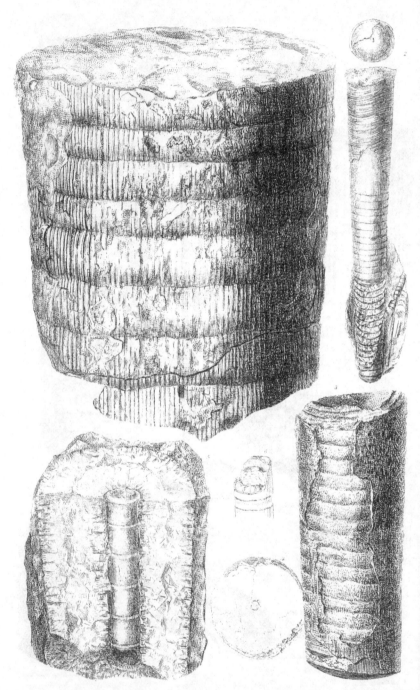

PLATE XXVII.

UPPER SILURIAN CEPHALOPODA.

Fig. 1. ORTHOCERAS FILOSUM, Sow. L. L., Leintwardine.

   2. O. GREGARIUM, Sow. L. L., near Ludlow.

3, 4. O. EXCENTRICUM, Sow. Woolhope Limest., Old Radnorshire.

   5. ORMOCERAS (Orthoceras) BRIGHTII, Sow. — broken open.
      W. L., Malverns.

   6. The siphuncle, slightly magnified.

PLATE XXVIII.

UPPER SILURIAN CEPHALOPODA.

Fig. 1, 2. ORTHOCERAS LUDENSE, Sow. L. L., Ludlow.

3. O. CANALICULATUM, Sow. W. S., Ledbury.

4. O. ANGULATUM (virgatum), Hisinger. L. L , Mocktree Forest.

5. O. DIMIDIATUM, Sow. L. L., Water-break-its-neck, Radnor Forest.

CEPHALOPODA.

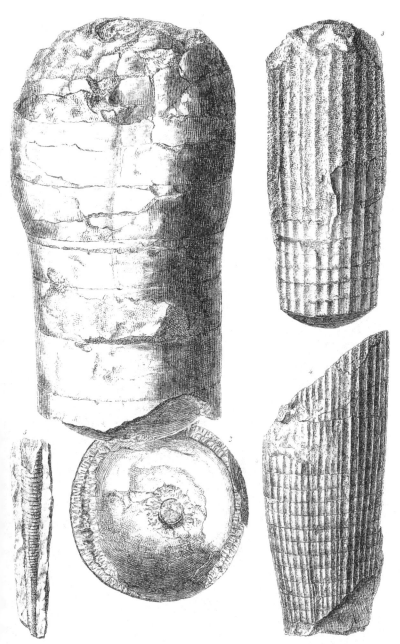

J. D. C. Sowerby.

Ford & West, Imp.t lith.t

CEPHALOPODA.

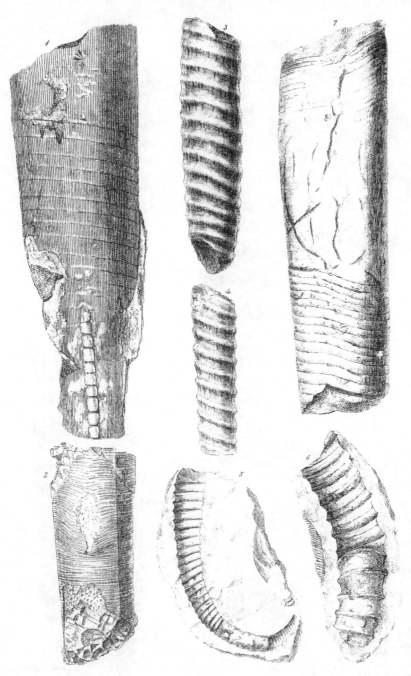

PLATE XXIX.

UPPER SILURIAN CEPHALOPODA.

Fig. 1. ORTHOCERAS BULLATUM (or striatum), Sow. U. L., Ludlow.

2. O. MOCKTREENSE, Sow. A. L., Mocktree Hays, Shropshire.

3, 4. O. IBEX, Sow. (Including O. articulatum.) Near Ludlow·

5, 6. O. PERELEGANS, Salter. (Lituites Ibex, and L. articulatus.)
L. L., Elton, fig. 5 ; Shelderton, fig. 7.

7. O. IMBRICATUM, Wahl. L. L., near Ludlow.

PLATE XXX.

UPPER SILURIAN CEPHALOPODA.

Figs 1, 2, 3. PHRAGMOCERAS (Orthoceras or Gomphoceras) PYRIFORME,
Sow. L. L., Leintwardine.

4   P. INTERMEDIUM (arcuatum β), McCoy. L. L., Shelderton.

CEPHALOPODA.

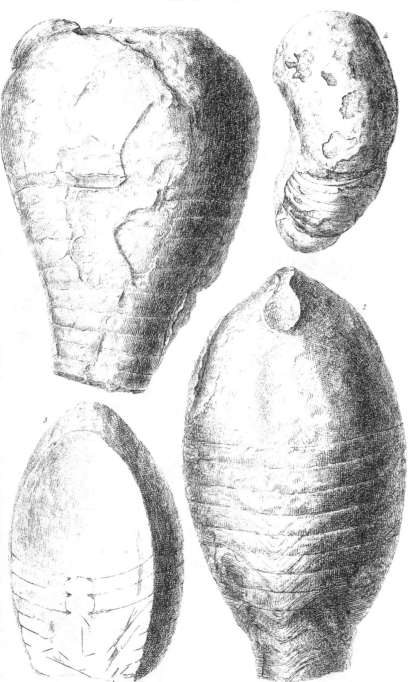

J.D.C. Sowerby

Ford & West Imp.t lith.

CEPHALOPODA.

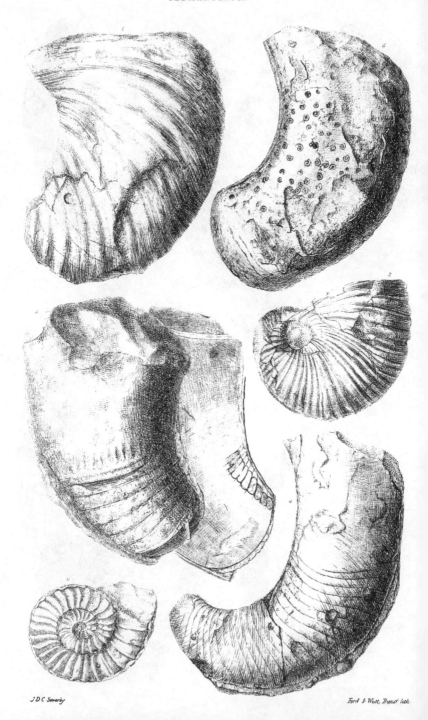

PLATE XXXI.

UPPER SILURIAN CEPHALOPODA.

Fig. 1, 2. PHRAGMOCERAS NAUTILEUM, Sow. W. S., Myddelton Hall, Caermarthen.

3. P. ARCUATUM, Sow. L. L., near Ledbury.

4. P. COMPRESSUM, Sow. L. L., near Aymestry.

5. LITUITES BIDDULPHII, Sow. W. L., Ledbury.

6. L. IBEX, Sow. L. L., near Ludlow.

PLATE XXXII.

UPPER SILURIAN CEPHALOPODA.

PHRAGMOCERAS VENTRICOSUM, Steininger.   L. L., Leint-
    wardine.

CEPHALOPODA.

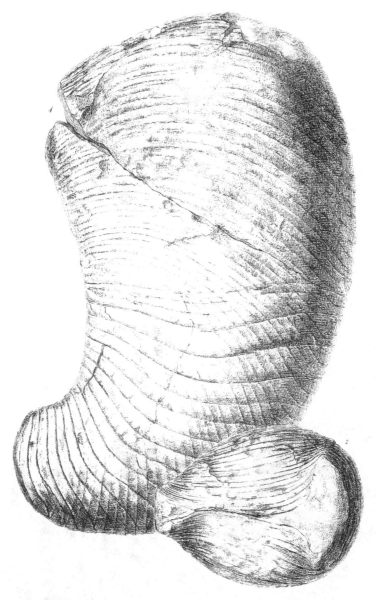

J.D.C.Sowerby.

Ford & West, Imp't lith.

CEPHALOPODA.

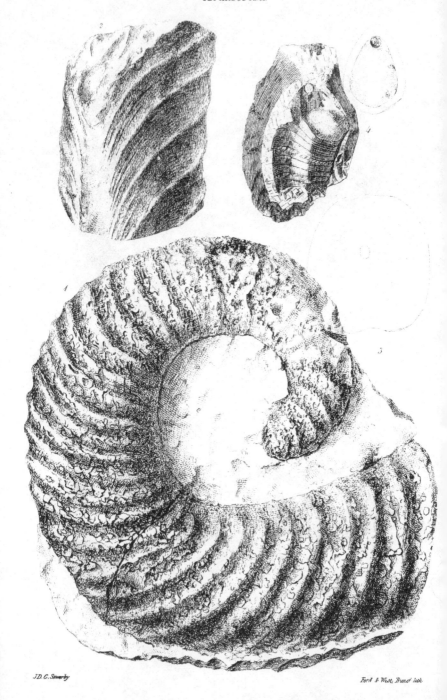

PLATE XXXIII.

UPPER SILURIAN CEPHALOPODA.

Figs. 1, 2, 3. LITUITES GIGANTEUS, Sow. L. L., Mocktree Hays.

   4. LITUITES? TORTUOSUS, Sow. L. L.? Between Welshpool and
      Berriew, Montgomeryshire.

d

PLATE XXXIV.

UPPER LUDLOW ROCK (TILESTONE).

Fig. 1. ONCHUS MURCHISONI, Agassiz. A true fish defence. Tin Mill, Downton.

2. LINGULA CORNEA, Sow. Same locality.

3. CUCULLELLA CAWDORI, Sow. in Upper? Ludlow Rock, Freshwater East, Pembrokeshire.

4. PTERINEA (Avicula) RECTANGULARIS, Sow. Horeb Chapel, Llandovery.

5. ORTHOCERAS SEMIPARTITUM, Sow. (perhaps a Diploceras). Same locality.

6. O. TRACHEALE, Sow. Same locality.

7. MODIOLOPSIS (Pullastra) LÆVIS, Sow. Horeb Chapel.

8. BELLEROPHON CARINATUS, Sow. do.

9. B. TRILOBATUS, Sow. Felindre, 10 m. W. of Knighton.

10. HOLOPELLA (Turritella) CONICA, Sow. Horeb Chapel.

10a. H. (Turritella) GREGARIA, Sow. Horeb Chapel.

11. H. (Turritella) OBSOLETA, Sow. Horeb Chapel; Felindre.

12, 13. TROCHUS? HELICITES, Sow. do.

14. TURBO WILLIAMSII, Sow. do.

15. GONIOPHORA (Cypricardia) CYMBÆFORMIS, Sow. Felindre.

16. CUCULLELLA (Cucullæa) ANTIQUA, Sow. Horeb Chapel.

17. C. (Cucullæa) OVATA, Sow. do.

18. CHONETES (Leptæna) LATA, Von Buch. do.

19. BELLEROPHON MURCHISONI, D'Orbigny? (B. striatus). do.

20. B. EXPANSUS, Sow. Young. (B. globatus). Felindre.

21. BEYRICHIA (Agnostus) TUBERCULATA, Klöden. Lodge Bank, Downton.

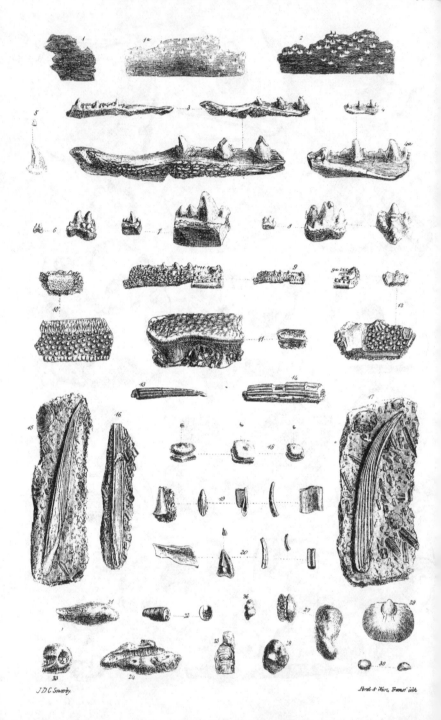

PLATE XXXV.

FOSSILS, CHIEFLY FISHES, OF THE UPPER LUDLOW BONE BED.

Fig. 1. Skin of SPHAGODUS, Agass.—1*a*, and 2, magnified portions.

   3—8. PLECTRODUS MIRABILIS, Agass. Jaws and teeth, natural size and magnified.

 9—12. P. (SCLERODUS) PUSTULIFERUS, Ag.

13, 14. ONCHUS MURCHISONI, Ag.

15—17. O. TENUISTRIATUS, Ag.

   18. Shagreen scales, probably of Onchus tenuistriatus, (Thelodus parvidens, Ag.) The stratum is chiefly made up of these.

19, 20. Indeterminable fragments, of bony texture.

23—28. Coprolites (probably of Plectrodus), and containing—Orthoceras semipartitum, fig. 22 ; Encrinite stems, fig. 23, 24; Lingula cornea, fig. 25 ; Holopella conica, fig. 26; Orbicula ruguta, fig. 27 ; Bellerophon expansus, fig 28.—all shells of the Upper Ludlow Rock.

   29. ORTHIS LUNATA, Sow. Frequent in this bed.

   30. Seeds or spores of some Cryptogamic land-plant, (Dr. Hooker).

   All the above are from the Bone bed at Ludford, Ludlow.

PLATE XXXVI.

FISHES OF THE OLD RED SANDSTONE.

Fig. 1, 2. ASTEROLEPIS. Interior and exterior view of a bony plate from the head. Elgin.

     3. Bony armour of the head of ASTEROLEPIS, from the same locality. Fig. 3* (wrongly numbered 2*), a portion magnified.

   4, 5. ONCHUS SEMISTRIATUS, Agass. Tenbury.

     6. CTENACANTHUS ORNATUS, Ag. The middle portion of a dorsal spine. Sapey, west of Worcester.

   7, 8. ONCHUS ARCUATUS, Ag. Bromyard.

     9. HOLOPTYCHIUS NOBILISSIMUS, Ag. One-third the natural length. Clashbinnie, Perthshire.

   10. Ventral scales, natural size.

   11. HOLOPTYCHIUS GIGANTEUS, Ag. Same locality.

   12. Under side of the scales.

   13. Tooth of HOLOPTYCHIUS, Elgin.

   14. Dorsal plate of COCCOSTEUS (called Cephalaspis' in the ' Sil. Syst.'), Herefordshire.

15, 15.* Serrated fish defence? PTYCHACANTHUS? DUBIUS, Ag. nat. size, and magnified. Herefordshire.

Pl. 36

The material originally positioned here is too large for reproduction in this reissue. A PDF can be downloaded from the web address given on page iv of this book, by clicking on 'Resources Available'.

PLATE XXXVII.

FISHES OF THE OLD RED SANDSTONE.

Figs. 1 to 3. CEPHALASPIS LYELLII, Ag. Forfarshire.

4. Side view of the head.

5. Exterior plates or scales of the head, highly magnified.

6. Scales from the sides of the body, magnified.

7. Scales from the flank.

8. Scales from the sides of the tail.

9. CEPH. LEWISII, Agass. Whitbatch, Shropshire.

10. CEPH. LLOYDII, Agass. Herefordshire.

10*. A portion of the surface, and the texture of the under layer, magnified.

11. CEPH. ROSTRATUS, Ag. Head with the outer surface removed. Whitbatch, Shropshire.

Printed in the United States
By Bookmasters